Advances in Experimental Medicine and Biology

Volume 1183

Advances in Microbiology, Infectious Diseases and Public Health

This book series focuses on current progress in the broad field of medical microbiology, and covers both basic and applied topics related to the study of microbes, their interactions with human and animals, and emerging issues relevant for public health. Original research and review articles present and discuss multidisciplinary findings and developments on various aspects of microbiology, infectious diseases, and their diagnosis, treatment and prevention.

The book series publishes review and original research contributions, short reports as well as guest edited thematic book volumes. All contributions will be published online first and collected in book volumes. There are no publication costs.

Advances in Microbiology, Infectious Diseases and Public Health is a subseries of *Advances in Experimental Medicine and Biology*, which has been publishing significant contributions in the field for over 30 years and is indexed in Medline, Scopus, EMBASE, BIOSIS, Biological Abstracts, CSA, Biological Sciences and Living Resources (ASFA-1), and Biological Sciences. 2018 Impact Factor: 2.126.

More information about this subseries at http://www.springer.com/series/13513

Giorgio Fedele · Clara Maria Ausiello
Editors

Pertussis Infection and Vaccines

Advances in Microbiology, Infectious
Diseases and Public Health
Volume 12

 Springer

Editors
Giorgio Fedele
Vaccine Preventable Diseases –
Reference Labs Unit, Department of
Infectious Diseases
Istituto Superiore di Sanità
Rome, Italy

Clara Maria Ausiello
Former Research Director in the
Department of Infectious Diseases
Istituto Superiore di Sanità
Rome, Italy

ISSN 0065-2598   ISSN 2214-8019   (electronic)
Advances in Experimental Medicine and Biology
ISSN 2365-2675   ISSN 2365-2683   (electronic)
Advances in Microbiology, Infectious Diseases and Public Health
ISBN 978-3-030-33248-8   ISBN 978-3-030-33249-5   (eBook)
https://doi.org/10.1007/978-3-030-33249-5

This Springer imprint is published by the registered company Springer Nature Switzerland AG.
The registered company address is: Gewerbestrasse 11, 6330 Cham, Switzerland

Preface

Pertussis or whooping cough remains one of the most poorly controlled vaccine-preventable diseases. Universal vaccination has dramatically reduced its incidence but has failed to bring it completely under control. In the last decades, pertussis epidemics have been registered in several countries, including those with high vaccination coverage. It is becoming apparent that immunity conferred by acellular pertussis vaccines wanes more rapidly than expected. Unlike old-generation whole-cell vaccines, acellular vaccines, while protecting against the disease, do not seem to prevent colonization. Increasing incidence among adolescents and adults makes them a reservoir for transmission to unimmunized infants, who are at risk of severe disease and death. The present volume has been conceived to give readers a comprehensive and up-to-date view on these topics.

The first two chapters, by N. Ring, J. S. Abrahams, S. Bagby, A. Preston, and I. MacArthur and by A. Barkoff and Q. He, respectively, discuss the evolution of *Bordetella pertussis*, including how vaccination is changing the circulating *B. pertussis* population. An insight on strain variation at gene and genome level is provided with the aim to elucidate how bacterial evolution affects the resurgence of pertussis and to provide crucial information for surveillance of the disease.

Research efforts tried to elucidate the mechanisms by which *B. pertussis* causes disease in humans, but several aspects remain poorly understood. In the third chapter, K. Scanlon, C. Skerry, and N. Carbonetti describe the roles played by three main virulence factors of *B. pertussis*, pertussis toxin, adenylate cyclase toxin-hemolysin, and tracheal cytotoxin, in infection and disease pathogenesis, elucidating their role in collectively suppressing protective immune responses while exacerbating damaging pathologies in the respiratory tract.

The host innate immune system plays a crucial role in response to infection and vaccination. In the fourth chapter, J. Gillard, E. van Schuppen, and D.A. Diavatopoulos describe recent understanding of the mechanisms through which innate immunity is programmed by *B. pertussis* and pertussis vaccines, how these innate responses prime adaptive immunity, as well as how this knowledge may be used for the development of new vaccines.

Immunological correlates of protection against pertussis are still elusive, but increasing evidence indicates that in animals, the hallmarks of protection are the priming of *B. pertussis*-specific T-helper (Th)-1 and Th-17 cells and their tissue residency. These features seem optimally primed by previous infection but insufficiently or only partially by current vaccines. In the fifth chapter, E. E. Lambert, A.M. Buisman, and C.A.C.M. van Els bring evidence that the infection drives such superior *B. pertussis*-specific CD4+ T-cell lineages also in humans and highlight key features of effector immunity downstream of Th-1 and Th-17 cell cytokines. The mismatch between infection- and vaccination-driven immunities appears to be a matter of fact. Resurgence of the disease could be attributed to suboptimal protection and waning of vaccine-induced immunity. The sixth chapter, by C.M. Ausiello, F. Mascart, V. Corbière, and G. Fedele, discusses the responses of the immune system to currently available whole-cell and acellular pertussis vaccines, with the aim to enlighten critical points needing further efforts to reach a good level of protection in vaccinated individuals. In particular, differences among currently available vaccines in terms of Th cells profiles induction and prevention of bacterial circulation are discussed. A review of results obtained by maternal immunization is also given.

New pertussis vaccines should be able to confer long-term protection against the disease and to prevent colonization of the respiratory tract. Several vaccine candidates are in preclinical development, and few others have recently completed phase I/phase II trials. These aspects are covered in the seventh chapter by D. Hozbor.

Public health systems need to confront against the resurgence of the disease. The eighth chapter, by P. Stefanelli, describes current knowledge on pertussis diagnosis, as well as optimal strategies for disease prevention and control. In this regard, the ninth chapter, by N. Guiso and F.Taieb, focuses on pragmatic approaches required to increase the control of pertussis in low- and middle-income countries where high numbers of death due to pertussis persist.

Many forms and clinical features of the disease, ranging from the most classical to atypical and much nuanced forms, have been reported. The tenth chapter, by I. Polinori and S. Esposito, focuses on typical and atypical presentation of clinical pertussis in infants and adolescents. A review of clinical management and supportive therapies is also provided.

The last chapter, a note by A. Cassone, critically re-examines the history of pertussis vaccines with their merits and limitations, particularly concerning the debated issue of waning immunity. A reflection on the complexity and apparent peculiarity of this field is put forward to the readers, to the final scope of discussing some aspects of the evolving strategies of disease control.

Rome, Italy   Giorgio Fedele
April 2019   Clara Maria Ausiello

Contents

How Genomics Is Changing What We Know About the Evolution and Genome of *Bordetella pertussis* 1
Natalie Ring, Jonathan S. Abrahams, Stefan Bagby, Andrew Preston, and Iain MacArthur

Molecular Epidemiology of *Bordetella pertussis* 19
Alex-Mikael Barkoff and Qiushui He

Role of Major Toxin Virulence Factors in Pertussis Infection and Disease Pathogenesis 35
Karen Scanlon, Ciaran Skerry, and Nicholas Carbonetti

Functional Programming of Innate Immune Cells in Response to *Bordetella pertussis* Infection and Vaccination 53
Joshua Gillard, Evi van Schuppen, and Dimitri A. Diavatopoulos

Superior *B. pertussis* Specific CD4+ T-Cell Immunity Imprinted by Natural Infection 81
Eleonora E. Lambert, Anne-Marie Buisman, and Cécile A. C. M. van Els

Human Immune Responses to Pertussis Vaccines 99
Clara M. Ausiello, Françoise Mascart, Véronique Corbière, and Giorgio Fedele

New Pertussis Vaccines: A Need and a Challenge 115
Daniela Hozbor

Pertussis: Identification, Prevention and Control 127
Paola Stefanelli

Pertussis in Low and Medium Income Countries: A Pragmatic Approach 137
Nicole Guiso and Fabien Taieb

Clinical Findings and Management of Pertussis 151
Ilaria Polinori and Susanna Esposito

Pertussis Vaccines and Vaccination Strategies.
An Ever-Challenging Health Problem 161
Antonio Cassone

Index 169

Adv Exp Med Biol - Advances in Microbiology, Infectious Diseases and Public Health (2019) 1183: 1–17
https://doi.org/10.1007/5584_2019_401
© Springer Nature Switzerland AG 2019
Published online: 19 July 2019

How Genomics Is Changing What We Know About the Evolution and Genome of *Bordetella pertussis*

Natalie Ring, Jonathan S. Abrahams, Stefan Bagby, Andrew Preston, and Iain MacArthur

Abstract

The evolution of *Bordetella pertussis* from a common ancestor similar to *Bordetella bronchiseptica* has occurred through large-scale gene loss, inactivation and rearrangements, largely driven by the spread of insertion sequence element repeats throughout the genome. *B. pertussis* is widely considered to be monomorphic, and recent evolution of the *B. pertussis* genome appears to, at least in part, be driven by vaccine-based selection. Given the recent global resurgence of whooping cough despite the wide-spread use of vaccination, a more thorough understanding of *B. pertussis* genomics could be highly informative. In this chapter we discuss the evolution of *B. pertussis*, including how vaccination is changing the circulating *B. pertussis* population at the gene-level, and how new sequencing technologies are revealing previously unknown levels of inter- and intra-strain variation at the genome-level.

Keywords

Bordetella pertussis · DNA sequencing · Evolution · Genomic variation · Whooping cough

N. Ring, J. S. Abrahams, S. Bagby, A. Preston, and I. MacArthur (✉)
Department of Biology and Biochemistry, and Milner Centre for Evolution, University of Bath, Bath, UK
e-mail: N.A.Ring@bath.ac.uk; J.S.Abrahams@bath.ac.uk; A.Preston@bath.ac.uk; I.MacArthur@bath.ac.uk

Abbreviations

ACT	adenylate cyclase toxin
ACV	acellular vaccine
CGH	comparative genomic hybridization
COG	Clusters of Orthologous Genes
FHA	Filamentous haemagglutinin
Fim(2/3)	Fimbrial protein (2/3)
HGT	horizontal gene transfer
INDEL	insertions and deletions
IS	insertion sequence
MLST	multilocus sequence typing
MLVA	multilocus variable-number tandem-repeat analysis
NGS	next generation sequencing
ONT	Oxford Nanopore Technologies
PacBio	Pacific Biosciences
PERISCOPE	PERtussIS Correlates Of Protection Europe (consortium)
PFGE	pulsed field gel electrophoresis
prn	Pertactin
ptx	Pertussis toxin
SMRT	single-molecule real-time
SNP	single nucleotide polymorphism
WCV	whole-cell vaccine
WGS	whole genome sequencing

1 The Ongoing Problem of *Bordetella pertussis*

Whooping cough, the infectious respiratory disease caused by the bacterium *Bordetella pertussis*, has been resurgent in many countries for the past two decades. This resurgence comes in spite of a global vaccination programme, with 90% of the target population receiving a single dose of *pertussis*-containing vaccine, and 85% receiving three doses (WHO 2018). In addition, there has been a shift in the epidemiological profile of the disease: whereas once most cases were reported in infants and unvaccinated children, the resurgence is also affecting vaccinated children, adolescents and adults, Fig. 1 (Strebel et al. 2001; Clark 2014). Data from the Centers for Disease Control, for example, show that from 1990–1997, the mean incidence of whooping cough per year in 11–19 year-olds was 2.54 per 100,000 people, whereas from 2010–2017, it was 20.43 (Centers for Disease Control 2019).

The original vaccination programme, introduced in the 1940s and 1950s, used a whole-cell vaccine (WCV). Initially, cases of the disease appeared to drop significantly. Due to perceived reactogenicity of the WCV (now largely discredited), it was replaced in many countries throughout the 1990s and early 2000s with an acellular vaccine (ACV) (for example: Cherry 1990, 1992, 1996; Cherry et al. 1993; Blumberg et al. 1993; Moore et al. 2004). The acellular vaccine contains one to five *B. pertussis* antigens, including Pertussis toxin (ptx), pertactin (prn), filamentous haemagglutinin (FHA) and the fimbrial proteins Fim2 and Fim3. These days, most developed countries use the ACV, although many developing countries continue to use the WCV. The recent use of the ACV has been strongly implicated in whooping cough's resurgence. However, concerns were raised about waning immunity conveyed by the WCV beginning in the early 1990s and, in many countries, the resurgence does seem to pre-date the switch to the ACV (De Serres et al. 1995; Cherry 1996; de Melker et al. 1997).

Three main potential causes are thought to have contributed to the recent observed increase in whooping cough cases: increased awareness of the disease coupled with improved diagnosis due to a switch from culture-based to PCR- and serology-based techniques; waning immunity,

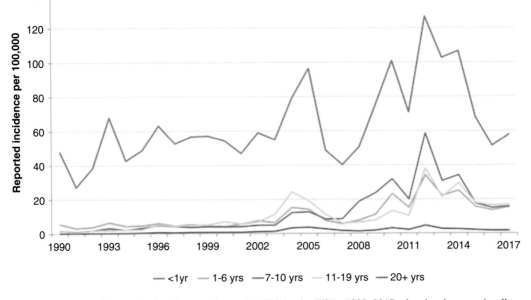

Fig. 1 Reported incidence of whooping cough per 100,000 in the USA, 1990–2017, showing increase in all age brackets. (Data source: CDC National notifiable diseases surveillance reports Centers for Disease Control (2019))

particularly that conveyed by the ACV compared to the WCV; and genetic variations in circulating *B. pertussis* strains, away from the vaccine strains (Sealey et al. 2015; Clark 2014; Ausiello and Cassone 2014). Here we focus on the latter from two different but highly interrelated perspectives: variation at the gene-level, and variation at the genome-level, with particular consideration of how recent developments in genomic research have contributed to our understanding of evolution and variation in a species which traditionally has been described as highly monomorphic (for example: Bart et al. 2010; Mooi 2010).

2   The Speciation of *Bordetella pertussis*

The study of the bordetellae has focussed largely on the three classical, pathogenic, *Bordetella* species: *B. bronchiseptica*, *B. pertussis* and *B. parapertussis*. However, the *Bordetella* genus contains many additional species which have been isolated from extremely diverse environments, including marine sponges, bioreactors, nitrifying sludge and mural paintings in ancient tombs (Wang et al. 2007; Bianchi et al. 2005; Sun et al. 2019; Tazato et al. 2015). Using previously published 16S sequence data derived from many *Bordetella* species to create a phylogenetic tree, Hamidou Soumana et al. (2017) demonstrated that eight out of the ten clades contained soil-dwelling bordetellae; the permeation of the environmental phenotype

throughout the phylogenetic tree hints at a soil-based origin for the *Bordetella* genus.

Key evolutionary milestones within the genus involve species that are capable of both environmental and pathogenic lifestyles, such as the key ancestor to the classical *Bordetella* species, a *B. bronchiseptica*-like bacterium. Multilocus sequence typing (MLST) studies of *B. bronchiseptica* have established two distinct complexes of isolates, complex I and IV. The majority of strains isolated from humans originate from Complex IV (Diavatopoulos et al. 2005; Park et al. 2012). Using the much higher discriminatory power of next generation sequencing (NGS) data, it has been demonstrated that *B. pertussis* and *B. parapertussis* evolved from different complexes, with *B. parapertussis* sharing a more recent ancestor with Complex I and *B. pertussis* with Complex IV (Linz et al. 2016).

Despite their apparently different evolutionary trajectories, *B. pertussis* and *B. parapertussis* cause remarkably similar pathologies in humans. In an example of convergent evolution, the genomes of both species have evolved primarily through genome reduction, mediated through homologous recombination of insertion sequences (IS). As seen in Table 1, the genome of *B. bronchiseptica* contains very few IS elements; in fact, the reference strain RB50 was originally believed to contain no IS elements at all (Parkhill et al. 2003; Preston et al. 2004). In contrast, *B. parapertussis* genomes contain around 30 IS elements, usually 22 copies of IS *1001* and 9 copies of IS *1002*. The *B. pertussis* genome contains the most IS elements: up to ten copies of IS *1002*,

Table 1 Characteristics of all classical *Bordetella* closed genomes available on RefSeq, March 2019

Characteristic	*B. pertussis*	*B. parapertussis*	*B. bronchiseptica*
Number of closed genomes	421	18	19
Genome size / Mb[a]	4.11 (4.04–4.39)	4.78 (4.77–4.90)	5.21 (5.08–5.34)
Number of predicted genes[a]	3,979 (3,856 – 4,239)	4,501 (4,490 – 4,574)	4,911 (4,774 – 5,090)
Number of proteins[a]	3,615 (3,425 – 3,866)	4,166 (4,157 – 4,184)	4,804 (4,663 – 4,993)
G + C %[a]	67.7 (67.7–67.8)	68.1 (67.8–68.1)	68.2 (68.1–68.4)
IS *481*[a]	256 (234–273)	–	1 (0–3)
IS *1001*[a]	–	22 (22–28)	1 (0–1)
IS *1002*[a]	8 (5–10)	9 (0–9)	0 (0–5)
IS *1663*[a]	17 (16–24)	–	2 (0–14)

[a]Figures shown are mean and (range)

around 20 copies of IS *1663*, and over 200 copies of IS *481*. The appearance and expansion of these IS elements is thought to have led to the speciation of *B. pertussis* and, separately, *B. parapertussis*, from their *B. bronchiseptica*-like ancestors (Parkhill et al. 2003; Preston et al. 2004; Diavatopoulos et al. 2005).

Genome reduction was key in the speciation of *B. pertussis*. The significant, IS-mediated, reduction in genome size of the *B. pertussis* genome (around 4.1 Mb) compared to the *B. bronchiseptica* genome (around 5.3 Mb) has led to a streamlined genome, depleted of many metabolic, membrane transport, surface structure synthesis and gene expression regulatory genes when compared to *B. bronchiseptica* genomes (Parkhill et al. 2003). Comparative genomic studies between *B. bronchiseptica* and *B. pertussis* reveal that the latter has around 1,200 fewer genes (Parkhill et al. 2003; Linz et al. 2016). In addition, insertions of IS elements into genes which are functional in *B. bronchiseptica* has resulted in the existence of over 350 pseudogenes in *B. pertussis*, compared to only around 20 in *B. bronchiseptica*. This sculpting of the B. *pertussis* genome via IS-mediated homologous recombination has produced a highly specialised pathogenic bacterium which is niche-restricted to the human nasopharynx. Traditionally, *B. pertussis* has been described as a monomorphic species (for example: Mooi 2010; Bart et al. 2010); however, since the advent of whole genome sequencing, genomics has been revealing that the bacterium may be less clonal than previously thought, with the introduction of vaccination and continued homologous recombination between IS elements driving gene- and genome-level variations respectively.

3   Vaccination Has Accelerated *B. pertussis* Gene-Level Evolution

3.1   Changes to Circulating Alleles

Over the last several decades, allele-typing of selected genes has shown a number of similar trends, characterised largely by the drift of genes away from the vaccine alleles (for example: Mooi et al. 1998; van der Zee et al. 1996; Mooi et al. 2001). One well-reviewed example involves *pertussis toxin*. Prior to the 1990s, the predominant ptx promoter allele was *ptxP1*. A new allele, *ptxP3*, was first observed in 1988 (Bart et al. 2010). In *ptxP3*, a SNP in the binding site for ptx's transcriptional regulator, BvgA, appears to increase the binding affinity between the promotor and regulator, thus increasing transcription and causing *ptxP3*-carrying strains to produce more ptx. The expression of other proteins involved in complement resistance is also altered and, together, these changes increase the transmissibility and severity of the disease caused by *ptxP3*-carrying strains (Mooi et al. 2009; Bart et al. 2010; King et al. 2013; de Gouw et al. 2014). This new allele spread rapidly throughout the 1990s, and *ptxP3* is now present in greater than 90% of recent isolates (Lam et al. 2012; Bart et al. 2010). A thorough screen of 343 strains representing 19 countries and six continents, spanning 90 years of *B. pertussis* isolation, showed that similarly rapid selective sweeps have also occurred in other antigen-related genes, including *ptxA*, *prn*, and *fim3* (Bart et al. 2014a).

In addition, analysis of 100 strains isolated during a 2012 whooping cough outbreak in the UK, after the introduction of the ACV, showed that the evolution of the antigens included in the ACV is occurring more rapidly than that of other *B. pertussis* surface proteins, which are presumably under similar levels of pressure from the human immune system (Sealey et al. 2015). Importantly, this analysis also showed that numerous different strains were circulating during the outbreak, rather than one particularly virulent strain or allelic profile being responsible. The same was also true for strains circulating during outbreaks in the USA, in California in 2010, and in Vermont and Washington in 2012 (Bowden et al. 2014; Bowden et al. 2016). This suggests that the strains circulating during outbreaks tend to be the same as those that circulate during non-outbreak periods, but that some unknown trigger causes an increase in whooping cough cases in regular four-yearly cycles.

Supporting the idea that the recent allelic changes we are seeing in *B. pertussis* are, in part, a response to the introduction of vaccination, Xu et al. (2015) and Du et al. (2016) showed that, in countries where vaccine uptake has been lower or delayed, the rate of change to the allelic profile of circulating strains has also been delayed. In the Philippines, for example, where the WCV is still in use, *prn2* has yet to appear, despite being the allele most frequently seen in ACV-adopting countries (Galit et al. 2015).

3.2 The Proliferation of Antigen-Deficient Strains

Another gene-level phenomenon which has been observed recently is the emergence of strains deficient in one or more of the antigens used in the ACV. During the pre-ACV era, antigen-deficient strains were occasionally isolated, albeit at very low frequencies. Individual strains with non-functional pertactin genes, for example, were isolated in Europe, North America and Japan in the 1990s (Mastrantonio et al. 1999; Weigand et al. 2018; Miyaji et al. 2013). The landmark

study by Bart et al. (2014a), in which all but around 20 strains were isolated prior to 2007, did not identify any strains which were prn-deficient (although one, BP310, has subsequently been resequenced by Zomer et al. (2018) and is likely to be deficient). Since the mid-2000s, however, the number of prn-deficient strains being isolated globally has increased rapidly. A study of Australian isolates from 2008–2012 showed an increase from 5% prn-deficient strains in 2008 to 78% prn-deficient strains in 2012 (Lam et al. 2014). Pertactin-deficiency appears to be polyclonal, affecting both dominant *prn* alleles, *prn1* and *prn2*, and arising through several different mechanisms including insertions of IS *481*, large deletions, and SNPs, with no single predominant causative mutation (Hegerle et al. 2012; Queenan et al. 2013; Barkoff et al. 2019).

As with changes to allelic profiles, countries with different vaccination strategies appear to be differently affected by the proliferation of antigen-deficiency. A longitudinal study of prn-deficiency in European countries between 1998 and 2015, Fig. 2, showed that the number of prn-deficient strains is increasing in all screened countries but,

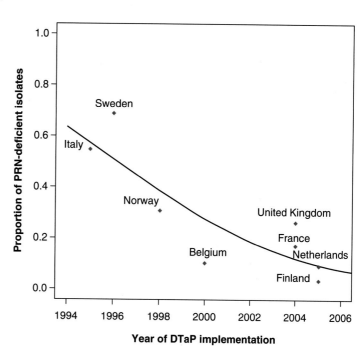

Fig. 2 Correlation between the introduction of a primary acellular pertussis vaccine containing pertactin (PRN) in a European country and the proportion of PRN-deficient isolates found in the study, 2012–2015. (Reproduced with permission from Barkoff et al. (2019))

the earlier a country introduced the ACV, the higher the percentage of strains currently found to be deficient (Barkoff et al. 2019). The rapid recent increase in prn-deficiency suggests that, although the deficiency may have always occurred in some strains by chance, it has been strongly selected for by the ACV compared to the WCV. Further supporting the idea of ACV-mediated selection pressure is the sustained decrease of prn-deficient strains circulating in Japan since pertactin was removed from the Japanese ACV in 2012 (Hiramatsu et al. 2017).

A smaller number of strains deficient in other antigens included in the ACV have also been identified. Bart et al. (2015) and Weigand et al. (2018) identified several geographically independent recent strains which were unable to produce FHA; the same strains were often also prn-deficient. In addition, a handful of strains have been isolated globally which are deficient in both prn and ptx (Bouchez et al. 2009; Williams et al. 2016; Weigand et al. 2018). However, both FHA and Pertussis toxin are thought to play more vital roles in whooping cough disease development than pertactin (Carbonetti 2010; Henderson et al. 2012; Serra et al. 2011). Hence, although occasional strains may develop deficiency in these antigens, the reduced ability of these deficient strains to cause disease may mean that this kind of antigen deficiency is unlikely to proliferate in the same way as prn-deficiency.

4 Recent Sequencing Advances Highlight Genome-Level Variation

4.1 The Limitations of Short-Read Sequencing

Whilst the wide availability of whole genome sequencing throughout the 2000s enabled a variety of high-throughput strain screens, the highly repetitive nature of the *B. pertussis* genome has made the assembly of closed, single-contig, *B. pertussis* genomes difficult. IS *481*, together with the smaller number of copies of other repeated regions, such as IS *1002*, IS *1663* and the rRNA operon, has confounded attempts to assemble closed genomes using short-read sequencing technologies, which produce sequencing reads shorter than the repeated section. The hundreds of *B. pertussis* genomes assembled using short-read sequencing alone have therefore tended to consist of several hundred contigs, ostensibly one per repeat in the genome. Thus, the majority of high-throughput screens throughout the 2000s were focussed on the gene-level differences between strains already discussed.

The presence of so many IS elements, however, means that assembly of closed *B. pertussis* genomes could be particularly informative: IS elements are able to move around the genome through homologous recombination, potentially causing genome-level structural changes which may be discernible only through single-contig assemblies (Bentley et al. 2008; Siguier et al. 2014). Despite the discovery that most whooping cough outbreaks tend to be polyclonal in nature, *B. pertussis* remains a relatively clonal species, with a low SNP rate compared to many other bacteria. In other species, in addition to gene-level variations, differences at a whole-genome level are known to contribute to altered gene expression and phenotypic diversity (Darch et al. 2014; Sousa et al. 1997). IS-mediated rearrangement may affect gene regulation and/or expression in *B. pertussis* by a number of mechanisms, including IS *481*'s inwards and outwards-facing promoters, as well as changing the distance of genes from the origin of replication (Amman et al. 2018). Limited evidence has already shown that certain genome-level differences can affect phenotype in *B. pertussis* in this way (Brinig et al. 2006).

The speciation of *B. pertussis* from *B. bronchiseptica* via IS element-mediated homologous recombination resulted in a variety of genomic arrangement differences between the two species alongside the reduction in genome size, and it is likely that IS-mediated genomic rearrangement in *B. pertussis* is an ongoing process (Parkhill et al. 2003). Indeed, prior to the advent of long-read sequencing, pulsed-field gel electrophoresis (PFGE) was one of the few methods able to discriminate between highly clonal *B. pertussis* strains: isolate screens using PFGE indicated that strains which seemed otherwise alike could vary significantly in terms of PFGE type (van Gent et al. 2015; Bisgard et al. 2001; Advani et al. 2004; Advani et al. 2013). Although the existence of numerous PFGE types was widely seen, however, it could not be confirmed how different PFGE types arose; they could represent different genomic arrangements, but could also have arisen due to mutagenesis at PFGE restriction sites, for example.

Thus, closed *B. pertussis* genome sequences may validate and further reveal genome-level differences between strains which otherwise appear to have highly similar or identical DNA content. The recent availability of long-read sequencing techniques, which can produce sequencing reads longer than 1,000 bp, has therefore revolutionised our ability to discover and investigate genome-level variations in *B. pertussis*.

4.2 Long-Read Sequencing Shows Extensive Inter-strain Genome Rearrangement

The first study to take advantage of long reads utilised Pacific Biosciences (PacBio) sequencing to produce closed, fully annotated, genomes for two *B. pertussis* strains: BP1917 and BP1920 (Bart et al. 2014b). The arrangement of these two strains differed significantly, with three large inversions and a variety of deletion and/or insertion events between the pair. Having proven the ability of long reads to close the genomes of BP1917 and BP1920, Bart et al. (2015) next sequenced 11 *B. pertussis* strains which represented the pandemic *ptxP3* lineage, again using PacBio sequencing to produce 10 kb-long reads. This cohort, which also included several strains deficient in prn and/or FHA, were characteristically similar in terms of SNPs but again showed significant differences in genome arrangement.

As is common for a developing technology, the cost of PacBio sequencing has rapidly decreased. Thus, higher-throughput strain screens have become increasingly feasible. Figure 3 shows the dramatic increase in closed genome sequences for the classical *Bordetella* species available from the NCBI's RefSeq database since 2014. Bowden et al. (2016) conducted the first whooping cough outbreak screen to utilise long-read sequencing alongside short-read sequencing in hybrid, sequencing 31 strains which had circulated during US whooping cough outbreaks in 2010 and 2012. The hybrid approach has been shown to improve the accuracy of assemblies produced using long-read sequencing, which still have an intrinsically higher error rate than short-read-only assemblies, particularly in homopolymeric tracts (Au et al. 2012; Koren et al. 2012). In the 31 genomes studied, 21 different arrangement profiles were observed; most consisted of inversions around the origin of replication. Bowden et al. also validated the arrangements using whole genome optical mapping and found that, in all cases, the boundaries between rearranged sections were composed of a repeated element: an insertion sequence, or the rRNA operon. The vast majority of the boundaries were IS *481* (89%), whilst the rest were composed of an rRNA operon, IS *1002* or a combination of IS *1002* and IS *481* together.

The most thorough investigation of *B. pertussis* genomic rearrangement to date also used a hybrid assembly strategy, combining PacBio long reads with Illumina short reads to close the genomes of 257 strains, dating from 1939 to 2014 (Weigand et al. 2017). When clustered based on their arrangement profiles, most isolates clustered according to allelic profile; for example, most *ptxP1* strains shared similar arrangements with other *ptxP1* strains. This clustering indicates that most structures are relatively stable, as supported by a clinical isolate which showed the same structure before and after 11 serial passages. Furthermore, these findings suggest that lineages are conserved not just in terms of SNPs, but also in genomic arrangement. Interestingly, Weigand et al. note that, on average, only half of their predicted IS *481* target sites

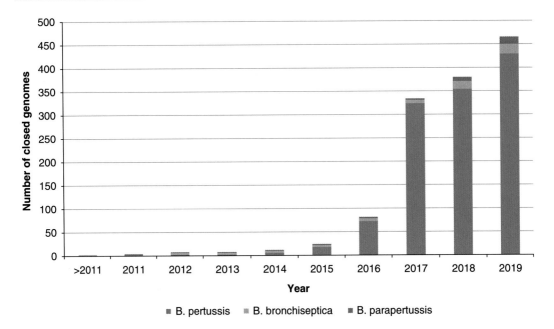

Fig. 3 Increase in numbers of closed classical *Bordetellae* genome sequences (available on RefSeq) since the commercial introduction of long-read sequencing technologies

are occupied in any given genome, suggesting a potential for further IS-mediated structural changes in future generations, assuming these sites are not non-permissive.

5   How Else Might *B. pertussis* Generate Diversity Through Genome-Level Variation?

The primary metric commonly used to assess diversity of bacterial species is SNPs. However, *B. pertussis* is a textbook example of a clonal bacterial pathogen: variation, when judged by SNPs alone, is extremely limited, even taking into account the accelerated mutation of *B. pertussis* genes since the introduction of vaccination. Bart et al. (2010) estimated the mutation rate between *B. pertussis* isolates to be 1 SNP per 8,675 bases, compared to, for example, 1 SNP per 3,000 bases in *Mycobacterium tuberculosis*, and 1 SNP per 6,700 bases in *Escherichia coli* O157: H7 (Fleischmann et al. 2002; Gutacker et al. 2002; Zhang et al. 2006). Diversity within a species is vital for its survival, in order to drive

adaptation; this is particularly true for pathogens, which are under pressure from the immune system (Mooi 2010). Therefore, a prominent question in *B. pertussis* genomics is: despite limited SNP diversity, how does *B. pertussis* generate diversity? The large numbers of closed genomes assembled using long-read sequencing have proven that rearrangements are a rich source of genome-level diversity, but can genomics also reveal other types of genome-level variation?

5.1   Harnessing Deletion as a Driver of Diversity

King et al. (2010) analysed the size of *B. pertussis* genomes of strains isolated over a 60-year period, demonstrating that genome streamlining has been an ongoing process, with recently isolated strains having smaller genomes and higher numbers of pseudogenes. Thus, *B. pertussis* is described as a species which is still undergoing genome reduction. Like genomic rearrangement, reduction is driven primarily by homologous recombination between insertion sequences. The large numbers of homologous IS elements in *B. pertussis* therefore produce a fertile mutational landscape capable of the generation of diversity.

Many bacterial species also generate diversity through the gain of genes, often by horizontal gene transfer (HGT), resulting in fluid gene content of the population, enabling the population to effectively respond to evolutionary bottlenecks that may arise over long or short timescales. In *B.* pertussis, however, HGT appears to occur very rarely (Linz et al. 2016). Gene content of a species or genus is often analysed using a "pangenome" approach, which consists of analysing which genes are consistently present (the core genome) in the population, and which genes are variably present (the accessory genome) (reviewed in Medini et al. 2005; Rouli et al. 2015). A number of studies have undertaken this analysis in *B. pertussis* using either comparative genomic hybridization (CGH) (for example: Zhang et al. 2006; Caro et al. 2006; Heikkinen et al. 2007; King et al. 2008) or NGS (for example: Park et al. 2012; Ding et al. 2017). These have shown that, despite extremely limited HGT and otherwise high levels of clonality between strains, *B. pertussis* maintains a moderate accessory genome, largely through gene loss rather than gene gain. For example, the most comprehensive pangenome study to date, using CGH on 171 *B. pertussis* strains, revealed that 15% of the genes present in the population appeared variably in the 171 strains studied. (King et al. 2010). By using a set of probes which included *B. bronchiseptica* and *B. parapertussis*, King et al. were able to avoid biasing their analysis towards genes that were present only in the *B. pertussis* reference strain, Tohama I, which has been shown to lack over 45 kb of the accessory genome of the population (Caro et al. 2008; Bouchez et al. 2008).

There is a lack of knowledge about the phenotypic impact of gene deletions in *B. pertussis*, however. As the cost of sequencing has plummeted, the frequency and ease with which genomes, and their constituent mutations, are published has far outpaced the publishing of their phenotypic impact. To cope with the deluge of data, ontology schemes strive to categorize genes into functional groups and estimate their function based on sequence homology. A variety of nucleotide polymorphisms in *B. pertussis*, such as those in *ptxP3*, have had their phenotypic impacts analysed, with some providing clear fitness advantages in the mouse model (Mooi et al. 2009). In contrast, many key gene deletions have yet to be experimentally characterized in *B. pertussis*. Using the ontology scheme Clusters of Orthologous Genes (COG), King et al. (2010) showed that, as expected, housekeeping genes were underrepresented in the deleted genes, whilst genes of unknown function were overrepresented by 25%. There therefore remains genetic "dark matter", genes with unknown function which are overrepresented in gene deletions.

Whilst *B. pertussis* is undisputedly undergoing genome reduction, evolution acts on phenotypes rather than genotypes. Therefore, it is likely that the *B. pertussis* genome is undergoing streamlining as an effect of certain phenotypes being selected against. It has been theorised that the transcriptional and translational cost of superfluous genes far outstrips the mere cost of DNA replication of such regions (Adler et al. 2014). *B. pertussis* maintains many seemingly functionless pseudogenes, despite the vast majority being shown to be transcriptionally inactive *in vitro* and in the mouse model (Bart et al. 2010; King et al. 2008; de Gouw et al. 2014). This supports the idea that the deletion of some genes provides a greater fitness benefit than the deletion of some others. Nonetheless, there is also evidence that the DNA content of the species is under selection, as pseudogenes have been shown to be mildly enriched in gene deletions, suggesting that streamlining of the DNA is also favoured to some extent (Kuo and Ochman 2010). Thus, the process of *B. pertussis* genome streamlining is likely to be a balance between entirely passive and entirely directed.

There have been five deleted regions, totalling over 50 genes, that have been deleted in all recently isolated clinical strains in comparison to the reference strain Tohama I (King et al. 2010; Heikkinen et al. 2007; Bouchez et al. 2008). In addition to clinical strains, Bart et al. (2014a) also investigated two strains that were used to make

WCVs. In one of the vaccine strains, the five deleted regions were present; if these regions impact cell surface antigens, the immunity conveyed by the WCV could therefore also be affected. This highlights the clinical importance of understanding the continual evolution of *B. pertussis*, which is in part driven by genome reduction.

5.2 Harnessing Duplication as a Driver of Diversity

Homologous recombination between IS elements has not only caused rearrangements and deletions in the *B. pertussis* genome, but also duplications ranging from single genes to large, multi-gene, regions. The general paradigm under which duplications occur is that a gene is duplicated, thus freeing the second copy from the purifying selection of the original copy of the gene, potentially allowing it over time to evolve a new function. However, the second copy of the gene may also maintain the same function of the first gene. These types of events are "canonical" duplications that are well documented in the bacterial kingdom (for example: Ohta 1989; Lynch 2002; Magadum et al. 2013). Taking the genes from Tohama I and clustering them based on 90% nucleotide homology using the tool CDHIT (available as a web server: http://weizhong-lab. ucsd.edu/cdhit_suite/cgi-bin/index.cgi?cmd=cd-hit-est), it can be seen that Tohama I maintains two copies of nine separate genes ranging from 97% to 100% homology (excluding IS elements and rRNA genes) The maintenance of these duplications provides further evidence that it is not the genome size of the bacterium itself that is the primary target of streamlining but certain phenotypic traits which are coded in the DNA.

Before the cost of NGS rapidly decreased in the late 2000s and early 2010s, duplications were primarily inferred by increased spot intensity in CGH or disturbances to southern blotting or PFGE patterns. Using these techniques, a number of multi-gene duplications were serendipitously discovered in the *B. pertussis* population. Using the power of long-read sequencing technologies, a number of studies had revealed further duplications, bringing the total to 13 serendipitously discovered mutations (Dalet et al. 2004; Caro et al. 2006; Heikkinen et al. 2007; Weigand et al. 2016; Dienstbier et al. 2018; Ring et al. 2018; Weigand et al. 2018). A recent study by Abrahams et al. (In preparation) systematically analysed the *B. pertussis* population in search of large multi-gene duplications. Previously published short-read sequencing data was utilised and read depth abnormalities were used to predict duplications. In the 473 strains analysed, over 400 duplications were found.

Abrahams et al. (In preparation) presents a deep description of duplications in *B. pertussis*. In addition to the quantity of duplications, over 90% of duplications were found at 11 "hotspot" loci but with varying gene contents, similar to a situation described previously in *M. tuberculosis* (Weiner et al. 2012). Interestingly, when the CNVs at each hotspot loci were mapped to a phylogenetic tree based on core genome SNPs, they appeared not to be vertically inherited and instead appeared to occur spontaneously many times at similar loci with subtly different gene content in each mutation, suggesting the existence of a potential phenotypic driver at those loci.

Large multi-gene duplications are known to be unstable in the bacterial kingdom, and this has also been demonstrated in *B. pertussis*. For instance, Dalet et al. (2004) noticed that subculturing a single isolate produced both high haemolytic and average haemolytic single colonies. Further analysis showed that colonies with high haemolysis had a duplication of the locus encoding adenylate cyclase, a key virulence factor. Dalet et al. further demonstrated that subculturing a strain with a duplication produced colonies with a single copy of the locus, thus indicating a mixed population, ostensibly caused by an unstable locus. This early study used PFGE and southern blotting to screen colonies for copy number of the locus. These "pre-genomics" tools provide high quality data, but largely answer very specific research hypotheses, in contrast to sequencing experiments, which shed light on a vast range of research questions. Abrahams et al.

used ultra-long nanopore sequencing reads (over 3,000 reads longer than 50 kb) to confirm the presence of between 1 and 5 copies of a single locus within an otherwise clonal population. This tentatively supports the findings of Dalet et al., showing that in a single sample there exists a variety of genetic configurations of a single locus. This study demonstrates the potential of long-read sequencing to not only confirm long-predicted genomic structural variations in *B. pertussis*, but also to play a key role in the discovery and further investigation of a variety of entirely unpredicted genomic phenomena. The next steps in understanding these new genomic phenomena in *B. pertussis* should aim to elucidate the existence and extent of any phenotypic effects stemming from large duplicated regions, as well as any contribution they make to whooping cough virulence.

6 What Is the Future for *B. pertussis* Genomic Research?

Changes to the allelic profile of circulating *B. pertussis* strains have been recorded for many decades, in the pre-genomics era and beyond. The wide-spread availability of whole genome sequencing since the early 2000s, though, has enabled the screening of larger numbers of strains isolated over the last hundred years, thus allowing us to understand, longitudinally, the extent to which *B. pertussis* has been evolving on the gene-level. As seen above, there is evidence that many of these gene-level changes, in terms of both allelic profile and antigenic deficiency, have been influenced by the introduction of first the WCV and then the ACV. Since the 2010s, long-read sequencing has allowed us to investigate *B. pertussis* on a new level, that of the whole genome. The existence of a wide variety of inter-strain genomic rearrangements is now well-established, and more recent evidence has begun to show that other types of genome-level differences, such as large tandem duplications, also exist. However, the contribution of these observed gene- and genome-level variations to

observed phenotypes is yet to be fully understood. In addition to informing our understanding of the continued evolution of the *B. pertussis* genome, a more thorough understanding of *B. pertussis* genomics could also contribute significantly to the future of whooping cough prevention strategies (Cherry 2019).

Long-read sequencing will have a major part to play in any future investigation of *B. pertussis* genomic variations. Until recently, high-throughput long-read sequencing was restricted to larger laboratories which could afford a PacBio sequencing system, thus all the early long-read studies described here took place at national health laboratories: the CDC in the US, and the Centre for Infectious Diseases Control in the Netherlands. Oxford Nanopore Technologies (ONT) sequencing may provide a more accessible alternative for smaller laboratories and, indeed, two studies from late 2018 have shown the feasibility of assembling multiple closed *B. pertussis* genomes using ONT sequencing in hybrid with Illumina sequencing (Ring et al. 2018; Bouchez et al. 2018). In addition, ultra-long ONT sequencing has recently revealed yet further *B. pertussis* structural complexity, in the form of highly mixed populations (Abrahams et al. In preparation). Thus, it is likely that future studies of *B. pertussis* genome structure will utilise both PacBio and ONT sequencing. For investigating the most complicated structural features, such as very long tandem duplications, there will likely be a preference for ONT sequencing, as there is theoretically no upper limit to the length of sequencing read which could be produced by nanopore sequencing (Schmid et al. 2018). It is also likely that any studies utilising long reads to investigate *B. pertussis* will use them in hybrid with a more accurate short-read technology, although improvements to both long-read technologies mean that highly accurate long-read-only assemblies are on the horizon, which could enable both base-level and genome-level interrogations using a single technology (Wenger et al. 2019; Oxford Nanopore Technologies 2018).

Alongside sequencing, there will still remain a place for other genomics tools, such as PFGE and optical mapping. The most recent survey of *B. pertussis* genomic diversity in the US, by

Weigand et al. (2019), demonstrates the potential for such a wholistic approach. Using a combination of short-read sequencing, long-read sequencing, multilocus variable-number tandem-repeat analysis (MLVA), PFGE and optical mapping, Weigand et al. were able in a single study to characterise the gene- and genome-level profiles, including allelic-profile, antigen deficiency, genome arrangement and the existence of several large tandem duplications, in 170 strains isolated between 2000 and 2013. Such detailed analyses will likely provide a springboard for future studies, for both the continued surveillance of *B. pertussis* evolution, and the investigation of any correlation between genotypic, genomic and phenotypic differences.

In summary, genomics has shown, and continues to show, that *B. pertussis* is not necessarily the entirely monomorphic species it is traditionally believed to be. Although the allelic profile of *B. pertussis* changed in response to the introduction of the WCV, and more rapidly since the switch to the ACV, diversity at the gene-level remains very limited when compared to many other bacteria. However, the wide availability of WGS, and particularly the more recent long-read sequencing technologies, have revealed dynamic and substantial genome-level variations, both between and within strains. Future work may utilise a wholistic approach, focussing on the further elucidation and phenotypic characterisation of both gene- and genome-level phenomena together, ultimately informing our understanding of how diversity is generated in species with limited base-level inter-strain variation and, perhaps, the role this has played in the resurgence of whooping cough.

References

Abrahams JS, Ring N, Quick J, Loman N, Peng Y, Weigand MA, Tondella ML, Williams MM, Bagby S, Gorringe AR, Preston A (In preparation) Duplications drive diversity in the monomorphic pathogen *Bordetella pertussis* on an underestimated scale

Adler M, Anjum M, Berg OG, Andersson DI, Sandegren L (2014) High fitness costs and instability of gene duplications reduce rates of evolution of new genes by duplication-divergence mechanisms. Mol Biol Evol 31(6):1526–1535. https://doi.org/10.1093/molbev/msu111

Advani A, Donnelly D, Hallander H (2004) Reference system for characterization of Bordetella pertussis pulsed-field gel electrophoresis profiles. J Clin Microbiol 42(7):2890–2897. https://doi.org/10.1128/jcm.42.7.2890-2897.2004

Advani A, Hallander HO, Dalby T, Krogfelt KA, Guiso N, Njamkepo E, von Konnig CH, Riffelmann M, Mooi FR, Sandven P, Lutynska A, Fry NK, Mertsola J, He Q (2013) Pulsed-field gel electrophoresis analysis of Bordetella pertussis isolates circulating in Europe from 1998 to 2009. J Clin Microbiol 51(2):422–428. https://doi.org/10.1128/jcm.02036-12

Amman F, D'Halluin A, Antoine R, Huot L, Bibova I, Keidel K, Slupek S, Bouquet P, Coutte L, Caboche S, Locht C, Vecerek B, Hot D (2018) Primary transcriptome analysis reveals importance of IS elements for the shaping of the transcriptional landscape of Bordetella pertussis. RNA Biol 15 (7):967–975. https://doi.org/10.1080/15476286.2018.1462655

Au KF, Underwood JG, Lee L, Wong WH (2012) Improving PacBio long read accuracy by Short read alignment. PLoS One 7(10):e46679. https://doi.org/10.1371/journal.pone.0046679

Ausiello CM, Cassone A (2014) Acellular pertussis vaccines and pertussis resurgence: revise or replace? MBio 5(3):e01339–e01314. https://doi.org/10.1128/mBio.01339-14

Barkoff AM, Mertsola J, Pierard D, Dalby T, Hoegh SV, Guillot S, Stefanelli P, van Gent M, Berbers G, Vestrheim D, Greve-Isdahl M, Wehlin L, Ljungman M, Fry NK, Markey K, He Q (2019) Pertactin-deficient Bordetella pertussis isolates: evidence of increased circulation in Europe, 1998 to 2015. Euro surveillance: bulletin Europeen sur les maladies transmissibles Eur Commun Dis Bull 24(7). https://doi.org/10.2807/1560-7917.es.2019.24.7.1700832

Bart MJ, van Gent M, van der Heide HG, Boekhorst J, Hermans P, Parkhill J, Mooi FR (2010) Comparative genomics of prevaccination and modern Bordetella pertussis strains. BMC Genomics 11:627. https://doi.org/10.1186/1471-2164-11-627

Bart MJ, Harris SR, Advani A, Arakawa Y, Bottero D, Bouchez V, Cassiday PK, Chiang CS, Dalby T, Fry NK, Gaillard ME, van Gent M, Guiso N, Hallander HO, Harvill ET, He Q, van der Heide HG, Heuvelman K, Hozbor DF, Kamachi K, Karataev GI, Lan R, Lutynska A, Maharjan RP, Mertsola J, Miyamura T, Octavia S, Preston A, Quail MA, Sintchenko V, Stefanelli P, Tondella ML, Tsang RS, Xu Y, Yao SM, Zhang S, Parkhill J, Mooi FR (2014a) Global population structure and evolution of Bordetella pertussis and their relationship with vaccination. MBio 5(2):e01074. https://doi.org/10.1128/mBio.01074-14

Bart MJ, Zeddeman A, van der Heide HGJ, Heuvelman K, van Gent M, Mooi FR (2014b) Complete genome sequences of Bordetella pertussis isolates B1917 and B1920, representing two predominant global lineages. Genome Announc 2(6). https://doi.org/10.1128/genomeA.01301-14

Bart MJ, van der Heide HGJ, Zeddeman A, Heuvelman K, van Gent M, Mooi FR (2015) Complete genome sequences of 11 Bordetella pertussis strains representing the pandemic ptxP3 lineage. Genome Announc 3(6). https://doi.org/10.1128/genomeA.01394-15

Bentley DR, Balasubramanian S, Swerdlow HP, Smith GP, Milton J, Brown CG, Hall KP, Evers DJ, Barnes CL, Bignell HR, Boutell JM, Bryant J, Carter RJ, Keira Cheetham R, Cox AJ, Ellis DJ, Flatbush MR, Gormley NA, Humphray SJ, Irving LJ, Karbelashvili MS, Kirk SM, Li H, Liu X, Maisinger KS, Murray LJ, Obradovic B, Ost T, Parkinson ML, Pratt MR, Rasolonjatovo IM, Reed MT, Rigatti R, Rodighiero C, Ross MT, Sabot A, Sankar SV, Scally A, Schroth GP, Smith ME, Smith VP, Spiridou A, Torrance PE, Tzonev SS, Vermaas EH, Walter K, Wu X, Zhang L, Alam MD, Anastasi C, Aniebo IC, Bailey DM, Bancarz IR, Banerjee S, Barbour SG, Baybayan PA, Benoit VA, Benson KF, Bevis C, Black PJ, Boodhun A, Brennan JS, Bridgham JA, Brown RC, Brown AA, Buermann DH, Bundu AA, Burrows JC, Carter NP, Castillo N, Chiara ECM, Chang S, Neil Cooley R, Crake NR, Dada OO, Diakoumakos KD, Dominguez-Fernandez B, Earnshaw DJ, Egbujor UC, Elmore DW, Etchin SS, Ewan MR, Fedurco M, Fraser LJ, Fuentes Fajardo KV, Scott Furey W, George D, Gietzen KJ, Goddard CP, Golda GS, Granieri PA, Green DE, Gustafson DL, Hansen NF, Harnish K, Haudenschild CD, Heyer NI, Hims MM, Ho JT, Horgan AM, Hoschler K, Hurwitz S, Ivanov DV, Johnson MQ, James T, Huw Jones TA, Kang GD, Kerelska TH, Kersey AD, Khrebtukova I, Kindwall AP, Kingsbury Z, Kokko-Gonzales PI, Kumar A, Laurent MA, Lawley CT, Lee SE, Lee X, Liao AK, Loch JA, Lok M, Luo S, Mammen RM, Martin JW, McCauley PG, McNitt P, Mehta P, Moon KW, Mullens JW, Newington T, Ning Z, Ling Ng B, Novo SM, O'Neill MJ, Osborne MA, Osnowski A, Ostadan O, Paraschos LL, Pickering L, Pike AC, Chris Pinkard D, Pliskin DP, Podhasky J, Quijano VJ, Raczy C, Rae VH, Rawlings SR, Chiva Rodriguez A, Roe PM, Rogers J, Rogert Bacigalupo MC, Romanov N, Romieu A, Roth RK, Rourke NJ, Ruediger ST, Rusman E, Sanches-Kuiper RM, Schenker MR, Seoane JM, Shaw RJ, Shiver MK, Short SW, Sizto NL, Sluis JP, Smith MA, Ernest Sohna Sohna J, Spence EJ, Stevens K, Sutton N, Szajkowski L, Tregidgo CL, Turcatti G, Vandevondele S, Verhovsky Y, Virk SM, Wakelin S, Walcott GC, Wang J, Worsley GJ, Yan J, Yau L, Zuerlein M, Mullikin JC, Hurles ME, McCooke NJ,

West JS, Oaks FL, Lundberg PL, Klenerman D, Durbin R, Smith AJ (2008) Accurate whole human genome sequencing using reversible terminator chemistry. Nature 456(7218):53–59. https://doi.org/10.1038/nature07517

Bianchi F, Careri M, Mustat L, Malcevschi A, Musci M (2005) Bioremediation of toluene and naphthalene: development and validation of a GC-FID method for their monitoring. Ann Chim 95(7–8):515–524

Bisgard KM, Christie CD, Reising SF, Sanden GN, Cassiday PK, Gomersall C, Wattigney WA, Roberts NE, Strebel PM (2001) Molecular epidemiology of Bordetella pertussis by pulsed-field gel electrophoresis profile: Cincinnati, 1989-1996. J Infect Dis 183 (9):1360–1367. https://doi.org/10.1086/319858

Blumberg DA, Lewis K, Mink CM, Christenson PD, Chatfield P, Cherry JD (1993) Severe reactions associated with diphtheria-tetanus-pertussis vaccine: detailed study of children with seizures, hypotonic-hyporesponsive episodes, high fevers, and persistent crying. Pediatrics 91(6):1158–1165

Bouchez V, Caro V, Levillain E, Guigon G, Guiso N (2008) Genomic content of bordetella pertussis clinical isolates circulating in areas of intensive children vaccination. PLoS One 3(6):e2437. https://doi.org/10.1371/journal.pone.0002437

Bouchez V, Brun D, Cantinelli T, Dore G, Njamkepo E, Guiso N (2009) First report and detailed characterization of B. pertussis isolates not expressing pertussis toxin or pertactin. Vaccine 27(43):6034–6041. https://doi.org/10.1016/j.vaccine.2009.07.074

Bouchez V, Baines SL, Guillot S, Brisse S (2018) Complete genome sequences of Bordetella pertussis clinical isolate FR5810 and reference strain Tohama from combined Oxford Nanopore and Illumina sequencing. Microbiol Resour Announc 7(19). https://doi.org/10.1128/mra.01207-18

Bowden KE, Williams MM, Cassiday PK, Milton A, Pawloski L, Harrison M, Martin SW, Meyer S, Qin X, DeBolt C, Tasslimi A, Syed N, Sorrell R, Tran M, Hiatt B, Tondella ML (2014) Molecular epidemiology of the pertussis epidemic in Washington state in 2012. J Clin Microbiol 52(10):3549–3557. https://doi.org/10.1128/jcm.01189-14

Bowden KE, Weigand MR, Peng Y, Cassiday PK, Sammons S, Knipe K, Rowe LA, Loparev V, Sheth M, Weening K, Tondella ML, Williams MM, Blokesch M (2016) Genome structural diversity among 31 Bordetella pertussis isolates from two recent U.S. whooping cough statewide epidemics. Mol Biol Physiol 1(3). https://doi.org/10.1128/mSphere.00036-16

Brinig MM, Cummings CA, Sanden GN, Stefanelli P, Lawrence A, Relman DA (2006) Significant gene order and expression differences in Bordetella pertussis despite limited gene content variation. J Bacteriol 188 (7):2375–2382. https://doi.org/10.1128/JB.188.7.2375-2382.2006

Carbonetti NH (2010) Pertussis toxin and adenylate cyclase toxin: key virulence factors of Bordetella pertussis and cell biology tools. Future Microbiol 5:455–469. https://doi.org/10.2217/fmb.09.133

Caro V, Hot D, Guigon G, Hubans C, Arrive M, Soubigou G, Renauld-Mongenie G, Antoine R, Locht C, Lemoine Y, Guiso N (2006) Temporal analysis of French Bordetella pertussis isolates by comparative whole-genome hybridization. Microbes Infect 8 (8):2228–2235. https://doi.org/10.1016/j.micinf.2006.04.014

Caro V, Bouchez V, Guiso N (2008) Is the sequenced Bordetella pertussis strain Tohama I representative of the species? J Clin Microbiol 46(6):2125–2128. https://doi.org/10.1128/jcm.02484-07

Centers for Disease Control (2019) Reported pertussis incidence by age group and year | CDC. https://www.cdc.gov/pertussis/surv-reporting/cases-by-age-group-and-year.html#modalIdString_CDCTable_0. Accessed 20 Feb 2019

Cherry JD (1990) 'Pertussis vaccine encephalopathy': it is time to recognize it as the myth that it is. JAMA 263 (12):1679–1680

Cherry JD (1992) Pertussis: the trials and tribulations of old and new pertussis vaccines. Vaccine 10 (14):1033–1038

Cherry JD (1996) Historical review of pertussis and the classical vaccine. J Infect Dis 174(Suppl 3):S259–S263

Cherry JD (2019) The 112-year odyssey of pertussis and pertussis vaccines—mistakes made and implications for the future. J Pediatr Infect Dis Soc. https://doi.org/10.1093/jpids/piz005

Cherry JD, Holtzman AE, Shields WD, Buch D, Nielsen C, Jacobsen V, Christenson PD, Zachau-Christiansen B (1993) Pertussis immunization and characteristics related to first seizures in infants and children. J Pediatr 122(6):900–903

Clark TA (2014) Changing pertussis epidemiology: everything old is new again. J Infect Dis 209(7):978–981. https://doi.org/10.1093/infdis/jiu001

Dalet K, Weber C, Guillemot L, Njamkepo E, Guiso N (2004) Characterization of adenylate cyclase-hemolysin gene duplication in a Bordetella pertussis isolate. Infect Immun 72(8):4874–4877. https://doi.org/10.1128/iai.72.8.4874-4877.2004

Darch SE, McNally A, Harrison F, Corander J, Barr HL, Paszkiewicz K, Holden S, Fogarty A, Crusz SA, Diggle SP (2014) Recombination is a key driver of genomic and phenotypic diversity in a Pseudomonas aeruginosa population during cystic fibrosis infection. Scientific Reports, Published online: 12 January 2015. https://doi.org/10.1038/srep07649

de Gouw D, Hermans PW, Bootsma HJ, Zomer A, Heuvelman K, Diavatopoulos DA, Mooi FR (2014) Differentially expressed genes in Bordetella pertussis strains belonging to a lineage which recently spread globally. PLoS One 9(1):e84523. https://doi.org/10.1371/journal.pone.0084523

de Melker HE, Conyn-van Spaendonck MA, Rumke HC, van Wijngaarden JK, Mooi FR, Schellekens JF (1997) Pertussis in the Netherlands: an outbreak despite high levels of immunization with whole-cell vaccine. Emerg Infect Dis 3(2):175–178. https://doi.org/10.3201/eid0302.970211

De Serres G, Boulianne N, Douville Fradet M, Duval B (1995) Pertussis in Quebec: ongoing epidemic since the late 1980s. Can Commun Dis Rep Releve des maladies transmissibles au Canada 21(5):45–48

Diavatopoulos DA, Cummings CA, Schouls LM, Brinig MM, Relman DA, Mooi FR (2005) Bordetella pertussis, the causative agent of whooping cough, evolved from a distinct, human-associated lineage of B. bronchiseptica. PLoS Pathog 1(4):e45. https://doi.org/10.1371/journal.ppat.0010045

Dienstbier A, Pouchnik D, Wildung M, Amman F, Hofacker IL, Parkhill J, Holubova J, Sebo P, Vecerek B (2018) Comparative genomics of Czech vaccine strains of Bordetella pertussis. Pathog Dis 76(7). https://doi.org/10.1093/femspd/fty071

Ding W, Max Planck Institute for Developmental Biology T, Germany, Baumdicker F, Mathematisches Institut A-LUoF, 79104 Freiburg, Germany, Neher RA, Max Planck Institute for Developmental Biology T, Germany, Biozentrum and SIB Swiss Institute of Bioinformatics UoB, 4056 Basel, Switzerland (2017) panX: pan-genome analysis and exploration. Nucleic Acids Res 46(1). https://doi.org/10.1093/nar/gkx977

Du Q, Wang X, Liu Y, Luan Y, Zhang J, Li Y, Liu X, Ma C, Li H, Wang Z, He Q (2016) Direct molecular typing of Bordetella pertussis from nasopharyngeal specimens in China in 2012-2013. Eur J Clin Microbiol Infect Dis 35(7):1211–1214. https://doi.org/10.1007/s10096-016-2655-3

Fleischmann RD, Alland D, Eisen JA, Carpenter L, White O, Peterson J, DeBoy R, Dodson R, Gwinn M, Haft D, Hickey E, Kolonay JF, Nelson WC, Umayam LA, Ermolaeva M, Salzberg SL, Delcher A, Utterback T, Weidman J, Khouri H, Gill J, Mikula A, Bishai W, Jacobs WR Jr, Venter JC, Fraser CM (2002) Whole-genome comparison of Mycobacterium tuberculosis clinical and laboratory strains. J Bacteriol 184 (19):5479–5490

Galit SR, Otsuka N, Furuse Y, Almonia DJ, Sombrero LT, Capeding RZ, Lupisan SP, Saito M, Oshitani H, Hiramatsu Y, Shibayama K, Kamachi K (2015) Molecular epidemiology of Bordetella pertussis in the Philippines in 2012-2014. Int J Infect Dis 35:24–26. https://doi.org/10.1016/j.ijid.2015.04.001

Gutacker MM, Smoot JC, Migliaccio CA, Ricklefs SM, Hua S, Cousins DV, Graviss EA, Shashkina E, Kreiswirth BN, Musser JM (2002) Genome-wide analysis of synonymous single nucleotide polymorphisms in Mycobacterium tuberculosis complex organisms: resolution of genetic relationships among closely related microbial strains. Genetics 162(4):1533–1543

Hamidou Soumana I, Linz B, Harvill ET (2017) Environmental origin of the genus Bordetella. Front Microbiol 8. https://doi.org/10.3389/fmicb.2017.00028

Hegerle N, Paris AS, Brun D, Dore G, Njamkepo E, Guillot S, Guiso N (2012) Evolution of French Bordetella pertussis and Bordetella parapertussis isolates: increase of Bordetellae not expressing pertactin. Clin Microbiol Infect 18(9):E340–E346. https://doi.org/10.1111/j.1469-0691.2012.03925.x

Heikkinen E, Kallonen T, Saarinen L, Sara R, King AJ, Mooi FR, Soini JT, Mertsola J, He Q (2007) Comparative genomics of Bordetella pertussis reveals progressive gene loss in Finnish strains. PLoS One 2(9):e904. https://doi.org/10.1371/journal.pone.0000904

Henderson MW, Inatsuka CS, Sheets AJ, Williams CL, Benaron DJ, Donato GM, Gray MC, Hewlett EL, Cotter PA (2012) Contribution of Bordetella filamentous hemagglutinin and adenylate cyclase toxin to suppression and evasion of interleukin-17-mediated inflammation. Infect Immun 80(6):2061–2075. https://doi.org/10.1128/iai.00148-12

Hiramatsu Y, Miyaji Y, Otsuka N, Arakawa Y, Shibayama K, Kamachi K (2017) Significant decrease in Pertactin-deficient Bordetella pertussis isolates, Japan. Emerg Infect Dis 23(4):699–701. https://doi.org/10.3201/eid2304.161575

King AJ, van Gorkom T, Pennings JL, van der Heide HG, He Q, Diavatopoulos D, Heuvelman K, van Gent M, van Leeuwen K, Mooi FR (2008) Comparative genomic profiling of Dutch clinical Bordetella pertussis isolates using DNA microarrays: identification of genes absent from epidemic strains. BMC Genomics 9:311. https://doi.org/10.1186/1471-2164-9-311

King AJ, van Gorkom T, van der Heide HG, Advani A, van der Lee S (2010) Changes in the genomic content of circulating Bordetella pertussis strains isolated from the Netherlands, Sweden, Japan and Australia: adaptive evolution or drift? BMC Genomics 11:64. https://doi.org/10.1186/1471-2164-11-64

King AJ, van der Lee S, Mohangoo A, van Gent M, van der Ark A, van de Waterbeemd B (2013) Genome-wide gene expression analysis of Bordetella pertussis isolates associated with a resurgence in pertussis: elucidation of factors involved in the increased fitness of epidemic strains. PLoS One 8(6):e66150. https://doi.org/10.1371/journal.pone.0066150

Koren S, Schatz MC, Walenz BP, Martin J, Howard J, Ganapathy G, Wang Z, Rasko DA, McCombie WR, Jarvis ED, Phillippy AM (2012) Hybrid error correction and de novo assembly of single-molecule sequencing reads. Nat Biotechnol 30(7):693–700. https://doi.org/10.1038/nbt.2280

Kuo CH, Ochman H (2010) The extinction dynamics of bacterial pseudogenes. PLoS Genet 6(8):e1001050. https://doi.org/10.1371/journal.pgen.1001050

Lam C, Octavia S, Bahrame Z, Sintchenko V, Gilbert GL, Lan R (2012) Selection and emergence of pertussis toxin promoter ptxP3 allele in the evolution of Bordetella pertussis. Infect Genet Evol 12(2):492–495. https://doi.org/10.1016/j.meegid.2012.01.001

Lam C, Octavia S, Ricafort L, Sintchenko V, Gilbert GL, Wood N, McIntyre P, Marshall H, Guiso N, Keil AD, Lawrence A, Robson J, Hogg G, Lan R (2014) Rapid increase in Pertactin-deficient Bordetella pertussis isolates, Australia. Emerg Infect Dis 20(4):626–633. https://doi.org/10.3201/eid2004.131478

Linz B, Ivanov YV, Preston A, Brinkac L, Parkhill J, Kim M, Harris SR, Goodfield LL, Fry NK, Gorringe AR, Nicholson TL, Register KB, Losada L, Harvill ET (2016) Acquisition and loss of virulence-associated factors during genome evolution and speciation in three clades of Bordetella species. BMC Genomics 17 (1):767. https://doi.org/10.1186/s12864-016-3112-5

Lynch M (2002) Genomics. Gene duplication and evolution. Science 297(5583):945–947. https://doi.org/10.1126/science.1075472

Magadum S, Banerjee U, Murugan P, Gangapur D, Ravikesavan R (2013) Gene duplication as a major force in evolution. J Genet 92(1):155–161

Mastrantonio P, Spigaglia P, van Oirschot H, van der Heide HG, Heuvelman K, Stefanelli P, Mooi FR (1999) Antigenic variants in Bordetella pertussis strains isolated from vaccinated and unvaccinated children. Microbiology 145(Pt 8):2069–2075. https://doi.org/10.1099/13500872-145-8-2069

Medini D, Donati C, Tettelin H, Masignani V, Rappuoli R (2005) The microbial pan-genome. Curr Opin Genet Dev 15(6):589–594. https://doi.org/10.1016/j.gde.2005.09.006

Miyaji Y, Otsuka N, Toyoizumi-Ajisaka H, Shibayama K, Kamachi K (2013) Genetic analysis of Bordetella pertussis isolates from the 2008–2010 pertussis epidemic in Japan. PLoS One 8(10):e77165. https://doi.org/10.1371/journal.pone.0077165

Mooi FR (2010) Bordetella pertussis and vaccination: the persistence of a genetically monomorphic pathogen. Infect Genet Evol 10(1):36–49. https://doi.org/10.1016/j.meegid.2009.10.007

Mooi FR, van Oirschot H, Heuvelman K, van der Heide HG, Gaastra W, Willems RJ (1998) Polymorphism in the Bordetella pertussis virulence factors P.69/pertactin and pertussis toxin in The Netherlands: temporal trends and evidence for vaccine-driven evolution. Infect Immun 66(2):670–675

Mooi FR, van Loo IH, King AJ (2001) Adaptation of Bordetella pertussis to vaccination: a cause for its reemergence? Emerg Infect Dis 7(3 Suppl):526–528. https://doi.org/10.3201/eid0707.010708

Mooi FR, van Loo IH, van Gent M, He Q, Bart MJ, Heuvelman KJ, de Greeff SC, Diavatopoulos D, Teunis P, Nagelkerke N, Mertsola J (2009) Bordetella pertussis strains with increased toxin production associated with pertussis resurgence. Emerg Infect Dis 15(8):1206–1213. https://doi.org/10.3201/eid1508.081511

Moore DL, Le Saux N, Scheifele D, Halperin SA (2004) Lack of evidence of encephalopathy related to pertussis vaccine: active surveillance by IMPACT, Canada, 1993–2002. Pediatr Infect Dis J 23(6):568–571

Ohta T (1989) Role of gene duplication in evolution. Genome 31(1):304–310

Oxford Nanopore Technologies (2018) Clive G Brown: nanopore community meeting 2018 talk. https://nanoporetech.com/about-us/news/clive-g-brown-nanopore-community-meeting-2018-talk. Accessed 15 Jan 2019

Park J, Zhang Y, Buboltz AM, Zhang X, Schuster SC, Ahuja U, Liu M, Miller JF, Sebaihia M, Bentley SD, Parkhill J, Harvill ET (2012) Comparative genomics of the classical Bordetella subspecies: the evolution and exchange of virulence-associated diversity amongst closely related pathogens. BMC Genomics 13:545. https://doi.org/10.1186/1471-2164-13-545

Parkhill J, Sebaihia M, Preston A, Murphy LD, Thomson N, Harris DE, Holden MTG, Churcher CM, Bentley SD, Mungall KL, Cerdeño-Tárraga AM, Temple L, James K, Harris B, Quail MA, Achtman M, Atkin R, Baker S, Basham D, Bason N, Cherevach I, Chillingworth T, Collins M, Cronin A, Davis P, Doggett J, Feltwell T, Goble A, Hamlin N, Hauser H, Holroyd S, Jagels K, Leather S, Moule S, Norberczak H, O'Neil S, Ormond D, Price C, Rabbinowitsch E, Rutter S, Sanders M, Saunders D, Seeger K, Sharp S, Simmonds M, Skelton J, Squares R, Squares S, Stevens K, Unwin L, Whitehead S, Barrell BG, Maskell DJ (2003) Comparative analysis of the genome sequences of Bordetella pertussis, Bordetella parapertussis and Bordetella bronchiseptica. Nat Genet 35(1):32–40. https://doi.org/10.1038/ng1227

Preston A, Parkhill J, Maskell DJ (2004) The bordetellae: lessons from genomics. Nat Rev Microbiol 2 (5):379–390. https://doi.org/10.1038/nrmicro886

Queenan AM, Cassiday PK, Evangelista A (2013) Pertactin-negative variants of Bordetella pertussis in the United States. N Engl J Med 368(6):583–584. https://doi.org/10.1056/NEJMc1209369

Ring N, Abrahams JS, Jain M, Olsen H, Preston A, Bagby S (2018) Resolving the complex Bordetella pertussis genome using barcoded nanopore sequencing. Microb Genom 4(11). https://doi.org/10.1099/mgen.0.000234

Rouli L, Merhej V, Fournier PE, Raoult D (2015) The bacterial pangenome as a new tool for analysing pathogenic bacteria. New Microbes New Infect 7:72–85. https://doi.org/10.1016/j.nmni.2015.06.005

Schmid M, Frei D, Patrignani A, Schlapbach R, Frey JE, Remus-Emsermann MNP, Ahrens CH (2018) Pushing the limits of de novo genome assembly for complex prokaryotic genomes harboring very long, near identical repeats. bioRxiv. https://doi.org/10.1101/300186

Sealey KL, Harris SR, Fry NK, Hurst LD, Gorringe AR, Parkhill J, Preston A (2015) Genomic analysis of isolates from the United Kingdom 2012 pertussis outbreak reveals that vaccine antigen genes are unusually fast evolving. J Infect Dis 212(2):294–301. https://doi.org/10.1093/infdis/jiu665

Serra DO, Conover MS, Arnal L, Sloan GP, Rodriguez ME, Yantorno OM, Deora R (2011) FHA-mediated cell-substrate and cell-cell adhesions are critical for Bordetella pertussis biofilm formation on abiotic surfaces and in the mouse nose and the trachea. PLoS One 6(12):e28811. https://doi.org/10.1371/journal.pone.0028811

Siguier P, Laboratoire de Microbiologie et Génétique Moléculaires UMdR, Centre National de Recherche Scientifique, Toulouse Cedex, France, Gourbeyre E, Laboratoire de Microbiologie et Génétique Moléculaires UMdR, Centre National de Recherche Scientifique, Toulouse Cedex, France, Chandler M, Laboratoire de Microbiologie et Génétique Moléculaires UMdR, Centre National de Recherche Scientifique, Toulouse Cedex, France (2014) Bacterial insertion sequences: their genomic impact and diversity. FEMS Microbiol Rev 38(5):865–891. https://doi.org/10.1111/1574-6976.12067

Sousa C, de Lorenzo V, Cebolla A (1997) Modulation of gene expression through chromosomal positioning in Escherichia coli. Microbiology 143(Pt 6):2071–2078. https://doi.org/10.1099/00221287-143-6-2071

Strebel P, Nordin J, Edwards K, Hunt J, Besser J, Burns S, Amundson G, Baughman A, Wattigney W (2001) Population-based incidence of pertussis among adolescents and adults, Minnesota, 1995-1996. J Infect Dis 183(9):1353–1359. https://doi.org/10.1086/319853

Sun H, Narihiro T, Ma X, Zhang XX, Ren H, Ye L (2019) Diverse aromatic-degrading bacteria present in a highly enriched autotrophic nitrifying sludge. Sci Total Environ 666:245–251. https://doi.org/10.1016/j.scitotenv.2019.02.172

Tazato N, Handa Y, Nishijima M, Kigawa R, Sano C, Sugiyama J (2015) Novel environmental species isolated from the plaster wall surface of mural paintings in the Takamatsuzuka tumulus: Bordetella muralis sp. nov., Bordetella tumulicola sp. nov. and Bordetella tumbae sp. nov. Int J Syst Evol Microbiol 65(12):4830–4838. https://doi.org/10.1099/ijsem.0.000655

van der Zee A, Vernooij S, Peeters M, van Embden J, Mooi FR (1996) Dynamics of the population structure of Bordetella pertussis as measured by IS1002-associated RFLP: comparison of pre- and post-vaccination strains and global distribution. Microbiology 142(Pt 12):3479–3485. https://doi.org/10.1099/13500872-142-12-3479

van Gent M, Heuvelman CJ, van der Heide HG, Hallander HO, Advani A, Guiso N, Wirsing von Konig CH, Vestrheim DF, Dalby T, Fry NK, Pierard D, Detemmerman L, Zavadilova J, Fabianova K, Logan C, Habington A, Byrne M, Lutynska A, Mosiej E, Pelaz C, Grondahl-Yli-Hannuksela K, Barkoff AM, Mertsola J, Economopoulou A, He Q, Mooi FR (2015) Analysis of Bordetella pertussis clinical isolates circulating in European countries during the period 1998-2012. Eur J Clin Microbiol Infect Dis 34(4):821–830. https://doi.org/10.1007/s10096-014-2297-2

Wang F, Grundmann S, Schmid M, Dorfler U, Roherer S, Charles Munch J, Hartmann A, Jiang X, Schroll R (2007) Isolation and characterization of 1,2,4-trichlorobenzene mineralizing Bordetella sp. and its bioremediation potential in soil. Chemosphere 67 (5):896–902. https://doi.org/10.1016/j.chemosphere.2006.11.019

Weigand MR, Peng Y, Loparev V, Johnson T, Juieng P, Gairola S, Kumar R, Shaligram U, Gowrishankar R, Moura H, Rees J, Schieltz DM, Williamson Y, Woolfitt A, Barr J, Tondella ML, Williams MM (2016) Complete genome sequences of four Bordetella pertussis vaccine reference strains from serum Institute of India. Genome Announc 4(6). https://doi.org/10.1128/genomeA.01404-16

Weigand MR, Peng Y, Loparev V, Batra D, Bowden KE, Burroughs M, Cassiday PK, Davis JK, Johnson T, Juieng P, Knipe K, Mathis MH, Pruitt AM, Rowe L, Sheth M, Tondella ML, Williams MM (2017) The history of Bordetella pertussis genome evolution includes structural rearrangement. J Bacteriol 199(8). https://doi.org/10.1128/jb.00806-16

Weigand MR, Pawloski LC, Peng Y, Ju H, Burroughs M, Cassiday PK, Davis JK, DuVall M, Johnson T, Juieng P, Knipe K, Loparev VN, Mathis MH, Rowe LA, Sheth M, Williams MM, Tondella ML (2018) Screening and genomic characterization of filamentous hemagglutinin-deficient Bordetella pertussis. Infect Immun 86(4). https://doi.org/10.1128/iai.00869-17

Weigand MR, Williams MM, Peng Y, Kania D, Pawloski LC, Tondella ML (2019) Genomic survey of Bordetella pertussis diversity, United States, 2000-2013. Emerg Infect Dis 25(4):780–783. https://doi.org/10.3201/eid2504.180812

Weiner B, Gomez J, Victor TC, Warren RM, Sloutsky A, Plikaytis BB, Posey JE, van Helden PD, Gey van Pittius NC, Koehrsen M, Sisk P, Stolte C, White J, Gagneux S, Birren B, Hung D, Murray M, Galagan J (2012) Independent large scale duplications in multiple M. tuberculosis lineages overlapping the same genomic region. PLoS One 7(2). https://doi.org/10.1371/journal.pone.0026038

Wenger AM, Peluso P, Rowell WJ, Chang P-C, Hall RJ, Concepcion GT, Ebler J, Fungtammasan A, Kolesnikov A, Olson ND, Toepfer A, Chin C-S, Alonge M, Mahmoud M, Qian Y, Phillippy AM, Schatz MC, Myers G, DePristo MA, Ruan J, Marschall T, Sedlazeck FJ, Zook JM, Li H, Koren S, Carroll A, Rank DR, Hunkapiller MW (2019) Highly-accurate long-read sequencing improves variant detection and assembly of a human genome. bioRxiv, 519025

WHO (2018) Global and regional immunization profile. WHO. https://www.who.int/immunization/monitoring_surveillance/data/gs_gloprofile.pdf?ua=1. Accessed 9 Jan 2019

Williams MM, Sen K, Weigand MR, Skoff TH, Cunningham VA, Halse TA, Tondella ML (2016) Bordetella pertussis strain lacking Pertactin and pertussis toxin. Emerg Infect Dis 22(2):319–322. https://doi.org/10.3201/eid2202.151332

Xu Y, Liu B, Grondahl-Yli-Hannuksila K, Tan Y, Feng L, Kallonen T, Wang L, Peng D, He Q, Zhang S (2015) Whole-genome sequencing reveals the effect of vaccination on the evolution of Bordetella pertussis. Sci Rep 5:12888. https://doi.org/10.1038/srep12888

Zhang W, Qi W, Albert TJ, Motiwala AS, Alland D, Hyytia-Trees EK, Ribot EM, Fields PI, Whittam TS, Swaminathan B (2006) Probing genomic diversity and evolution of Escherichia coli O157 by single nucleotide polymorphisms. Genome Res 16(6):757–767. https://doi.org/10.1101/gr.4759706

Zomer A, Otsuka N, Hiramatsu Y, Kamachi K, Nishimura N, Ozaki T, Poolman J, Geurtsen J (2018) Bordetella pertussis population dynamics and phylogeny in Japan after adoption of acellular pertussis vaccines. Microb Genom. https://doi.org/10.1099/mgen.0.000180

Adv Exp Med Biol - Advances in Microbiology, Infectious Diseases and Public Health (2019) 1183: 19–33
https://doi.org/10.1007/5584_2019_402
© Springer Nature Switzerland AG 2019
Published online: 25 July 2019

Molecular Epidemiology of *Bordetella pertussis*

Alex-Mikael Barkoff and Qiushui He

Abstract

Although vaccination has been effective, *Bordetella pertussis* is increasingly causing epidemics, especially in industrialized countries using acellular vaccines (aPs). One factor behind the increased circulation is the molecular changes on the pathogen level. After pertussis vaccinations were introduced, changes in the fimbrial (Fim) serotype of the circulating strains was observed. When bacterial typing methods improved, further changes between the vaccine and circulating strains, especially among the common virulence genes including pertussis toxin (PT) and pertactin (PRN) were noticed. Moreover, development of genome based techniques including pulsed-field gel electrophoresis (PFGE), multiple-locus variable number tandem repeat analysis (MLVA) and whole-genome sequencing (WGS) have offered a better resolution to monitor *B. pertussis* strains. After the introduction of aP vaccines, *B. pertussis* strains that are deficient to vaccine antigens, especially PRN, have appeared widely. On the other hand, antimicrobial resistance to first line drugs (macrolides) against *B. pertussis* is still low in many countries and therefore no globally evaluated antimicrobial susceptibility test values have been recommended. In this review, we focus on the molecular changes in the bacteria, which have or may have affected the past and current epidemiology of pertussis.

Keywords

Bordetella pertussis · Genetic variation · Pertactin · Vaccination

A.-M. Barkoff
Institute of Biomedicine, Department of Microbiology, Virology and Immunology, University of Turku, Turku, Finland
e-mail: ambark@utu.fi

Q. He (✉)
Institute of Biomedicine, Department of Microbiology, Virology and Immunology, University of Turku, Turku, Finland

Department of Medical Microbiology, Capital Medical University, Beijing, China
e-mail: qiushui.he@utu.fi

1   Introduction

Whooping cough or pertussis is mainly caused by Gram-negative bacterium *Bordetella pertussis,* which is strictly a human pathogen. Vaccination against the disease started with whole-cell vaccines (wP) earliest in the 1940s, but mostly in the 1950–60s. Currently, wP vaccines have been mostly replaced by acellular vaccines (aP), containing pertussis toxin (PT), pertactin (PRN), filamentous hemagglutinin (FHA) and Fimbriae 2 or 3 (Fim2/3) in industrialized countries with the change taking place during the late 1990s and in the early twenty-first century (Cherry 1996; Watanabe and Nagai 2005; Kallonen and He 2009; Campbell et al. 2012; Barkoff et al. 2015). In Europe, the only exception is Poland where the wP vaccine is still used for primary vaccinations, although aP is commercially available and used by approximately 60% of the Polish population in 2013 (Polak et al. 2018). However, many developing countries are further using primary wP vaccinations due to cost-effectiveness (Mahmood et al. 2013). Despite extensive immunisations, pertussis is still one of the world's worst controlled vaccine preventable disease.

The disease has known to be cyclic with epidemic periods every 3–5 years (Mooi et al. 2000). However, a large cycle of nationwide pertussis epidemics started in 2008 from Australia, and within the next few years, epidemics were noted in the Netherlands, the UK and the US (Campbell et al. 2012; Winter et al. 2012; van der Maas et al. 2013; Sealey et al. 2014; Winter et al. 2014). In addition, seroprevalence studies have shown that the disease is clearly under reported, being highly circulating in many countries, and not only being restricted to certain continents or areas (Barkoff et al. 2015). Resurgence of this vaccine preventable disease has raised questions on the contributing factors causing the increase. Speculated key reasons include a decrease in vaccine efficacy, waning immunity, better surveillance and reporting, newer laboratory tests and genetic changes within the *B. pertussis* (Cherry 2013). The latter of these include many factors, and is affected by vaccination and natural boosting (Mooi et al. 2013).

Several typing methods have been developed to study bacterial changes. Serotyping is the oldest of these techniques going back to the 1960s. The first notifications of a serotype change after wP vaccinations was noted in England, and serotype analysis was recommended to be used to recognize emerging epidemiological patterns in the US (Eldering et al. 1969). When bacterial typing methods further improved, the genotyping of *prn*, PT subunit S1 (*ptxA*) and PT promoter (*ptxP*) genes showed even more changes between the vaccine and circulating strains (Mooi et al. 1998; Mastrantonio et al. 1999; Mooi et al. 1999). Furthermore, development of genome based techniques including pulsed-field gel electrophoresis (PFGE), multiple-locus variable number tandem repeat analysis (MLVA) and whole-genome

sequencing (WGS) has offered a more dynamic separation and monitoring of *B. pertussis* strains (Parkhill et al. 2003; Advani et al. 2004; Schouls et al. 2004). Changes in the production of the main vaccine antigens has been noticed after the introduction of aP vaccines. Especially, *B. pertussis* strains not expressing vaccine antigen PRN have appeared globally (Bouchez et al. 2009; Stefanelli et al. 2009; Barkoff et al. 2012; Otsuka et al. 2012). Many mechanisms causing the non-expression have been found by gene sequencing and specific PCR, targeting the most variant areas with insertions and deletions in the *prn* gene (Lam et al. 2014; Pawloski et al. 2014; Zeddeman et al. 2014). Antimicrobial resistance of *B. pertussis* against the first line drugs (macrolides), used for prophylaxis and treatment of the infected patients, has been low in many countries. Therefore, no standardized method for antimicrobial susceptibility testing (AST) of *B. pertussis* is available from EUCAST (Lonnqvist et al. 2018).

This review chapter will focus on the genetic changes of the *B. pertussis* during the vaccination era. The aim is to offer a view of the methodology, to show the current key factors in the evolution of the bacteria, to provide some future insights for surveillance and to show the impact of molecular changes of the bacteria on vaccine effectiveness.

2 Genotyping and Serotyping

Genotyping of the main virulence genes (*ptxA*, *prn*, *fim2*, *fim3*, *ptxP*, *ptxC*) of *B. pertussis* has become less discriminative. Commonly, genotyping is performed by PCR and sequencing of the genes of interest (van Loo et al. 2002; van Gent et al. 2015). Many alleles have been found among the virulence genes: 13 alleles for *ptxA*, 18 for *prn*, 14 for *ptxP*, two for *fim2* and six to *fim3*, to our knowledge (van Loo et al. 2002; Miyaji et al. 2013; Bart et al. 2014; Simmonds et al. 2014; van Gent et al. 2015; Barkoff et al. 2018; Polak et al. 2018). Moreover, the *ptxP3* allele is believed to confer enhanced virulence (Mooi et al. 2009). A study done among five

European countries with different vaccines used, demonstrated how vaccination drives *B. pertussis* isolates to express genotypes other than that of the vaccine antigens (*ptxA2/4*, *prn1*) (van Amersfoorth et al. 2005). Novel European studies have shown that currently circulating *B. pertussis* populations carry an increasing frequency (>90%) of the prevalent *ptxA1*, *prn2* and *ptxP3* genotypes (van Gent et al. 2015; Barkoff et al. 2018; Petridou et al. 2018). However, changes among these genotypes are noted in Austria, where types other than *prn2* were found, and in Poland where the wP vaccine is used (Wagner et al. 2015; Polak et al. 2018). In Poland, the *ptxP1* and *prn1* alleles have been dominant for decades, but within the latest isolates (2010–2016) a change in the prevalence was noticed. Now, more than 90% of the strains carry the *prn2* and *ptxP3* alleles. In addition, the frequency of *fim3–1* genotype (approx. 70%) is also in line with other European countries, excluding France where both, *fim3–1* and *fim3–2* alleles, are equally presented (van Gent et al. 2015; Barkoff et al. 2018; Polak et al. 2018). Similar findings to Europe can be found in other continents. In the US, during the wP era, strains carried *prn1*, *ptxP1*, *ptxA1/2 and fim3–1* genotypes (Schmidtke et al. 2012). However, a shift from allele *prn1* to *prn2/3* occurred in the 1970s when wP was still in use (Cassiday et al. 2000). Genotypes of *ptxP3*, *ptxA1* and *fim3–1/2* started to increase during the aP era and are now dominant in the US (Schmidtke et al. 2012; Bowden et al. 2014). In Australia, studies also indicate how diversity has decreased and *ptxA1*, *prn2* and *ptxP3* genotypes are now prevalent (Octavia et al. 2012; Lam et al. 2014). In Japan, the frequency of prevalent *ptxA1*, *prn2* and *ptxP3* alleles is increasing, but *ptxA2*, *prn1* and *ptxP1* alleles were still seen among 40% of the isolates in 2011–2012. Moreover, two separate groups have formed among the isolates (Miyaji et al. 2013; Zomer et al. 2018). Similar strain profiles and frequencies to Japan were noticed in Cambodia (Moriuchi et al. 2017). Like Japan and Cambodia, China also carries isolates with distinct profiles compared to other countries. Two studies showed that circulating isolates harbored *prn1*, *ptxP1* and *fim3–1* genotypes among isolates collected in 1953–2005 and 2012–2013 (Zhang et al. 2010; Du et al. 2016). However, in a recent publication, two groups of *B. pertussis* isolates were noticed in the Shanghai area, 1) the prevalent profile in "western" countries with *prn2* and *ptxP3* (41%) and 2) Chinese/Japanese profile with *prn1* and *ptxP1* (59%) (Fu et al. 2018). It should be kept in mind that aP primary vaccinations replaced wP vaccinations in China in 2013, and it remains to be seen whether the "western" profile increases in this country. An additional study from the Philippines observed strains harboring *ptxP1*, *ptxA1*, *prn1* and *fim3–1* alleles (Galit et al. 2015). Interestingly in Peru and Iran, many strains collected in 2012 carried the "western" profile of *ptxA1*, *prn2* and *ptxP3* alleles (Bailon et al. 2016; Sadeghpour Heravi et al. 2018). It seems that *B. pertussis* populations, globally, are moving towards *ptxA1*, *prn2*, *ptxP3* and *fim3–1* genotypes and so far both wP and aP vaccinations have influenced this trend.

Serotyping discriminates *B. pertussis* isolates into three sub-groups: Fim2, Fim3 and Fim2,3. Two methods are used for serotyping (1) a side agglutination test with specific antisera against Fim2 or 3, and (2) a specific monoclonal based ELISA method (Eldering et al. 1969; Heikkinen et al. 2008). The serotype of the strains included in the vaccines (wP & aP) has shifted the serotypes of circulating strains throughout the years (Elomaa et al. 2005; Gorringe and Vaughan 2014). A study by Hallander et al. combines data from wP, aP and the non-vaccine era to demonstrate this phenomenon. Study results showed that Fim3 was circulating during the wP era (Fim2 in the vaccine), Fim2 during the non-vaccine era and after the change to aP vaccines, Fim3 became prevalent again (Hallander et al. 2005). Altogether, during the aP vaccine era, the Fim3 serotype has dominated in Europe (van Gent et al. 2015). However, in a recent study, isolates harboring Fim2 were found in almost half of the strains collected (Barkoff et al. 2018). In Poland, where the wP vaccine has been regularly used, the Fim2 serotype has been dominant during the 2000s (van Gent et al. 2015). This is in contrast to the study by Hallander (Hallander et al. 2005), and is most likely due to the different vaccine strains containing Fim3 serotype. In Japan, similar dominance of Fim3 is seen like in Europe (Miyaji et al. 2013). It is also known that outbreaks modify the serotype of circulating strains, as natural immunity in population increases (Gorringe and Vaughan 2014). This, however, does not self-explain why Fim3 is dominating in countries where aP vaccines are used. As Fim3 seems to be more expressed on the bacterial surface than Fim2, it may give an advantage against a highly immunized population (Heikkinen et al. 2008). It is also known that the expression of fimbriae depends on a run of C residues in the promoter regions of *fim2* and *fim3* genes, which can make genes capable of phase variation (Willems et al. 1990; Chen et al. 2010). Studies have also shown that patients infected by Fim2 strains developed antibodies to Fim3, suggesting a difference in fimbrial expression *in vivo* and *in vitro* (Heikkinen et al. 2008; Alexander et al. 2012).

3 PFGE and MLVA Used for Genomic Analyses

PFGE has been used for the typing of *B. pertussis* since the 1990s. As a method it is laborious, but has a better discriminative power than genotyping of the virulence genes (Advani et al. 2004; Cassiday et al. 2016). In addition, the recommended method and two reference systems for nomenclature of PFGE profiles (BpSR-profiles and clusters) have been described and are compared in Fig. 1 (Mooi et al. 2000; Advani et al. 2004). Although recommendations are made, they are used by a minority of laboratories, which makes the global comparison of results difficult. However, published studies are descriptive, and the change between intervals of time and profiles can be seen. In Europe, the main PFGE profiles have been BpSR3, BpSR5, BpSR10, BpSR11 and BpSR12. European wide studies have shown how BpSR11 (29%), BpSR10 (10%) and other profiles (36%) dominated during 1998–2006, whereas, in 2002–2012, the BpSR3, BpSR5 and BpSR10 profiles began to increase

(Advani et al. 2013; van Gent et al. 2015). In a recent study (2012–2015), a clear shift within the PFGE-profiles was noticed among the European *B. pertussis* populations. BpSR3 (29.4%) and BpSR10 (27.2%) became prevalent, whereas the previously dominant BpSR11 and the number and frequency of other profiles decreased. Furthermore, country specific differences were noticed, e.g. BpSR3 was not found among any of the Swedish isolates (Barkoff et al. 2018). A study in the US by Cassiday et al. summarizes PFGE findings during 2000–2012. The results from this study suggest a similar change among the US and European *B. pertussis* populations. They noticed that CDC013 (BpSR11) was prevalent (41%) in 2000–2009, but decreased during 2010–2012 (9%), whereas CDC002 (BpSR3) increased from 4% to 25%. Although, the number of other profiles increased, it mainly consisted of three profiles (CDC217, CDC237, CDC253). The major difference was the profile CDC010 (BpSR10), which declined significantly between these time periods in the US (Cassiday et al. 2016). Nevertheless, it was noticed in a European study that some profiles were dominant in some countries and some in others (Barkoff et al. 2018). In addition, the time-frame of this comparison between these two studies is not exactly the same (Fig. 2). Therefore, a direct comparison in PFGE profiles between Europe and the US would help with the global surveillance of pertussis. Moreover in China, during the period of 2012–2013, 16 strains were tested by PFGE, and the prevalent profiles detected were BpSR23 and BpFINR9. These profiles were prevalent in the 1990s and are rarely seen in Europe today (van Gent et al. 2015; Wang and He 2015).

So far, seven PFGE clusters have been defined, and there are three subgroups in cluster IV (IVα, β and Υ) (Caro et al. 2006; Shuel et al. 2013; Cassiday et al. 2016; Barkoff et al. 2018). For cluster analysis, *B. pertussis* isolates in Europe, in the US and also in Canada mainly belong to group IV (Shuel et al. 2013, Cassiday et al. 2016, Barkoff et al. 2018). Studies from wP vaccine using countries like Poland, Turkey and China show different results when compared to

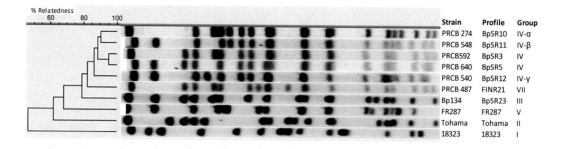

Fig. 1 PFGE-profiles and clusters obtained from reference strains and Finnish *B. pertussis* strain collection *For PFGE profiles, the unweighted pair group with arithmetic clustering (UPGMA) dendrogram type with the Dice similarity coefficient, 1% optimization, and 1% tolerance were used, whereas for clusters 1.5% optimization and 2% band tolerance were applied

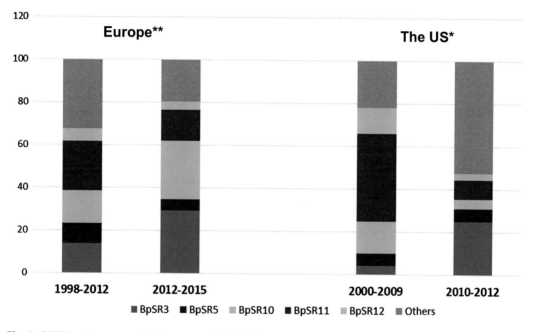

Fig. 2 PFGE-profiles reported in Europe and in the US during two different eras*** *In the US in 2010–2012, other profiles included three major profiles of CDC217 (4%), CDC237 (24%) and CDC253 (6%). These profiles accounted <1% of other profiles in 2000–2009. CDC253 is closely related to BpSR10

**In Europe, during 2012–2015, no BpSR3 was seen in Sweden ***Picture composed of data obtained from van Gent et al. (2015), Cassiday et al. (2016) and Barkoff et al. (2018)

Europe and the US. A large study made in Turkey showed that most of the strains during 2001–2009 belonged to PFGE cluster II (Nar Otgun et al. 2011). In Poland, strains mainly belonged to cluster III during 2000–2007, but have totally shifted to cluster IV during 2010–2016. This may be due to the high use of commercial aP vaccines (Polak et al. 2018).

Multiple-locus variable-number tandem repeat analysis (MLVA) was first described by Schouls et al. (Schouls et al. 2004). It is based on the variable-number of tandem repeats (VNTR) within the *B. pertussis* genome and it is performed by PCR and sequencing. This method is proven to have better discriminative power than common genotyping. However, studies in Europe have shown how the number and frequency of MLVA types (MT), diminish among the aP vaccinated populations. Results from Schouls et al. indicated that MT27 and MT29 were the dominant profiles in the Netherlands during 1992–2000. Nevertheless, they found many MTs, but the number of strains among each MTs was rather low, excluding MT27 and MT29. (Schouls et al. 2004). Later, Swedish and British studies were concluded with similar results (Advani et al. 2009; Litt et al. 2009). After this, the frequency of MT27 increased and reached more than 75% in Europe during 2002–2012. At the same time the second most common type MT29 almost vanished. Interestingly, MT78 was dominant in Finland during this time frame (van Gent et al. 2015). Furthermore, in a recent European study, the frequency of MT27 among the 265 strains collected in nine countries during 2012–2015 was over 80% and the Simpson's diversity index [calculated based on the formula $D = 1 - \Sigma n(n - 1)/N(N - 1)$, where n indicates the number of individual profiles and N is the number of all profiles] was only 0.34, indicating a poor discriminative power of this method (Barkoff et al. 2018). In addition, MT27 now became dominant in Finland (>85%). Interestingly, in Denmark, the number of profiles other than MT27 was more than 50% and MT28 represented almost 30% of the strains (Barkoff et al. 2018). A change to MT27 has also been seen in Japan, where MT186 was dominant in 2002–2007, but after 2008, MT27 became prevalent. In the US, similar dominance (>75%) of MT27 is seen (Bowden et al. 2014; Bowden et al. 2016). However, different MTs are further seen in countries with wP vaccination. In China, no clear dominant MT type was found, MT types such as MT55, MT91, MT136 and MT152 were presented in similar frequencies (Wang and He 2015; Xu et al. 2015b). A similarly low frequency of MT27 was seen in Poland (1998–2006), where MT70 and MT29 have been prevalent (van Gent et al. 2015). In short, it is clear that long-term aP vaccination has had an effect on the high frequency of MT27. However, the situation in Denmark is interesting as this country uses aP containing only PT, and profiles other than MT27

are dominant. More discriminative approaches such as WGS should be used to reveal whether this is due to the small number of collected study strains or a beneficial genomic arrangement for *B. pertussis* to escape immunity acquired from the monocomponent vaccine in this country. Moreover, there are variations among MTs between countries using aP and wP vaccines.

4   Whole Genome Sequencing

Parkhill et al. sequenced and annotated the whole genome of the *B. pertussis* reference strain Tohama I, and showed e.g. the amount of *IS481* elements, which mediates deletions and interrupts in the *B. pertussis* genes (Parkhill et al. 2003). Since then, many genomes of *B. pertussis* strains have been sequenced. Bart et al. performed comparative genomic analyses to a worldwide collection of 343 strains isolated between 1920 and 2010 (Bart et al. 2014). They identified two different lineages from which the currently circulating isolates have evolved. They showed how vaccination has driven the strains genetically more close to each other's in time, and found a similar average density of SNPs (0.0013 SNPs/bp) to that reported previously by Parkhill et al.. However, the highest density was found in virulence-associated and transport/binding related genes (0.0016 and 0.0015 SNPs/bp, respectively) (Bart et al. 2014). The study clearly indicates that global transmission of new strains is very rapid and that the worldwide population of *B. pertussis* is evolving in response to vaccine introduction. Subsequently, Sealey et al. reported that especially genes that encode vaccine antigens such as PT, FHA and PRN are unusually fast to evolve when compared to other genes, with significantly higher SNP density (0.00173 SNPs/bp, the UK and 0.00291 SNPs/bp, global) in the aP era. They also proposed that only a few SNPs are related to the increased virulence of *ptxP3* strains (Sealey et al. 2014). Xu et al. sequenced and analyzed the complete genomes of 40 strains isolated from China and Finland during 1956–2008, as well as 11 previously sequenced strains from the Netherlands, where different vaccination strategies have been used over the past 50 years (Xu et al. 2015a). They found that evolution of the *B. pertussis* populations is closely associated with the country vaccination coverage. Furthermore, studies from the US have shown that genomic rearrangements have occurred in the form of large inversions, which were frequently flanked by *IS481* insertions (Weigand et al. 2017b), which are in line with the previous studies. Furthermore, this group has detected novel mutations among vaccine antigen genes leading to deficiency of PRN and FHA (Weigand et al. 2017a, 2018). In the latter, they found that mutations in the *fhaB* gene and in other genes affected FHA production. Double mutations in different genes within the same strain were also found to cause FHA deficiency. By using WGS, Bouchez et al. have defined a core genome multilocus sequence typing scheme (cgMLST) comprising of 2038 loci (Bouchez et al. 2018). They demonstrated its congruence with whole genome single nucleotide polymorphism variation. As speculated by authors the cgMLST method has the potential to study transmission of particular *B. pertussis* lineages such as those that evolve towards a lack of expression of vaccine antigens. It is clear that WGS has offered a more wide perspective to study the evolution and transmission of *B. pertussis*. Since culture is performed less and less in clinical microbiology laboratories, WGS of DNAs isolated directly from clinical samples should be considered and compared in the future. Further standardization and harmonization of these epidemiological typing methods including WGS are needed for global surveillance of pertussis.

5 Vaccine Antigen Deficient *B. pertussis* Isolates

The expression of vaccine antigens by *B. pertussis* has greatly changed during the last 10 years, and is under constant transition. So far, the main pressure has been on PRN, which is a member of the autotransporter family and plays a role in the adhesion to cell membrane of the host (Mattoo and Cherry 2005). To detect vaccine antigen production of *B. pertussis* isolates, two methods have been used, (1) a classical western blot to target the protein of interest and (2) a more rapid ELISA method based on specific monoclonal antibodies (Bouchez et al. 2009; Barkoff et al. 2014). The appearance of PRN deficient *B. pertussis* isolates began mainly after the introduction of aP vaccines. The first reports describing these strains were from France and Italy in 2009 (Bouchez et al. 2009; Stefanelli et al. 2009). Soon after, PRN-negative isolates were noted in countries with the change from wP to aP vaccination, including Australia, Finland, Japan, Norway, Sweden, the Netherlands and the US (Barkoff et al. 2012; Otsuka et al. 2012; Lam et al. 2013; Queenan et al. 2013; Zeddeman et al. 2014). A novel study from Europe showed an increased frequency of PRN deficient isolates, when four *B. pertussis* strain panels (N = 661) were collected from several European countries in 1998–2015. During this period, the frequency of PRN deficient isolates increased from 1.0% to 24.9% (Barkoff et al. 2019). Furthermore, a similar increase has been detected in Australia, where the frequency of PRN deficient strains was 5% in 2008 (N = 39), from where it rapidly increased to 78% in 2012 (N = 36), and also in the US, where isolates from two outbreaks (California and Washington state) indicated an increase in the frequency from 12.1% in 2010 (N = 33) to 73.1% in 2012 (N = 216) (Pawloski et al. 2014). Moreover, a study combining isolates from eight states in the US (N = 753) showed that 85% of the strains were PRN deficient (Martin et al. 2015). Recently, PRN deficient isolates have also been found in Slovenia, where the frequency increased from 0% (2006–2011) to 89.5% (2014–2017), and from Canada (Tsang et al. 2014; Kastrin et al. 2019). The appearance of these strains seems to depend on the primary aP vaccination. However, a recent study by Polak et al. showed that 4/188 isolates collected in 1959–2016 were PRN deficient in Poland, a country where the wP vaccine is used. These strains were found after 2010 (4/26, 15.4%) and according to their estimation, commercial aP vaccines have replaced the intake of wP vaccine by 60% in 2013 (Polak et al. 2018). In addition, in

Denmark where the PT-vaccine is used, PRN deficient isolates (N = 4) were found among isolates collected in 2012–2015 (Barkoff et al. 2019). These findings indicate that only "a minor" pressure from aP vaccinations are needed for the transformation, and that these strains also translocate with travel as Denmark is a passage between Central Europe and Scandinavia. Indeed, "high" prevalence of PRN deficient strains have been found in Scandinavian countries Sweden and Norway (Barkoff et al. 2019). The pressure from aP booster vaccines is difficult to determine, as booster schedules vary largely among different countries (Barkoff et al. 2015). Some research have also studied "historical" isolates, and it seems that only two PRN deficient isolates have been found before the introduction of aP vaccines, one in Finland and one in the US (Hegerle et al. 2012; Pawloski et al. 2014; Zomer et al. 2018). Moreover, these two strains may have evolved during large outbreaks, when multiple strain profiles are usually detected for a short period of time. In contrast, in Japan, the frequency of PRN deficient isolates has even decreased, after a change to an aP vaccine not containing PRN (Hiramatsu et al. 2017).

Many mechanisms are causing PRN deficiency. Common methods to identify these mechanisms are either specific PCRs flanking the area of mutations, or sequencing of the whole *prn* gene (Lam et al. 2014; Pawloski et al. 2014; Zeddeman et al. 2014). The earliest findings included insertion of the *IS481* element in the region II of the *prn* gene and a deletion of 84 bp (26^109) in the signal sequence (ΔSS) (Bouchez et al. 2009; Barkoff et al. 2012; Otsuka et al. 2012). When PRN deficient isolates become more prevalent, a number of causing mechanisms identified increased, and the *IS481* element was found either in a forward or a reverse formation in two different positions (1613^1614; 2735^2736) of the *prn* gene. Other mechanisms include a deletion of 49 bp in the ΔSS (32^80), deletions in the promoter and signal region (-283^-40; -292^1340; -1513^145)a 22 kb inversion of the promotor region (-20892^-75), an insertion of *IS1002* element in position 1613^1614, point mutations

leading to stop codons (223STOP:C > T, 1273STOP:C > T, 2077STOP:G > T), disruption of the promoter (-162:G > A), addition of single nucleotide (1185:G), deletion of single nucleotide (631^632:T), a deletion of 1612 bp in the promotor region (-2090^-478) and the deletion of the whole *prn* gene. However, for some strains no mutations within the *prn* gene or promoter, causing the PRN deficiency, has been found (Hegerle et al. 2012; Bowden et al. 2014; Lam et al. 2014; Tsang et al. 2014; Zeddeman et al. 2014; Weigand et al. 2017a; Barkoff et al. 2019). The *IS481* insertion is still the main factor for PRN deficiency, but other mechanisms are increasing except in Japan, where only two mechanisms (*IS481* and ΔSS) are prevalent (Hiramatsu et al. 2017; Zomer et al. 2018). All mechanisms included in this study are presented in Table 1.

Molecular typing of PRN deficient strains shows different patterns. Strains isolated in Australia, Canada, Europe and the US are mainly carrying the prevalent genotypes of *ptxA1*, *prn2*, *ptxP3* and MT27 profile (Bouchez et al. 2009; Stefanelli et al. 2009; Bowden et al. 2014; Lam et al. 2014; Tsang et al. 2014; Barkoff et al. 2018), whereas in Japan PRN deficient isolates are still carrying moderate numbers of *ptxP1*, *prn1* and *ptxA2* alleles related to the MT186 profile, although the MT27 profile with *ptxP3*, *prn2* and *ptxA2* alleles (as in Australia, Europe and the US) is dominating the Japanese PRN deficient isolates (Otsuka et al. 2012; Zomer et al. 2018). For PFGE, variations among the strains can be seen with individual profiles, although the number of studies describing PRN deficient isolates with PFGE profiles is very limited. In the study by Pawloski et al., PFGE-profiles of PRN deficient (mainly CDC013 and CDC217) and PRN producing (mainly CDC002 and CDC237) strains differed among the Washington state epidemy (Pawloski et al. 2014). Tsang et al. found five PFGE profiles (076 and 021 dominant) among the 12 PRN deficient isolates in Canada (Tsang et al. 2014). In Europe, a study by Hegerle et al. showed that PRN deficient isolates belonged to cluster IVα or IVβ, whereas PRN producing isolates mostly

Table 1 Mechanisms causing PRN deficiency, position in the *prn* gene and appearance by country

Mechanism	Position in *prn* gene	Countries
IS481 insertion[a,b,c,d,e,f,g,h]	1,613^1,614 & 2,735^2,736	Australia, Canada, France, Belgium, Denmark, Italy, the NL, Norway, Sweden, the UK, the US
84 bp deletion[g,i,j]	26^109	Finland, Japan, the US
49 bp deletion[h]	32^80	Finland
22 kb inversion in the prn promoter[a,g,h]	−20,892^-75	Belgium, France, Italy, the NL, Sweden, the UK, the US
IS1002 insertion[c]	1,613^1,614	Australia
STOP:C > T[a,e,h]	223 & 1,273	Canada, the NL, Norway, Sweden, the UK
STOP:G>T[a]	2077	Finland
Addition of G[c]	1,185	Australia
Deletion of T[a]	631^632	Denmark, Italy, Sweden
deletions in the promoter[j]	−2090^-478; −283^-40; −292^1,340; −1,513^145	the US
Deletion of prn gene[a,c,f]	Δprn	Australia, Denmark, France, Norway
Disruption G > A[e]	−162	Canada
Unknown[a,c,e]	N/A	Australia, Canada, the UK

[a]Barkoff et al. (2019); [b]Bouchez et al. (2009); [c]Lam et al. (2014); [d]Zomer et al. (2018); [e]Tsang et al. (2014); [f]Hegerle et al. (2012); [g]Bowden et al. (2014); [h]Zeddeman et al. (2014); [i]Barkoff et al. (2012); [j]Octavia et al. (2012); [k]Weigand et al. (2017a)

harbored the IVγ (Hegerle et al. 2012). Barkoff et al. showed that PRN deficient isolates carried different PFGE profiles with dominance on BpSR3, BpSR10 and BpSR11 profiles (Barkoff et al. 2018, 2019). To summarize these findings, it seems that PRN deficient isolates are evolving towards uniformity, which can be seen especially within the Japanese isolates as described above. The strains with a dominant MT27-*ptxA1*, *prn2*, *ptxP3* profile may have an increased fitness among the aP immunized population.

Many studies have shown that PRN-deficient isolates cause classical pertussis symptoms with cascades and prolonged cough (Bouchez et al. 2009; Barkoff et al. 2012; Martin et al. 2015; Williams et al. 2016). In addition, PRN deficient isolates may have a selective advantage in aP immunized population (Martin et al. 2015). However, the latter still needs further investigation as a recent study, in which strains were collected and compared from several European countries, showed that these strains were found equally among vaccinated and unvaccinated individuals (Barkoff et al. 2019). Furthermore, when the time of introduction of primary vaccination with PRN-containing aP was compared in these European countries, the results suggested that the longer the period since the introduction of primary aP vaccination, the higher the frequency of circulating PRN-deficient strains.

Other vaccine antigen protein deficient *B. pertussis* isolates has been low. The first PT deficient isolate was reported in France, and to date two of these isolates have been found in this country. The mechanism causing deficiency of these two strains differed, in one the whole *ptx* operon was deleted and for the second the mechanism was unknown. The molecular background of the strains was similar (*prn2*, Fim3) and both belonged to PFGE group IVγ (Hegerle et al. 2012). In addition, one PT negative isolate has been found in the US with similar characteristics and mechanisms to the French strain where the whole *ptx* operon was deleted and both harbored MT27. Interestingly the US strain was also PRN deficient (Williams et al. 2016). FHA deficient isolates have appeared more frequently than PT deficient, although their number is low. These strains have been found from France (N = 2), Sweden (N = 2) and the US (N = 5) (Hegerle et al. 2012; Bart et al. 2015; Weigand et al. 2018). Interestingly, four of these isolates were found during the pre-vaccine era in the US (N = 3) and France (N = 1). (Bouchez et al. 2015;

Weigand et al. 2018). Several mechanisms were found to stop the production of FHA among the nine isolates, these include *IS481* insertion and single nucleotide deletion/insertion among the FHA coding gene *fhaB*, and an *IS481* insertion within the *Bordetella* Bvg-intermediate-phase gene (*bipA*). For one isolate the mechanism was unknown. In addition, four of the FHA deficient strains were also PRN deficient (Bart et al. 2015; Bouchez et al. 2015; Weigand et al. 2018). Most of the contemporary isolates carried *ptxP3*, *ptxA1* and *prn2* alleles, whereas their PFGE-profiles varied. According to a French study, FHA deficient strains had an elevated expression of other virulence genes of *ptxA*, *cyaA*, and *prn* (Bouchez et al. 2015). For Fim2 or 3 deficiency, three isolates have been found so far, one in Canada, one in Japan and one in Norway (Miyaji et al. 2013; Shuel et al. 2013; Barkoff et al. 2019). In addition, one tracheal colonization factor A deficient isolate was recently found in Poland (Polak et al. 2018). Altogether, PT, FHA and Fim deficient isolates are sporadic and it remains to be seen whether there will be an increase among these isolates.

6 Genetic Change and Antimicrobial Resistance

Antimicrobial therapy is immediately needed, when pertussis is diagnosed, for bacterial clearance and to prevent further transmission of the disease to other individuals. It is also recommended for prophylaxis. Commonly, macrolides [erythromycin (ERY) and azithromycin (AZT)] are the first-line drugs to treat *B. pertussis* infection (Kilgore et al. 2016; Lonnqvist et al. 2018). So far, macrolide resistant isolates have mainly been found in China, but isolates have also been found sporadically in Europe, the Middle East and in North and South America (Torres et al. 2015; Centers for Disease Control and Prevention (CDC) 1994; Wilson et al. 2002; Guillot et al. 2012; Shahcheraghi et al. 2014; Wang et al. 2014; Wang et al. 2014). However, the frequency of these strains is increasing in China and the latest studies from

Beijing (2013–2014), Zhejiang province (2016) and Shanghai (2016–2017) show a high frequency of these isolates varying between 60–92% (Yang et al. 2015; Fu et al. 2018). It also seems that these strains are found from different areas in China, indicating that these strains are not only clonally expanded in one location.

Currently, the only mechanism to cause the macrolide resistance is a point mutation changing nucleotide A to G at the position 2047 (A2047G) in the domain V of the 23S rRNA gene of *B. pertussis*, as described previously (Bartkus et al. 2003; Guillot et al. 2012; Wang et al. 2015; Lonnqvist et al. 2018). However, *B. pertussis* carries 3 copies of the 23S rRNA gene and so far, only one study has shown that the mutation occurred in all copies (Guillot et al. 2012). It remains to be shown if other mutations or even molecular changes are involved in the resistance. Therefore, culture-based AST is required for the determination of the macrolide resistance. So far, the resistant *B. pertussis* isolates detected have been highly resistant (>256 µg/mL) to macrolides (both ERY and AZT) when measured by the minimum inhibition concentration (MIC) test, whereas sensitive strains show almost no tolerance against macrolides (both ERY and AZT <0.250 µg/mL) (Guillot et al. 2012; Stefanelli et al. 2017; Fu et al. 2018; Lonnqvist et al. 2018). The effect of A2047G mutation to macrolide resistance is shown in Fig. 3. The macrolide resistant *B. pertussis* isolate found in France carried *ptxP3*, *ptxA1* and *prn2* allele and expressed all vaccine antigens (PT, FHA, PRN and Fim3) (Guillot et al. 2012). However, a recent study by Fu et al. showed that isolates (N = 141) from Shanghai had mainly two different characteristics based on the resistance and genetic profiles of the bacteria: sensitive (41.1%) and resistant (58.9%) groups. Sensitive strains carried genotypes of *ptxP3/prn2/ptxC2*, like those prevalent in Europe and in the US, whereas almost all resistant strains had the profile *ptxP1/prn1/ptxC1* (81/83, 97.6%). Other differences between these two groups of strains were not significant. However, the resistant rate of 57.5% in Shanghai was much lower than those reported in Xi'an (87.5%) and Beijing

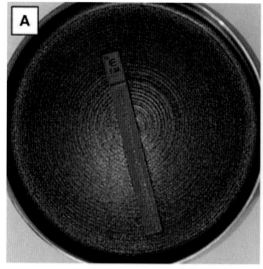

Fig. 3 Effect of A2047G mutation in the 23S rRNA of *Bordetella pertussis* to MIC

(**a**) Macrolide resistant *B. pertussis* isolate; (**b**) Macrolide sensitive *B. pertussis* isolate. Photographed by Barkoff and He, modified from Lonnqvist et al. (2018)

(91.9%) during 2013–2014 (Wang et al. 2014; Yang et al. 2015).

Other alternative antimicrobial agents to treat pertussis patients, if macrolides cannot be used, are trimethoprim-sulfamethoxazole (TMS) and quinolones (e.g. Nalidixic acid). For TMS, no studies have described resistance among the *B. pertussis* strains (Jakubu et al. 2017; Lonnqvist et al. 2018; Hua et al. 2019). However, one study from Japan has described *B. pertussis* strains resistant to Nalicidic acid. All six resistant strains carried a mutation A260G in the *gyrA* gene. Further investigation showed that these strains had different PFGE-profiles, although four were closely related (Ohtsuka et al. 2009). However, as there is no resistance for TMS, which is the second line of antibiotics to treat pertussis patients, it remains to be seen whether the quinolone resistance will be a matter of importance as they are rarely used.

7 Conclusions

In industrialized countries using aP vaccines, common genotypes, MLVA types and PFGE profiles are identified in *B. pertussis* populations.

With PFGE many strain profiles can be seen and it is evident that PFGE profiles vary among populations immunized by aP or wP vaccines. Furthermore, the increased use of aP vaccines in Poland has clearly harbored the Polish strains towards those circulating in industrialized countries where aP vaccines have been used for 10–20 years, and as there has been a change to aP vaccination in China after 2013, it remains to be seen how this will influence the *B. pertussis* population in this country. Moreover, the number of all other, than the prevalent PFGE profiles, is decreasing indicating that the shift results are caused by the use of aP vaccines. Vaccine antigen deficient strains, especially PRN-deficient ones, have emerged and increased especially in the aP vaccinated populations, and so far they have not been reported in countries where wP vaccines are used. The circulating strains in developing countries using wP vaccines harbor different geno-, MLVA- and PFGE-types than those found in countries using the aP vaccine. So far, macrolide resistance of *B. pertussis* is associated with a point mutation in 23S rRNA gene of *B. pertussis*, and is widely detected only in China. WGS has offered a more wide perspective to study the evolution and transmission of

B. pertussis. Since culture is performed less and less in clinical microbiology laboratories, WGS of DNAs isolated directly from clinical samples should be considered in the future. This review also underscores the importance of long-term surveillance of *B. pertussis* because of its impact on the effectiveness of vaccines.

References

Advani A, Donnelly D, Hallander H (2004) Reference system for characterization of Bordetella pertussis pulsed-field gel electrophoresis profiles. J Clin Microbiol 42(7):2890–2897

Advani A, Van der Heide HG, Hallander HO et al (2009) Analysis of Swedish Bordetella pertussis isolates with three typing methods: characterization of an epidemic lineage. J Microbiol Methods 78(3):297–301

Advani A, Hallander HO, Dalby T et al (2013) Pulsed-field gel electrophoresis analysis of Bordetella pertussis isolates circulating in Europe from 1998 to 2009. J Clin Microbiol 51(2):422–428

Alexander F, Matheson M, Fry NK et al (2012) Antibody responses to individual Bordetella pertussis fimbrial antigen Fim2 or Fim3 following immunization with the five-component acellular pertussis vaccine or to pertussis disease. Clin Vaccine Immunol 19 (11):1776–1783

Bailon H, Leon-Janampa N, Padilla C et al (2016) Increase in pertussis cases along with high prevalence of two emerging genotypes of Bordetella pertussis in Peru, 2012. BMC Infect Dis 16:422-016-1700-2

Barkoff AM, Mertsola J, Guillot S et al (2012) Appearance of Bordetella pertussis strains not expressing the vaccine antigen pertactin in Finland. Clin Vaccine Immunol 19(10):1703–1704

Barkoff AM, Guiso N, Guillot S et al (2014) A rapid ELISA-based method for screening Bordetella pertussis strain production of antigens included in current acellular pertussis vaccines. J Immunol Methods 408:142–148

Barkoff AM, Grondahl-Yli-Hannuksela K, He Q (2015) Seroprevalence studies of pertussis: what have we learned from different immunized populations. Pathog Dis 73(7):10

Barkoff AM, Mertsola J, Pierard D et al (2018) Surveillance of circulating Bordetella pertussis strains in Europe during 1998–2015. J Clin Microbiol 56:5

Barkoff AM, Mertsola J, Pierard D et al (2019) Pertactin-deficient Bordetella pertussis isolates: evidence of increased circulation in Europe, 1998 to 2015. Euro Surveill 24(7):10

Bart MJ, Harris SR, Advani A et al (2014) Global population structure and evolution of Bordetella pertussis and their relationship with vaccination. MBio 5(2): e01074–e01014

Bart MJ, van der Heide HG, Zeddeman A et al (2015) Complete Genome Sequences of 11 Bordetella pertussis Strains Representing the Pandemic ptxP3 Lineage. Genome Announc 3(6):10

Bartkus JM, Juni BA, Ehresmann K et al (2003) Identification of a mutation associated with erythromycin resistance in Bordetella pertussis: implications for surveillance of antimicrobial resistance. J Clin Microbiol 41(3):1167–1172

Bouchez V, Brun D, Cantinelli T et al (2009) First report and detailed characterization of B. pertussis isolates not expressing Pertussis Toxin or Pertactin. Vaccine 27 (43):6034–6041

Bouchez V, Hegerle N, Strati F et al (2015) New data on vaccine antigen deficient bordetella pertussis isolates. Vaccines (Basel) 3(3):751–770

Bouchez V, Baines SL, Guillot S et al (2018) Complete genome sequences of bordetella pertussis clinical isolate FR5810 and reference strain tohama from combined Oxford nanopore and illumina sequencing. Microbiol Resour Announc 7(19):10

Bowden KE, Williams MM, Cassiday PK et al (2014) Molecular epidemiology of the pertussis epidemic in Washington State in 2012. J Clin Microbiol 52 (10):3549–3557

Bowden KE, Weigand MR, Peng Y et al (2016) Genome structural diversity among 31 bordetella pertussis Isolates from two recent U.S. Whooping cough statewide epidemics. mSphere 1(3):10

Campbell P, McIntyre P, Quinn H et al (2012) Increased population prevalence of low pertussis toxin antibody levels in young children preceding a record pertussis epidemic in Australia. PLoS One 7(4):e35874

Caro V, Elomaa A, Brun D et al (2006) Bordetella pertussis, Finland and France. Emerg Infect Dis 12 (6):987–989

Cassiday P, Sanden G, Heuvelman K et al (2000) Polymorphism in Bordetella pertussis pertactin and pertussis toxin virulence factors in the United States, 1935–1999. J Infect Dis 182(5):1402–1408

Cassiday PK, Skoff TH, Jawahir S et al (2016) Changes in predominance of pulsed-field gel electrophoresis profiles of bordetella pertussis isolates, United States, 2000–2012. Emerg Infect Dis 22(3):442–448

Centers for Disease Control and Prevention (CDC) (1994) Erythromycin-resistant Bordetella pertussis--Yuma County, Arizona, May-October 1994. MMWR Morb Mortal Wkly Rep 43(44):807–810

Chen Q, Decker KB, Boucher PE et al (2010) Novel architectural features of Bordetella pertussis fimbrial subunit promoters and their activation by the global virulence regulator BvgA. Mol Microbiol 77 (5):1326–1340

Cherry JD (1996) Historical review of pertussis and the classical vaccine. J Infect Dis 174(3):S259–S263

Cherry JD (2013) Pertussis: challenges today and for the future. PLoS Pathog 9(7):e1003418

Du Q, Wang X, Liu Y et al (2016) Direct molecular typing of Bordetella pertussis from nasopharyngeal specimens in China in 2012–2013. Eur J Clin Microbiol Infect Dis 35(7):1211–1214

Eldering G, Holwerda J, Davis A et al (1969) Bordetella pertussis serotypes in the United States. Appl Microbiol 18(4):618–621

Elomaa A, Advani A, Donnelly D et al (2005) Strain variation among Bordetella pertussis isolates in finland, where the whole-cell pertussis vaccine has been used for 50 years. J Clin Microbiol 43 (8):3681–3687

Fu P, Wang C, Tian H et al (2018) Bordetella pertussis Infection in Infants and Young Children in Shanghai, China, 2016–2017: clinical features, genotype variations of antigenic genes and macrolides resistance. Pediatr Infect Dis J 38:370–376. https://doi.org/10.1097/INF.0000000000002160

Galit SR, Otsuka N, Furuse Y et al (2015) Molecular epidemiology of Bordetella pertussis in the Philippines in 2012–2014. Int J Infect Dis 35:24–26

Gorringe AR, Vaughan TE (2014) Bordetella pertussis fimbriae (Fim): relevance for vaccines. Expert Rev Vaccines 13(10):1205–1214

Guillot S, Descours G, Gillet Y et al (2012) Macrolide-resistant Bordetella pertussis infection in newborn girl, France. Emerg Infect Dis 18(6):966–968

Hallander HO, Advani A, Donnelly D et al (2005) Shifts of Bordetella pertussis variants in Sweden from 1970 to 2003, during three periods marked by different vaccination programs. J Clin Microbiol 43(6):2856–2865

Hegerle N, Paris AS, Brun D et al (2012) Evolution of French Bordetella pertussis and Bordetella parapertussis isolates: increase of Bordetellae not expressing pertactin. Clin Microbiol Infect 18(9): E340–E346

Heikkinen E, Xing DK, Olander RM et al (2008) Bordetella pertussis isolates in Finland: serotype and fimbrial expression. BMC Microbiol 8:162-2180-8-162

Hiramatsu Y, Miyaji Y, Otsuka N et al (2017) Significant decrease in pertactin-deficient bordetella pertussis Isolates, Japan. Emerg Infect Dis 23(4):699–701

Hua CZ, Wang HJ, Zhang Z et al (2019) In vitro activity and clinical efficacy of macrolides, cefoperazone-sulbactam and piperacillin/piperacillin-tazobactam against Bordetella pertussis and the clinical manifestations in pertussis patients due to these isolatesA single-center study in Zhejiang Province, China. J Glob Antimicrob Resist. https://doi.org/10.1016/j.jgar.2019.01.029

Jakubu V, Zavadilova J, Fabianova K et al (2017) Trends in the minimum inhibitory concentrations of erythromycin, clarithromycin, azithromycin, ciprofloxacin, and trimethoprim/sulfamethoxazole for strains of bordetella pertussis isolated in the Czech Republic in 1967–2015. Cent Eur J Public Health 25(4):282–286

Kallonen T, He Q (2009) Bordetella pertussis strain variation and evolution postvaccination. Expert Rev Vaccines 8(7):863–875

Kastrin T, Barkoff AM, Paragi M et al (2019) High prevalence of currently circulating Bordetella pertussis isolates not producing vaccine antigen pertactin in Slovenia. Clin Microbiol Infect 25(2):258–260

Kilgore PE, Salim AM, Zervos MJ et al (2016) Pertussis: microbiology, disease, treatment, and prevention. Clin Microbiol Rev 29(3):449–486

Lam C, Octavia S, Ricafort L et al (2013) Emergence of pertactin deficient Bordetella pertussis in Australia is due to independent events. In: 10th international symposium on bordetella, Dublin, Ireland, Poster(P11)

Lam C, Octavia S, Ricafort L et al (2014) Rapid Increase in Pertactin-deficient Bordetella pertussis Isolates, Australia. Emerg Infect Dis 20(4):626–633

Litt DJ, Neal SE, Fry NK (2009) Changes in genetic diversity of the Bordetella pertussis population in the United Kingdom between 1920 and 2006 reflect vaccination coverage and emergence of a single dominant clonal type. J Clin Microbiol 47(3):680–688

Lonnqvist E, Barkoff AM, Mertsola J et al (2018) Antimicrobial susceptibility testing of Finnish Bordetella pertussis isolates collected during 2006–2017. J Glob Antimicrob Resist 14:12–16

Mahmood K, Pelkowski S, Atherly D et al (2013) Hexavalent IPV-based combination vaccines for public-sector markets of low-resource countries. Hum Vaccin Immunother 9(9):1894–1902

Martin SW, Pawloski L, Williams M et al (2015) Pertactin-negative Bordetella pertussis strains: evidence for a possible selective advantage. Clin Infect Dis 60(2):223–227

Mastrantonio P, Spigaglia P, van Oirschot H et al (1999) Antigenic variants in Bordetella pertussis strains isolated from vaccinated and unvaccinated children. Microbiology 145(Pt 8):2069–2075

Mattoo S, Cherry JD (2005) Molecular pathogenesis, epidemiology, and clinical manifestations of respiratory infections due to Bordetella pertussis and other Bordetella subspecies. Clin Microbiol Rev 18 (2):326–382

Miyaji Y, Otsuka N, Toyoizumi-Ajisaka H et al (2013) Genetic Analysis of Bordetella pertussis Isolates from the 2008–2010 pertussis epidemic in Japan. PLoS One 8(10):e77165

Mooi FR, van Oirschot H, Heuvelman K et al (1998) Polymorphism in the Bordetella pertussis virulence factors P.69/pertactin and pertussis toxin in The Netherlands: temporal trends and evidence for vaccine-driven evolution. Infect Immun 66 (2):670–675

Mooi FR, He Q, van Oirschot H et al (1999) Variation in the Bordetella pertussis virulence factors pertussis toxin and pertactin in vaccine strains and clinical isolates in Finland. Infect Immun 67(6):3133–3134

Mooi FR, Hallander H, Wirsing von Konig CH et al (2000) Epidemiological typing of Bordetella pertussis isolates: recommendations for a standard methodology. Eur J Clin Microbiol Infect Dis 19(3):174–181

Mooi FR, van Loo IH, van Gent M et al (2009) Bordetella pertussis strains with increased toxin production associated with pertussis resurgence. Emerg Infect Dis 15(8):1206–1213

Mooi FR, van der Maas NA, De Melker HE (2013) Pertussis resurgence: waning immunity and pathogen adaptation – two sides of the same coin. Epidemiol Infect 142:685–694. https://doi.org/10.1017/S0950268813000071

Moriuchi T, Vichit O, Vutthikol Y et al (2017) Molecular epidemiology of Bordetella pertussis in Cambodia determined by direct genotyping of clinical specimens. Int J Infect Dis 62:56–58

Nar Otgun S, Durmaz R, Karagoz A, Ertek M et al (2011) Pulsed-field gel electrophoresis characterization of Bordetella pertussis clinical isolates circulating in Turkey in 2001–2009. Eur J Clin Microbiol Infect Dis 30 (10):1229–1236

Octavia S, Sintchenko V, Gilbert GL et al (2012) Newly emerging clones of Bordetella pertussis carrying prn2 and ptxP3 alleles implicated in Australian pertussis epidemic in 2008–2010. J Infect Dis 205 (8):1220–1224

Ohtsuka M, Kikuchi K, Shimizu K et al (2009) Emergence of quinolone-resistant Bordetella pertussis in Japan. Antimicrob Agents Chemother 53(7):3147–3149

Otsuka N, Han HJ, Toyoizumi-Ajisaka H et al (2012) Prevalence and genetic characterization of pertactin-deficient Bordetella pertussis in Japan. PLoS One 7 (2):e31985

Parkhill J, Sebaihia M, Preston A et al (2003) Comparative analysis of the genome sequences of Bordetella pertussis, Bordetella parapertussis and Bordetella bronchiseptica. Nat Genet 35(1):32–40

Pawloski LC, Queenan AM, Cassiday PK et al (2014) Prevalence and molecular characterization of pertactin-deficient bordetella pertussis in the United States. Clin Vaccine Immunol 21(2):119–125

Petridou E, Jensen CB, Arvanitidis A et al (2018) Molecular epidemiology of Bordetella pertussis in Greece, 2010–2015. J Med Microbiol 67(3):400–407

Polak M, Zasada AA, Mosiej E et al (2018) Pertactin-deficient Bordetella pertussis isolates in Poland-a country with whole-cell pertussis primary vaccination. Microbes Infect. https://doi.org/10.1016/j.micinf.2018.12.001

Queenan AM, Cassiday PK, Evangelista A (2013) Pertactin-negative variants of Bordetella pertussis in the United States. N Engl J Med 368(6):583–584

Sadeghpour Heravi F, Nikbin VS, Nakhost Lotfi M et al (2018) Strain variation and antigenic divergence among Bordetella pertussis circulating strains isolated from patients in Iran. Eur J Clin Microbiol Infect Dis 37(10):1893–1900

Schmidtke AJ, Boney KO, Martin SW et al (2012) Population diversity among Bordetella pertussis isolates, United States, 1935–2009. Emerg Infect Dis 18 (8):1248–1255

Schouls LM, van der Heide HG, Vauterin L et al (2004) Multiple-locus variable-number tandem repeat analysis of Dutch Bordetella pertussis strains reveals rapid genetic changes with clonal expansion during the late 1990s. J Bacteriol 186(16):5496–5505

Sealey KL, Harris SR, Fry NK et al (2014) Genomic analysis of isolates from the UK 2012 pertussis outbreak reveals that vaccine antigen genes are unusually fast evolving. J Infect Dis 212(2):294–301

Shahcheraghi F, Nakhost Lotfi M, Nikbin VS et al (2014) The first macrolide-resistant bordetella pertussis strains isolated from Iranian patients. Jundishapur J Microbiol 7(6):e10880

Shuel M, Jamieson FB, Tang P et al (2013) Genetic analysis of Bordetella pertussis in Ontario, Canada reveals one predominant clone. Int J Infect Dis 17(6): e413–e417

Simmonds K, Fathima S, Chui L et al (2014) Dominance of two genotypes of Bordetella pertussis during a period of increased pertussis activity in Alberta, Canada: January to August 2012. Int J Infect Dis 29:223–225

Stefanelli P, Fazio C, Fedele G et al (2009) A natural pertactin deficient strain of Bordetella pertussis shows improved entry in human monocyte-derived dendritic cells. New Microbiol 32(2):159–166

Stefanelli P, Buttinelli G, Vacca P et al (2017) Severe pertussis infection in infants less than 6 months of age: clinical manifestations and molecular characterization. Hum Vaccin Immunother 13(5):1073–1077

Torres R, Moraes P, Oliveira K (2015, October 18–22) Bordetella pertussis antibiotic resistance in Southern Brazil: a 3-year surveillance study. XXVIII Brazilian congress of microbiology. Accessed 1 June 2018

Tsang RS, Shuel M, Jamieson FB et al (2014) Pertactin-negative Bordetella pertussis strains in Canada: characterization of a dozen isolates based on a survey of 224 samples collected in different parts of the country over the last 20 years. Int J Infect Dis 28:65–69

van Amersfoorth SC, Schouls LM, van der Heide HG et al (2005) Analysis of Bordetella pertussis populations in European countries with different vaccination policies. J Clin Microbiol 43(6):2837–2843

van der Maas NA, Mooi FR, de Greeff SC et al (2013) Pertussis in the Netherlands, is the current vaccination strategy sufficient to reduce disease burden in young infants? Vaccine 31(41):4541–4547

van Gent M, Heuvelman CJ, van der Heide HG et al (2015) Analysis of Bordetella pertussis clinical isolates circulating in European countries during the period 1998–2012. Eur J Clin Microbiol Infect Dis 34 (4):821–830

van Loo IH, Heuvelman KJ, King AJ et al (2002) Multilocus sequence typing of Bordetella pertussis based on surface protein genes. J Clin Microbiol 40 (6):1994–2001

Wagner B, Melzer H, Freymuller G et al (2015) Genetic Variation of Bordetella pertussis in Austria. PLoS One 10(7):e0132623

Wang Z, He Q (2015) Bordetella pertussis Isolates circulating in China where whole cell Vaccines have been used for 50 years. Clin Infect Dis 61 (6):1028–1029

Wang Z, Cui Z, Li Y et al (2014) High prevalence of erythromycin-resistant Bordetella pertussis in Xi'an, China. Clin Microbiol Infect 20(11):O825–O830

Wang Z, Han R, Liu Y et al (2015) Direct detection of erythromycin-resistant Bordetella pertussis in clinical specimens by PCR. J Clin Microbiol 53 (11):3418–3422

Watanabe M, Nagai M (2005) Acellular pertussis vaccines in Japan: past, present and future. Expert Rev Vaccines 4(2):173–184

Weigand MR, Peng Y, Cassiday PK et al (2017a) Complete genome sequences of bordetella pertussis isolates with novel pertactin-deficient deletions. Genome Announc 5(37). https://doi.org/10.1128/genomeA.00973-17

Weigand MR, Peng Y, Loparev V et al (2017b) The history of bordetella pertussis genome evolution includes structural rearrangement. J Bacteriol 199 (8):10

Weigand MR, Pawloski LC, Peng Y et al (2018) Screening and genomic characterization of filamentous hemagglutinin-deficient bordetella pertussis. Infect Immun 86(4):10

Willems R, Paul A, van der Heide HG et al (1990) Fimbrial phase variation in Bordetella pertussis: a novel mechanism for transcriptional regulation. EMBO J 9 (9):2803–2809

Williams MM, Sen K, Weigand MR et al (2016) Bordetella pertussis strain lacking pertactin and pertussis toxin. Emerg Infect Dis 22(2):319–322

Wilson KE, Cassiday PK, Popovic T et al (2002) Bordetella pertussis isolates with a heterogeneous phenotype for erythromycin resistance. J Clin Microbiol 40(8):2942–2944

Winter K, Harriman K, Zipprich J et al (2012) California pertussis epidemic, 2010. J Pediatr 161(6):1091–1096

Winter K, Glaser C, Watt J et al (2014) Pertussis epidemic--California, 2014. MMWR Morb Mortal Wkly Rep 63(48):1129–1132

Xu Y, Liu B, Grondahl-Yli-Hannuksila K et al (2015a) Whole-genome sequencing reveals the effect of vaccination on the evolution of Bordetella pertussis. Sci Rep 5:12888

Xu Y, Zhang L, Tan Y et al (2015b) Genetic diversity and population dynamics of Bordetella pertussis in China between 1950–2007. Vaccine 33(46):6327–6331

Yang Y, Yao K, Ma X et al (2015) Variation in bordetella pertussis susceptibility to erythromycin and virulence-related genotype changes in China (1970–2014). PLoS One 10(9):e0138941

Zeddeman A, van Gent M, Heuvelman CJ et al (2014) Investigations into the emergence of pertactin-deficient Bordetella pertussis isolates in six European countries, 1996 to 2012. Euro Surveill 19(33):20881

Zhang L, Xu Y, Zhao J et al (2010) Effect of vaccination on Bordetella pertussis strains, China. Emerg Infect Dis 16(11):1695–1701

Zomer A, Otsuka N, Hiramatsu Y et al (2018) Bordetella pertussis population dynamics and phylogeny in Japan after adoption of acellular pertussis vaccines. Microb Genom 4. https://doi.org/10.1099/mgen.0.000180

Adv Exp Med Biol - Advances in Microbiology, Infectious Diseases and Public Health (2019) 1183: 35–51
https://doi.org/10.1007/5584_2019_403
© Springer Nature Switzerland AG 2019
Published online: 3 August 2019

Role of Major Toxin Virulence Factors in Pertussis Infection and Disease Pathogenesis

Karen Scanlon, Ciaran Skerry, and Nicholas Carbonetti

Abstract

Bordetella pertussis produces several toxins that affect host-pathogen interactions. Of these, the major toxins that contribute to pertussis infection and disease are pertussis toxin, adenylate cyclase toxin-hemolysin and tracheal cytotoxin. Pertussis toxin is a multisubunit protein toxin that inhibits host G protein-coupled receptor signaling, causing a wide array of effects on the host. Adenylate cyclase toxin-hemolysin is a single polypeptide, containing an adenylate cyclase enzymatic domain coupled to a hemolysin domain, that primarily targets phagocytic cells to inhibit their antibacterial activities. Tracheal cytotoxin is a fragment of peptidoglycan released by *B. pertussis* that elicits damaging inflammatory responses in host cells. This chapter describes these three virulence factors of *B. pertussis*, summarizing background information and focusing on the role of each toxin in infection and disease pathogenesis, as well as their role in pertussis vaccination.

Keywords

Adenylate cyclase toxin-hemolysin · Bordetella toxins · Pertussis pathogenesis · Pertussis toxin · Tracheal Cytotoxin

1 Pertussis Toxin

1.1 Background

Pertussis toxin (Agarwal et al. 2009) is a multisubunit (AB_5) protein toxin secreted by *B. pertussis*. PT binds mammalian cell surface glycosylated molecules (Witvliet et al. 1989) in a non-saturable and non-specific manner (Finck-Barbancon and Barbieri 1996), indicating lack of a specific receptor. PT is endocytosed and transported by the retrograde pathway to the endoplasmic reticulum (el Baya et al. 1997; Plaut and Carbonetti 2008; Plaut et al. 2016), from where the A subunit (S1) translocates to the cytosol (Hazes et al. 1996; Pande et al. 2006; Worthington and Carbonetti 2007). In the cytosol, S1 ADP-ribosylates the alpha subunit of heterotrimeric G proteins of the $G_{i/o}$ class in mammalian cells, inhibiting activation of these G proteins by ligand-bound G protein-coupled receptors (GPCR) (Katada 2012). This modification has multiple effects on host cell activities, since many different GPCRs couple to G_i proteins. Binding of the PT B pentamer to mammalian cells elicits various signaling effects independently of the enzymatic activity of S1, but

K. Scanlon, C. Skerry, and N. Carbonetti (✉)
Department of Microbiology & Immunology, University of Maryland School of Medicine, Baltimore, MD, USA
e-mail: kscanlon@som.umaryland.edu; cskerry@som.umaryland.edu; ncarbonetti@som.umaryland.edu

these effects are relatively transient and concentration-dependent (Carbonetti 2010; Mangmool and Kurose 2011; Wong and Rosoff 1996; Schneider et al. 2009; Zocchi et al. 2005) and it is unclear whether this binding and signaling activity of PT is relevant in vivo. PT is an important virulence factor for *B. pertussis* and is central to the pathogenesis of pertussis infection and disease (as described below) and a detoxified form of PT is an important component of currently used acellular pertussis vaccines (aPV) (Coutte and Locht 2015).

1.2 Role in Infection and Disease Pathogenesis

1.2.1 Systemic and Local Effects of PT

Systemic administration of purified PT to experimental animals has a variety of biological and toxic effects (Munoz et al. 1981). The most important of these is leukocytosis (Morse and Morse 1976; Munoz et al. 1981; Hinds et al. 1996; Nogimori et al. 1984), a large increase in the number of circulating white blood cells that is also a prominent feature in human infants with pertussis and high levels of which correlate with fatal outcome (Winter et al. 2015). PT likely induces leukocytosis through a number of mechanisms (Carbonetti 2016), including reduced expression of leukocyte surface adhesion molecules such as LFA-1 (Schenkel and Pauza 1999) and CD62L (Hodge et al. 2003; Hudnall and Molina 2000), inhibition of LFA-1-dependent lymphocyte arrest on lymph node high endothelial venules (Bargatze and Butcher 1993; Warnock et al. 1998), and inhibition of chemokine receptor signaling affecting leukocyte migration (Beck et al. 2014; Pham et al. 2008). PT treatment of experimental animals also has adverse effects on control of heart rate and other cardiac functions, independently of leukocytosis (Grimm et al. 1998; Wainford et al. 2008; Adamson et al. 1993; Zheng et al. 2005). Another effect of PT is reduction of

vascular barrier integrity (Dudek et al. 2007), which may contribute to the pathogenesis of experimental autoimmune encephalitis (EAE) in animal models (Bennett et al. 2010), several of which require PT as an adjuvant to stimulate disease (Munoz et al. 1984; Arimoto et al. 2000; Zhao et al. 2008). This led to the recent speculation that PT effects from sub-clinical pertussis infections may be a contributor to exacerbations of multiple sclerosis in humans (Rubin and Glazer 2016). Other effects of systemic administration of PT include vasoactive sensitization to histamine (Munoz et al. 1981) and induction of insulinemia/hypoglycemia (Yajima et al. 1978), although whether these are relevant to pertussis disease in humans is not clear.

The extent to which PT contributes to the respiratory pathology of pertussis is less clear. Experimental animals, such as mice and guinea pigs, do not cough when administered purified PT (Hewitt and Canning 2010) but baboons experimentally-infected with *B. pertussis* suffer the typical severe paroxysmal pertussis cough (Warfel et al. 2012). However, baboons infected with PT-deficient strains of *B. pertussis* show no cough symptoms at all, despite being colonized for the same duration as wild type *B. pertussis*-infected animals (Merkel, unpublished data). This may be because PT is necessary to induce robust respiratory inflammation, as demonstrated in mouse models (Connelly et al. 2012; Khelef et al. 1994), which may be an important contributor to the cough pathology (see more below). In addition, data from natural and volunteer *B. pertussis* infections indicate that PT plays a role in respiratory pathology. For instance, a PT-deficient strain isolated from a 3-month-old unvaccinated infant in France was associated with a relatively mild and short time course of disease (Bouchez et al. 2009). Also, human volunteers intranasally inoculated with the candidate pertussis vaccine strain BPZE1, which expresses a genetically inactivated form of PT (Mielcarek et al. 2006), did not cough (Thorstensson et al.

2014), although this strain contains two other modifications in potential virulence factors. The paroxysmal nature of pertussis cough may also be an effect of PT activity, since PT can inhibit desensitization of receptors stimulated by tussive agents (Maher et al. 2011), thereby preventing the cessation of the coughing response. In addition, PT effects in experimental animals are long-lived (Carbonetti et al. 2003; Carbonetti et al. 2007) and so the longevity of pertussis cough may also be an effect of PT activity.

Another important activity of PT is likely in promoting fatal pertussis infection in young infants. PT induces leukocytosis and there is a significant correlation between high levels of leukocytosis and fatal outcome in these infants (Rowlands et al. 2010; Surridge et al. 2007; Pierce et al. 2000; Winter et al. 2015). Animal models provide additional evidence supporting the role of PT in promoting fatal outcome in pertussis. For example, PT is required for lethality of *B. pertussis* infection in neonatal Balb/c (Weiss and Goodwin 1989) and C57BL/6 mice (Scanlon et al. 2017). However, the specific mechanisms by which PT promotes lethality in pertussis remain to be determined and probably involve pathologies beyond leukocytosis and respiratory effects.

1.2.2 Effects of PT on B. pertussis Colonization and Host Immune Responses

Most of our understanding of the effects of PT on colonization and immune responses is derived from mouse model experiments, with some recent data from baboon studies, by comparing infection with isogenic strains differing only in PT production. In mice, PT clearly promotes colonization of the respiratory tract, presumably through effects on immune responses (Carbonetti et al. 2003; Connelly et al. 2012). PT inhibits early innate immune defenses, including recruitment of neutrophils

to *B. pertussis*-infected lungs (Carbonetti et al. 2003, 2005; Kirimanjeswara et al. 2005; Andreasen and Carbonetti 2008) and antibacterial activity of resident airway macrophages (Carbonetti et al. 2007). PT also inhibits some aspects of adaptive immunity, including migration of human dendritic cells in response to lymphatic chemokines (Fedele et al. 2011) and generation of serum antibody responses to *B. pertussis* during infection (Carbonetti et al. 2004; Mielcarek et al. 1998). However, at the peak of infection PT promotes lung inflammation (Connelly et al. 2012; Khelef et al. 1994; Andreasen et al. 2009), probably through a variety of mechanisms. One such mechanism may involve the PT-dependent increased expression of the epithelial anion exchanger pendrin, which promotes lung inflammatory pathology during *B. pertussis* infection (Scanlon et al. 2014). Another mechanism involves PT-dependent inhibition of the resolution of lung inflammation (Connelly et al. 2012), possibly by inhibiting activity of specialized pro-resolving lipid mediators such as resolvins and lipoxins (Levy and Serhan 2014), that signal through PT-sensitive GPCRs (Maddox et al. 1997; Krishnamoorthy et al. 2010; Chattopadhyay et al. 2018; Jo et al. 2016).

It is unclear whether lung inflammatory pathology correlates with the severe cough of human pertussis, but the recently developed baboon model of pertussis may shed some light on this. Baboons cough paroxysmally when infected with a clinical isolate of *B. pertussis* (Warfel et al. 2012; Warfel and Merkel 2014) although, as stated above, they do not develop cough when infected with an isogenic PT-deficient strain. In addition, PT production was associated with cough in rats experimentally infected with *B. pertussis* (Parton et al. 1994). Even though PT may not be the direct cause of pertussis cough, these findings indicate that there is a strong association between PT production and cough responses, which may be due to the exacerbated inflammatory responses at the peak of infection that are promoted by PT.

1.3 PT as a Pertussis Vaccine Component and Therapeutic Target

Since the move to aPV in most countries in recent years, a detoxified form of PT has been a component of all of these vaccines, in recognition of its role as a protective antigen against severe disease in infants (Coutte and Locht 2015). Although most of these vaccines contain other pertussis antigens, a monocomponent detoxified PT vaccine is used in Denmark (Thierry-Carstensen et al. 2013). Since pertussis has remained under control in Denmark (Dalby et al. 2016), this indicates that immune responses to this single component can protect a population from pertussis disease effectively. In addition, in the baboon model a monocomponent PT vaccine was sufficient to protect infant animals born to vaccinated mothers from pertussis disease (Kapil et al. 2018).

PT may also be an important target for therapy against pertussis. In one trial, administration of intravenous pertussis immunoglobulin (P-IGIV), which contains high levels of anti-PT antibodies, resulted in significant reduction in leukocytosis in infants with pertussis, and reduced paroxysmal coughing and bradycardic episodes (Bruss et al. 1999). In animal models, treatment with humanized murine monoclonal antibodies specific for PT was more effective than P-IGIV in preventing leukocytosis in mice and reduced leukocytosis when administered therapeutically to infected baboons (Nguyen et al. 2015), highlighting the therapeutic potential of this approach. Very recently, inhibitors of cyclophilins, such as cyclosporine A, were found to inhibit PT activity in cultured cells (Ernst et al. 2018), representing an additional therapeutic strategy targeting PT.

2 Adenylate Cyclase Toxin-Hemolysin

2.1 Background

Adenylate cyclase toxin-hemolysin (CyaA, ACT or AC-Hly) is a repeat-in-toxin (RTX) cytotoxin expressed by three closely related *Bordetella* species, *B. pertussis*, *B. parapertussis* and *B. bronchiseptica* (Endoh et al. 1980; Hewlett et al. 1976; Glaser et al. 1988a). This 1706 amino acid polypeptide is encoded by *cyaA* in an operon containing the type I secretion apparatus genes *cyaB*, *cyaD* and *cyaE* (Glaser et al. 1988b). CyaA contains an N-terminal adenylate cyclase (AC) enzyme domain of 364 amino acid residues and a C-terminal RTX hemolysin moiety of ~1300 residues (Glaser et al. 1988b). The hemolysin moiety is comprised of a hydrophobic pore-forming domain (Benz et al. 1994), an active domain containing two posttranslationally acylated lysine residues (Lys 860 and Lys 983) (Hackett et al. 1994, 1995), a receptor binding RTX domain with characteristic calcium-binding glycine- and aspartate-rich nonapeptide repeats (Rose et al. 1995) and a C-terminal secretion signal for the type I secretion system (Bumba et al. 2016; Sebo and Ladant 1993). This toxin plays a key role in establishing *B. pertussis* infection. Activities mediated by both the AC and hemolysin domains of CyaA function to subvert host innate immunity and thereby facilitate initial colonization (as described below). Current formulations of aPV do not contain CyaA. However, recently a focus on research elucidating the structure-function relationship of CyaA has emerged, with the aim of using this new understanding to rationally develop CyaA vaccine candidates (Cheung et al. 2006; Osickova et al. 2010; Boehm et al. 2018).

2.2   Role in Infection and Disease Pathogenesis

2.2.1   Host Cell Subversion by CyaA

First described in 1976 as both a soluble and bacterial cell-associated enzyme (Hewlett et al. 1976), CyaA physically associates with filamentous haemagglutinin (FHA) to mediate toxin retention on the bacterial surface (Zaretzky et al. 2002). Interestingly, target cell intoxication is not dependent on surface-associated CyaA and instead newly secreted CyaA is necessary (Gray et al. 2004). However, unlike PT, which acts conventionally as a soluble factor (described above), exogenous administration of recombinant CyaA fails to rescue a "wild type" *B. pertussis* phenotype in mice infected with a *cyaA*-deficient strain (Carbonetti et al. 2005). In vitro studies have demonstrated rapid aggregation of CyaA in solution (Rogel et al. 1991). Hence, CyaA is hypothesized to be a short-lived toxin acting in close proximity to the bacterium (Vojtova et al. 2006) but the exact role of soluble vs bacterial-associated CyaA remains to be determined. Given that CyaA functions at the site of the bacterial-host interface, this toxin does not contribute to systemic pathologies during *B. pertussis* infection.

At the site of infection, locally secreted CyaA performs a number of functions to subvert host cell biology and potentiate cell death. Upon insertion into the target cell membrane, CyaA takes on one of two conformations (Osickova et al. 1999). It is proposed that a monomeric form of CyaA acts to translocate the catalytic AC enzyme domain into the host cell cytosol, whilst oligomerization of CyaA potentiates the formation of a hemolytic pore and that actions performed by the CyaA monomer improve the efficacy of the hemolysin moiety (Osickova et al. 1999; Fiser et al. 2012). The exact mechanism by which CyaA translocates its AC domain into the host cytosol is still under investigation. Recently, recombinant CyaA has been shown to display calcium-dependent phospholipase A (PLA)

activity and this mechanism has been associated with CyaA-induced cytotoxicity in macrophages and macrophage release of free fatty acids (Gonzalez-Bullon et al. 2017). Ostolaza et al. suggest that CyaA-PLA acts to remodel the host cell membrane (releasing membrane fatty acids and lysophospholipids), generating a "toroidal pore" that facilitates transport of the AC domain (Ostolaza et al. 2017). However, whether CyaA itself, and not an *E. coli*-derived contaminant, possesses PLA activity has been contested (Masin et al. 2018; Ostolaza 2018), hence more studies are required to elucidate the contribution of PLA activity to CyaA biology. Translocation of the AC domain is concomitant with Ca^{2+} influx in the host cell (Fiser et al. 2007). This process is required to activate calpain-mediated processing of CyaA and liberation of the AC domain into the cytosol (Bumba et al. 2010; Uribe et al. 2013). In addition, CyaA-driven Ca^{2+} influx alters endocytic trafficking in the host cell membrane and limits macropinocytic removal of pore-forming oligomerized CyaA (Fiser et al. 2012). In the cytosol, AC binds host calmodulin and catalyzes the rapid, unregulated conversion of cytoplasmic ATP into cAMP (Wolff et al. 1980; Hanski and Farfel 1985; Guo et al. 2005). CyaA-induced accumulation of cAMP prevents bactericidal activities of phagocytes by inhibiting oxidative burst and phagocytosis (Confer and Eaton 1982; Pearson et al. 1987; Kamanova et al. 2008; Friedman et al. 1987; Cerny et al. 2015), and modulates innate immune cell activation by inhibiting phagocyte maturation and suppressing the expression of proinflammatory cytokines and chemokines (Boyd et al. 2005; Njamkepo et al. 2000; Ross et al. 2004; Fedele et al. 2010; Spensieri et al. 2006). In addition, elevated cAMP levels inhibit the formation of neutrophil extracellular traps (NETs) and neutrophil apoptosis (Eby et al. 2014), whilst also promoting *B. pertussis* intracellular survival in macrophages and macrophage apoptosis (Valdez et al. 2016; Hewlett et al. 2006; Ahmad et al. 2016). In parallel, the oligomerized conformation of CyaA generates a cation-selective pore that induces potassium ion efflux from nucleated cells (Gray et al. 1998; Osickova et al. 1999). This activity promotes cell death caused by both apoptosis and necrosis (Basler et al. 2006; Khelef et al. 1993; Hewlett et al. 2006). In addition, CyaA-promoted potassium efflux induces IL-1β production by dendritic cells via activation of caspase-I and NALP3-containing inflammasome complex (Dunne et al. 2010) and activates mitogen-activated protein kinase (MAPK) and N-terminal protein kinase (JNK) signaling (Masin et al. 2015; Svedova et al. 2016). Hence, taken together CyaA uses both its AC domain and hemolysin moiety to induce cellular dysfunctions that promote bacterial survival and inhibit phagocyte-mediated bacterial clearance.

2.2.2 Effects of CyaA on B. pertussis Colonization and Host Immune Responses

CyaA was characterized as a toxin in 1982, when it was shown that CyaA inhibited phagocytosis and oxidative burst by human neutrophils (Confer and Eaton 1982). Since that time, CyaA has been found to specifically target myeloid phagocytic cells, using CD11b/CD18 integrin (known as $\alpha_M\beta_2$ integrin or complement receptor 3, CR3) on the host cell surface as a receptor (Guermonprez et al. 2001). CyaA can also penetrate lipid bilayers in the absence of this receptor (Martin et al. 2004; Szabo et al. 1994) and intoxicate most cells but with reduced efficacy (Gordon et al. 1989; Gray et al. 1999; Eby et al. 2010; Bassinet et al. 2000; Hanski and Farfel 1985). In vivo, CyaA acts primarily on phagocytic cells to inhibit clearance and promote B. pertussis colonization (Gueirard et al. 1998; Harvill et al. 1999). However, CyaA-induced apoptotic cell death of bronchopulmonary cells is also described following administration of airway-isolated bacteria (Gueirard et al. 1998).

A role for CyaA in bacterial colonization was first described by Weiss et al. (Weiss et al. 1983). In that study, B. pertussis mutants not expressing CyaA were found to be non-lethal in a histamine-sensitizing in vivo mouse assay (Weiss et al. 1983). Further to this, expression of CyaA was found to be required for colonization and lethality of B. pertussis in infant mice (Goodwin and Weiss 1990; Weiss and Goodwin 1989). Both AC activity and the hemolysin moiety have been shown to be required for competent colonization and virulence by B. pertussis (Gross et al. 1992; Khelef et al. 1992). Indeed, in a later study using B. pertussis strains that express CyaA deficient in AC or hemolysin activity, it was determined that actions performed by the AC domain were required for bacterial colonization and persistence in mouse lungs, whereas the hemolysin moiety was not involved in colonization (Skopova et al. 2017). However, the hemolysin moiety did contribute to B. pertussis-induced lethality, penetration of B. pertussis across the epithelial lining and recruitment of myeloid phagocytic cells into B. pertussis-infected tissue (Skopova et al. 2017). Hence, the pore-forming ability of CyaA is not required for bacterial persistence but contributes to B. pertussis-associated pathology.

2.3 CyaA as a Pertussis Vaccine Component

Whole-cell formulations of pertussis vaccines (wPV) displayed AC activity (Hewlett et al. 1977) and mass spectrometry analysis of wPV detected the presence of CyaA (Boehm et al. 2018). However, the aPV does not contain CyaA antigens. In studies using monoclonal antibodies and serum from convalescent individuals, it was found that antibodies against CyaA promote B. pertussis phagocytosis by neutrophils, validating one potentially beneficial effect of including a CyaA antigen in an aPV (Weingart and Weiss 2000; Weingart et al.

2000; Mobberley-Schuman et al. 2003). In addition, CyaA antibody-mediated neutrophil clearance of *B. pertussis* was shown to be important in an immune mouse challenge model (Andreasen and Carbonetti 2009). Given the potential for CyaA-mediated toxicity, current studies on developing CyaA as a vaccine antigen have focused of delineating the minimal essential regions of CyaA that confer protective immunity. Studies by Wang et al. show that the RTX domain of CyaA was sufficient to generate neutralizing antibodies and may represent an alternative to the use of full length CyaA (Wang et al. 2015). Indeed, in a mouse model, inclusion of the RTX domain in aP resulted in enhanced bacterial clearance after *B. pertussis* challenge, increased production of anti-PT antibodies, decreased production of proinflammatory cytokines and decreased recruitment of total macrophages (Boehm et al. 2018).

CyaA also displays an adjuvant effect in mice during immunization, with AC enzymatically inactive-CyaA generating greater and more potent adaptive immune responses than active CyaA (Orr et al. 2007; Cheung et al. 2006). When used as an adjuvant, CyaA induces the generation of antigen-specific Th17 cells by a pore-forming dependent mechanism (Dunne et al. 2010). Th17 cells mediate protective immunity to *B. pertussis* (Ross et al. 2013), hence generation of vaccine antigens that include specific functional regions of CyaA may prove beneficial in an aPV formulation.

3 Tracheal Cytotoxin

3.1 Background

Tracheal cytotoxin (TCT) is a peptidoglycan (PGN) fragment released by *B. pertussis*. PGN recycling is a process utilized by bacteria during cell division. First PGN is converted into its constituent parts, which are then available to the bacteria to be utilized in the synthesis of more PGN or for use as an energy source (Uehara and

Park 2008). Bacteria producing PGN molecules containing diaminopimelic acid (DAP), or DAP-type PGNs, such as *Bacillus subtilis, Neisseria gonorrhoeae, Lactobacillus acidophilus* and *B. pertussis*, lose large amounts (25–50%) of the DAP from their cell wall to their growth media compared to *E. coli* (6%) (Mauck et al. 1971; Boothby et al. 1973; Goodell 1985; Goodell et al. 1978; Chaloupka and Strnadova 1972; Hebeler and Young 1976). The inner membrane permease AmpG is a key component in this process (Jacobs et al. 1994). In *Bordetella* however, this transporter is defective, resulting in accumulation of extracellular fragments of PGN (Cookson et al. 1989a; Rosenthal et al. 1987; Mielcarek et al. 2006). Over 95% of these PGN fragments released from *Bordetella* are of the structure N-acetylglucosaminyl-1,6-anhydro-N-acetylmuramyl-L-alanine-D-glutamyl-mesoDAP-D-alanine. This released monomeric fragment of PGN constitutes the virulence factor TCT. TCT is a 921 dalton disaccharide tetrapeptide PGN fragment. The anhydrous nature of the acetylmuramyl saccharide indicates that TCT is not just an accidentally released PGN fragment (which would contain reducing as opposed to anhydrous muramic acid residues) (Goldman and Cookson 1988). This finding led to the hypothesis that TCT is processed via an as yet unidentified murein transglycosylase.

3.2 TCT in Pathogenesis

Host pattern recognition receptors trigger innate responses to PGN fragments. Following its release from the bacterial cell wall, non-ciliated cells of the airway epithelium internalize TCT (Flak and Goldman 1999; Flak et al. 2000). TCT is trafficked to the cytosol via PGN transporter Slc46A2 (Paik et al. 2017) where it is recognized by the pattern recognition receptor Nod1 (nucleotide-binding oligomerization domain protein 1) (Magalhaes et al. 2005). This leads to the activation of downstream NF-kB signaling (Paik et al. 2017) and

IL-1α production by non-ciliated cells of the murine airway epithelium, in a synergistic manner with LPS (Flak et al. 2000). Interestingly, neither TCT nor *Bordetella* lipooligosaccharide (LOS) elicit strong IL-1α production alone, but the combination of the two is a potent producer (Flak et al. 2000). Downstream effects of this activity include the production of nitric oxide (NO), which diffuses to neighboring ciliated cells causing cell death (Flak and Goldman 1999). In addition to promoting NO and IL-1α mediated inflammation, TCT inhibits neutrophil chemotaxis (Cundell et al. 1994) preventing optimal immune responses.

In Drosophila, TCT triggers the activation of the IMD pathway following cytosolic recognition by PGN recognition protein PGRP-LE (Lim et al. 2006). Interestingly, mammalian PGN recognition protein 4 (PGLYRP4) has a protective role against TCT-mediated pathogenesis (Skerry et al. 2019). Hypersecretion of TCT, using mutant strains lacking PGN recycling molecule AmpG, or deficiency in PGLYRP4 causes increased early lung pathology in mice, suggesting that PGLYRP4 limits early TCT-induced inflammation (Skerry et al. 2019).

Each member of the pathogenic *Bordetellae* produce a chemically identical TCT molecule (Cookson et al. 1989b; Folkening et al. 1987; Gentry-Weeks et al. 1988; Goldman and Cookson 1988). Additionally, all of the pathogenic *Bordetellae* induce a similar respiratory pathology typified by the destruction of the ciliated cells of the upper airways, in hosts ranging from humans (Paddock et al. 2008) to dogs (Oskouizadeh et al. 2011; Bemis 1992) and turkeys (Arp and Fagerland 1987) and TCT is the only factor capable of replicating this ciliated cell extrusion that is the hallmark of the bordetelloses (Cookson et al. 1989a; Endoh et al. 1986; Goldman et al. 1982; Goldman and Cookson 1988).

TCT analogues in other bacteria are associated with similar pathologies, e.g. *Neisseria gonorrhoeae*-released PGN fragments have been associated with damage to the mucosa of human fallopian tubes (Melly et al. 1984). PGN monomers released from *Vibrio fischeri* induce the regression of ciliated fields of *Euprymna scolopes* (Doino and McFall-Ngai 1995; Montgomery and McFall-Ngai 1994).

3.3 TCT in Vaccination

The transition from the use of the wPV to aPV has been associated with a reduction in the duration of immunity (Chen and He 2017). The wPV, derived from intact *B. pertussis* organisms, likely contained TCT, even if only in the form of intact PGN. The aPV consists of recombinant subunit proteins of *B. pertussis* but does not contain TCT or PGN. Freund's complete adjuvant, a potent contributor to vaccine responses, contains muramyl peptides like TCT (Kotani et al. 1986). It has therefore been speculated that the loss of TCT in aPV formulations contributed to their waning immunogenicity (Goldman and Cookson 1988). Candidate live attenuated *B. pertussis* vaccine BPZE1 utilizes the *E. coli* PGN recycling machinery, AmpG, resulting in <1% TCT activity (Mielcarek et al. 2006). However, it contains the full complement of *Bordetella* PGN and therefore the potential adjuvanticity. Loss of PT, TCT and dermonecrotic toxin (DNT) resulted in a vaccine strain which successfully colonized hosts and elicited a protective immune response, without the associated pathophysiology of "wild-type" infection (Skerry and Mahon 2011; Skerry et al. 2009; Feunou et al. 2010).

TCT is identical in structure to slow wave sleep promoting factor FSu (Martin et al. 1984). In rabbits, FSu has been shown to induce excess slow-wave sleep following intraventricular infusion of picomolar concentrations (Krueger et al. 1982). The sleep-promoting effects of this factor are separate from its immunomodulatory potential (Krueger and Karnovsky 1987). These somnogenic qualities, along with the small size of TCT, led to the untested hypothesis that TCT may have been the reason behind the wPV-associated drowsiness (Goldman and Cookson 1988).

4   Conclusion

As we describe here, much of the pathogenesis of pertussis is attributable to the major secreted toxins, PT, CyaA and TCT, produced during *B. pertussis* infection (their activities and effects are summarized in Fig. 1 and Table 1). Although there is still much to be learned about their activities and roles, it is clear that these toxins collectively suppress protective immune responses while exacerbating damaging pathologies in the respiratory tract and (in the case of PT) other organ systems. Further understanding of these toxin activities and the affected host responses will be important in the development of optimal therapeutics and vaccines to treat and prevent pertussis.

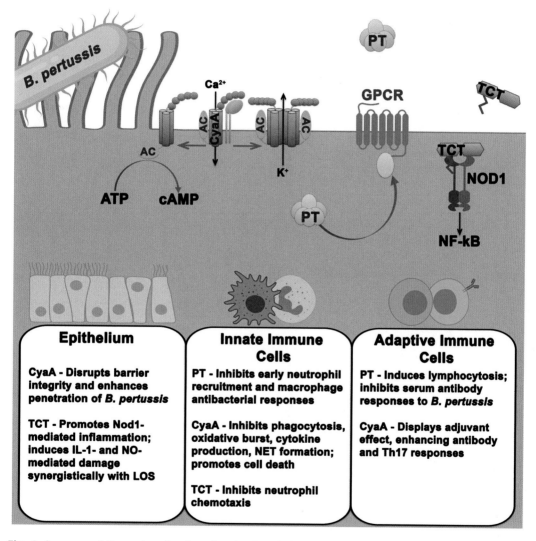

Fig. 1 Summary of the modes of action of major *B. pertussis* virulence factors. *B. pertussis* colonizes ciliated epithelial cells and releases adenylate cyclase toxin-hemolysin (CyaA), pertussis toxin (PT) and tracheal cytotoxin (TCT) to modulate host cell functions (schematically depicted) and subvert host immunity (listed by immune system). *AC* adenylate cyclase, *GPCR* G protein-coupled receptor, *Nod1* nucleotidebinding oligomerization domain protein 1, *NO* nitric oxide, *LOS* lipooligosaccharide, *NET* neutrophil extracellular traps

Table 1 Effect of *B. pertussis* virulence factors on host responses

Virulence Factor	Effect	References
Pertussis toxin	ADP-Ribosylation of Gi/α class GPCRs	Katada (2012)
	Induction of leukocytosis	Morse and Morse (1976), Munoz et al. (1981), Hinds et al. 1996 and Nogimori et al. (1984)
		Pham et al. (2008) and Beck et al. (2014)
	Inhibition of chemokine signaling	Grimm et al. (1998), Wainford et al. (2008), Adamson et al. (1993) and Zheng et al. (2005)
	Cardiac issues	Dudek et al. (2007)
		Munoz et al. (1981)
	Reduction of vascular barrier integrity	Yajima et al. (1978)
	Histamine sensitization	Maher et al. (2011)
	Insulinemia/hypoglycemia	Carbonetti et al. (2003) and Connelly et al. (2012)
	Prevention of cough cessation	Carbonetti et al. (2003), Carbonetti et al. (2005), Kirimanjeswara et al. (2005) and Andreasen and Carbonetti (2008)
	Promotes colonization	Carbonetti et al. (2007)
	Inhibits early defense	Connelly et al. (2012), Khelef et al. (1994) and Andreasen et al. (2009)
	Inhibits macrophage antibacterial responses	
	Promotes lung inflammation	
Adenylate cyclase toxin	Hemolytic pore formation	Osickova et al. (1999) and Fiser et al. (2012)
	cAMP intoxication	Wolff et al.(1980), Hanski and Furfel (1985) and Guo et al. (2005)
	Inhibition of oxidative burst and phagocytosis	Confer and Eaton (1982), Pearson et al. (1987), Kamanova et al. (2008), Friedman et al. (1987) and Cerny et al. (2015)
	Suppression of pro-inflammatory cytokines and chemokines	Boyd et al. (2005), Njamkepo et al. (2000) and Ross et al. (2004)
	Inhibition of NET formation	
	Induction of potassium ion efflux	Eby et al. (2014)
	Promotes cell death	Gray et al. (1998) and Osickova et al. (1999)
	Activation of MAP and JNK signaling	Basler et al. (2006), Khelef et al. (1993) and Hewlett et al. (2006)
	Promotes Colonization	Masin et al.(2015) and Svedova et al. (2016)
	Inhibits clearance	Weiss and Goodwin (1989)
		Gueirard et al. (1998)
Tracheal cytotoxin	Stimulation of Nod1	Magalhaes et al. (2005)
	Nf-kB activation	Paik et al. (2017)
	IL-1a production	Flak et al. (2000)
	Ciliated cell death	Flak and Goldman (1999)
	Inhibition of neutrophil chemotaxis	Cundell et al. (1994)

References

Adamson PB, Hull SS Jr, Vanoli E, De Ferrari GM, Wisler P, Foreman RD, Watanabe AM, Schwartz PJ (1993) Pertussis toxin-induced ADP ribosylation of inhibitor G proteins alters vagal control of heart rate in vivo. Am J Phys 265(2 Pt 2):H734–H740

Agarwal N, Lamichhane G, Gupta R, Nolan S, Bishai WR (2009) Cyclic AMP intoxication of macrophages by a Mycobacterium tuberculosis adenylate cyclase. Nature 460(7251):98–102

Ahmad JN, Cerny O, Linhartova I, Masin J, Osicka R, Sebo P (2016) cAMP signalling of Bordetella adenylate cyclase toxin through the SHP-1 phosphatase activates the BimEL-Bax pro-apoptotic cascade in phagocytes. Cell Microbiol 18(3):384–398

Andreasen C, Carbonetti NH (2008) Pertussis toxin inhibits early chemokine production to delay neutrophil recruitment in response to Bordetella pertussis respiratory tract infection in mice. Infect Immun 76 (11):5139–5148

Andreasen C, Carbonetti NH (2009) Role of neutrophils in response to Bordetella pertussis infection in mice. Infect Immun 77(3):1182–1188

Andreasen C, Powell DA, Carbonetti NH (2009) Pertussis toxin stimulates IL-17 production in response to Bordetella pertussis infection in mice. PLoS One 4 (9):e7079

Arimoto H, Tanuma N, Jee Y, Miyazawa T, Shima K, Matsumoto Y (2000) Analysis of experimental autoimmune encephalomyelitis induced in F344 rats by pertussis toxin administration. J Neuroimmunol 104 (1):15–21

Arp LH, Fagerland JA (1987) Ultrastructural pathology of Bordetella avium infection in turkeys. Vet Pathol 24 (5):411–418

Bargatze RF, Butcher EC (1993) Rapid G protein-regulated activation event involved in lymphocyte binding to high endothelial venules. J Exp Med 178 (1):367–372

Basler M, Masin J, Osicka R, Sebo P (2006) Pore-forming and enzymatic activities of Bordetella pertussis adenylate cyclase toxin synergize in promoting lysis of monocytes. Infect Immun 74(4):2207–2214

Bassinet L, Gueirard P, Maitre B, Housset B, Gounon P, Guiso N (2000) Role of adhesins and toxins in invasion of human tracheal epithelial cells by Bordetella pertussis. Infect Immun 68(4):1934–1941

Beck TC, Gomes AC, Cyster JG, Pereira JP (2014) CXCR4 and a cell-extrinsic mechanism control immature B lymphocyte egress from bone marrow. J Exp Med 211(13):2567–2581

Bemis DA (1992) Bordetella and Mycoplasma respiratory infections in dogs and cats. Vet Clin North Am Small Anim Pract 22(5):1173–1186

Bennett J, Basivireddy J, Kollar A, Biron KE, Reickmann P, Jefferies WA, McQuaid S (2010) Blood-brain barrier disruption and enhanced vascular permeability in the multiple sclerosis model EAE. J Neuroimmunol 229(1–2):180–191

Benz R, Maier E, Ladant D, Ullmann A, Sebo P (1994) Adenylate cyclase toxin (CyaA) of Bordetella pertussis. Evidence for the formation of small ion-permeable channels and comparison with HlyA of Escherichia coli. J Biol Chem 269(44):27231–27239

Boehm DT, Hall JM, Wong TY, DiVenere AM, Sen-Kilic E, Bevere JR, Bradford SD, Blackwood CB, Elkins CM, DeRoos KA, Gray MC, Cooper CG, Varney ME, Maynard JA, Hewlett EL, Barbier M, Damron FH (2018) Evaluation of adenylate cyclase toxoid antigen in acellular pertussis vaccines by using a Bordetella pertussis challenge model in mice. Infect Immun 86(10)

Boothby D, Daneo-Moore L, Higgins ML, Coyette J, Shockman GD (1973) Turnover of bacterial cell wall peptidoglycans. J Biol Chem 248(6):2161–2169

Bouchez V, Brun D, Cantinelli T, Dore G, Njamkepo E, Guiso N (2009) First report and detailed characterization of B. pertussis isolates not expressing pertussis toxin or pertactin. Vaccine 27(43):6034–6041

Boyd AP, Ross PJ, Conroy H, Mahon N, Lavelle EC, Mills KH (2005) Bordetella pertussis adenylate cyclase toxin modulates innate and adaptive immune responses: distinct roles for acylation and enzymatic activity in immunomodulation and cell death. J Immunol 175(2):730–738

Bruss JB, Malley R, Halperin S, Dobson S, Dhalla M, McIver J, Siber GR (1999) Treatment of severe pertussis: a study of the safety and pharmacology of intravenous pertussis immunoglobulin. Pediatr Infect Dis J 18 (6):505–511

Bumba L, Masin J, Fiser R, Sebo P (2010) Bordetella adenylate cyclase toxin mobilizes its beta2 integrin receptor into lipid rafts to accomplish translocation across target cell membrane in two steps. PLoS Pathog 6(5):e1000901

Bumba L, Masin J, Macek P, Wald T, Motlova L, Bibova I, Klimova N, Bednarova L, Veverka V, Kachala M, Svergun DI, Barinka C, Sebo P (2016) Calcium-driven folding of RTX domain beta-rolls ratchets translocation of RTX proteins through type I secretion ducts. Mol Cell 62(1):47–62

Carbonetti NH (2010) Pertussis toxin and adenylate cyclase toxin: key virulence factors of Bordetella pertussis and cell biology tools. Future Microbiol 5 (3):455–469

Carbonetti NH (2016) Pertussis leukocytosis: mechanisms, clinical relevance and treatment. Pathog Dis 74:ftw087. https://doi.org/10.1093/femspd/ftw087

Carbonetti NH, Artamonova GV, Mays RM, Worthington ZE (2003) Pertussis toxin plays an early role in respiratory tract colonization by Bordetella pertussis. Infect Immun 71(11):6358–6366

Carbonetti NH, Artamonova GV, Andreasen C, Dudley E, Mays RM, Worthington ZE (2004) Suppression of serum antibody responses by pertussis toxin after respiratory tract colonization by Bordetella pertussis and identification of an immunodominant lipoprotein. Infect Immun 72(6):3350–3358

Carbonetti NH, Artamonova GV, Andreasen C, Bushar N (2005) Pertussis toxin and adenylate cyclase toxin provide a one-two punch for establishment of Bordetella pertussis infection of the respiratory tract. Infect Immun 73(5):2698–2703

Carbonetti NH, Artamonova GV, Van Rooijen N, Ayala VI (2007) Pertussis toxin targets airway macrophages to promote Bordetella pertussis infection of the respiratory tract. Infect Immun 75(4):1713–1720

Cerny O, Kamanova J, Masin J, Bibova I, Skopova K, Sebo P (2015) Bordetella pertussis adenylate cyclase toxin blocks induction of bactericidal nitric oxide in macrophages through cAMP-dependent activation of the SHP-1 Phosphatase. J Immunol 194 (10):4901–4913

Chaloupka J, Strnadova M (1972) Turnover of murein in a diaminopimelic acid dependent mutant of Escherichia coli. Folia Microbiol (Praha) 17(6):446–455

Chattopadhyay R, Mani AM, Singh NK, Rao GN (2018) Resolvin D1 blocks H2O2-mediated inhibitory crosstalk between SHP2 and PP2A and suppresses endothelial-monocyte interactions. Free Radic Biol Med 117:119–131

Chen Z, He Q (2017) Immune persistence after pertussis vaccination. Hum Vaccin Immunother 13(4):744–756. https://doi.org/10.1080/21645515.2016.1259780

Cheung GY, Xing D, Prior S, Corbel MJ, Parton R, Coote JG (2006) Effect of different forms of adenylate cyclase toxin of Bordetella pertussis on protection afforded by an acellular pertussis vaccine in a murine model. Infect Immun 74(12):6797–6805

Confer DL, Eaton JW (1982) Phagocyte impotence caused by an invasive bacterial adenylate cyclase. Science 217 (4563):948–950

Connelly CE, Sun Y, Carbonetti NH (2012) Pertussis toxin exacerbates and prolongs airway inflammatory responses during Bordetella pertussis infection. Infect Immun 80(12):4317–4332

Cookson BT, Cho HL, Herwaldt LA, Goldman WE (1989a) Biological activities and chemical composition of purified tracheal cytotoxin of Bordetella pertussis. Infect Immun 57(7):2223–2229

Cookson BT, Tyler AN, Goldman WE (1989b) Primary structure of the peptidoglycan-derived tracheal cytotoxin of Bordetella pertussis. Biochemistry 28 (4):1744–1749

Coutte L, Locht C (2015) Investigating pertussis toxin and its impact on vaccination. Future Microbiol 10:241–254

Cundell DR, Kanthakumar K, Taylor GW, Goldman WE, Flak T, Cole PJ, Wilson R (1994) Effect of tracheal cytotoxin from Bordetella pertussis on human neutrophil function in vitro. Infect Immun 62(2):639–643

Dalby T, Andersen PH, Hoffmann S (2016) Epidemiology of pertussis in Denmark, 1995 to 2013. Euro Surveill 21(36)

Doino JA, McFall-Ngai MJ (1995) A transient exposure to symbiosis-competent bacteria induces light organ morphogenesis in the host squid. Biol Bull 189 (3):347–355

Dudek SM, Camp SM, Chiang ET, Singleton PA, Usatyuk PV, Zhao Y, Natarajan V, Garcia JG (2007) Pulmonary endothelial cell barrier enhancement by FTY720 does not require the S1P1 receptor. Cell Signal 19 (8):1754–1764

Dunne A, Ross PJ, Pospisilova E, Masin J, Meaney A, Sutton CE, Iwakura Y, Tschopp J, Sebo P, Mills KH (2010) Inflammasome activation by adenylate cyclase toxin directs Th17 responses and protection against Bordetella pertussis. J Immunol 185(3):1711–1719

Eby JC, Ciesla WP, Hamman W, Donato GM, Pickles RJ, Hewlett EL, Lencer WI (2010) Selective translocation of the Bordetella pertussis adenylate cyclase toxin across the basolateral membranes of polarized epithelial cells. J Biol Chem 285(14):10662–10670

Eby JC, Gray MC, Hewlett EL (2014) Cyclic AMP-mediated suppression of neutrophil extracellular trap formation and apoptosis by the Bordetella pertussis adenylate cyclase toxin. Infect Immun 82 (12):5256–5269

el Baya A, Linnemann R, von Olleschik-Elbheim L, Robenek H, Schmidt MA (1997) Endocytosis and retrograde transport of pertussis toxin to the Golgi complex as a prerequisite for cellular intoxication. Eur J Cell Biol 73(1):40–48

Endoh M, Takezawa T, Nakase Y (1980) Adenylate cyclase activity of Bordetella organisms. I. Its production in liquid medium. Microbiol Immunol 24 (2):95–104

Endoh M, Amitani M, Nakase Y (1986) Purification and characterization of heat-labile toxin from Bordetella bronchiseptica. Microbiol Immunol 30(7):659–673

Ernst K, Eberhardt N, Mittler AK, Sonnabend M, Anastasia A, Freisinger S, Schiene-Fischer C, Malesevic M, Barth H (2018) Pharmacological cyclophilin inhibitors prevent intoxication of mammalian cells with Bordetella pertussis toxin. Toxins 10(5)

Fedele G, Spensieri F, Palazzo R, Nasso M, Cheung GY, Coote JG, Ausiello CM (2010) Bordetella pertussis commits human dendritic cells to promote a Th1/Th17 response through the activity of adenylate cyclase toxin and MAPK-pathways. PLoS One 5(1): e8734

Fedele G, Bianco M, Debrie AS, Locht C, Ausiello CM (2011) Attenuated Bordetella pertussis vaccine candidate BPZE1 promotes human dendritic cell CCL21-induced migration and drives a Th1/Th17 response. J Immunol 186(9):5388–5396

Feunou PF, Kammoun H, Debrie AS, Mielcarek N, Locht C (2010) Long-term immunity against pertussis induced by a single nasal administration of live attenuated B. pertussis BPZE1. Vaccine 28 (43):7047–7053

Finck-Barbancon V, Barbieri JT (1996) Preferential processing of the S1 subunit of pertussis toxin that is bound to eukaryotic cells. Mol Microbiol 22(1):87–95

Fiser R, Masin J, Basler M, Krusek J, Spulakova V, Konopasek I, Sebo P (2007) Third activity of Bordetella adenylate cyclase (AC) toxin-hemolysin. Membrane translocation of AC domain polypeptide promotes calcium influx into CD11b+ monocytes independently of the catalytic and hemolytic activities. J Biol Chem 282(5):2808–2820

Fiser R, Masin J, Bumba L, Pospisilova E, Fayolle C, Basler M, Sadilkova L, Adkins I, Kamanova J, Cerny J, Konopasek I, Osicka R, Leclerc C, Sebo P (2012) Calcium influx rescues adenylate cyclase-hemolysin from rapid cell membrane removal and enables phagocyte permeabilization by toxin pores. PLoS Pathog 8(4):e1002580

Flak TA, Goldman WE (1999) Signalling and cellular specificity of airway nitric oxide production in pertussis. Cell Microbiol 1(1):51–60

Flak TA, Heiss LN, Engle JT, Goldman WE (2000) Synergistic epithelial responses to endotoxin and a naturally occurring muramyl peptide. Infect Immun 68 (3):1235–1242

Folkening WJ, Nogami W, Martin SA, Rosenthal RS (1987) Structure of Bordetella pertussis peptidoglycan. J Bacteriol 169(9):4223–4227

Friedman RL, Fiederlein RL, Glasser L, Galgiani JN (1987) Bordetella pertussis adenylate cyclase: effects of affinity-purified adenylate cyclase on human polymorphonuclear leukocyte functions. Infect Immun 55 (1):135–140

Gentry-Weeks CR, Cookson BT, Goldman WE, Rimler RB, Porter SB, Curtiss R 3rd (1988) Dermonecrotic toxin and tracheal cytotoxin, putative virulence factors of Bordetella avium. Infect Immun 56(7):1698–1707

Glaser P, Ladant D, Sezer O, Pichot F, Ullmann A, Danchin A (1988a) The calmodulin-sensitive adenylate cyclase of Bordetella pertussis: cloning and expression in Escherichia coli. Mol Microbiol 2 (1):19–30

Glaser P, Sakamoto H, Bellalou J, Ullmann A, Danchin A (1988b) Secretion of cyclolysin, the calmodulin-sensitive adenylate cyclase-haemolysin bifunctional protein of Bordetella pertussis. EMBO J 7 (12):3997–4004

Goldman WE, Cookson BT (1988) Structure and functions of the Bordetella tracheal cytotoxin. Tokai J Exp Clin Med 13(Suppl):187–191

Goldman WE, Klapper DG, Baseman JB (1982) Detection, isolation, and analysis of a released Bordetella pertussis product toxic to cultured tracheal cells. Infect Immun 36(2):782–794

Gonzalez-Bullon D, Uribe KB, Martin C, Ostolaza H (2017) Phospholipase A activity of adenylate cyclase toxin mediates translocation of its adenylate cyclase domain. Proc Natl Acad Sci U S A 114(33):E6784–E6793

Goodell EW (1985) Recycling of murein by Escherichia coli. J Bacteriol 163(1):305–310

Goodell EW, Fazio M, Tomasz A (1978) Effect of benzylpenicillin on the synthesis and structure of the cell envelope of Neisseria gonorrhoeae. Antimicrob Agents Chemother 13(3):514–526

Goodwin MS, Weiss AA (1990) Adenylate cyclase toxin is critical for colonization and pertussis toxin is critical for lethal infection by Bordetella pertussis in infant mice. Infect Immun 58(10):3445–3447

Gordon VM, Young WW Jr, Lechler SM, Gray MC, Leppla SH, Hewlett EL (1989) Adenylate cyclase toxins from Bacillus anthracis and Bordetella pertussis. Different processes for interaction with and entry into target cells. J Biol Chem 264(25):14792–14796

Gray M, Szabo G, Otero AS, Gray L, Hewlett E (1998) Distinct mechanisms for K+ efflux, intoxication, and hemolysis by Bordetella pertussis AC toxin. J Biol Chem 273(29):18260–18267

Gray MC, Ross W, Kim K, Hewlett EL (1999) Characterization of binding of adenylate cyclase toxin to target cells by flow cytometry. Infect Immun 67 (9):4393–4399

Gray MC, Donato GM, Jones FR, Kim T, Hewlett EL (2004) Newly secreted adenylate cyclase toxin is responsible for intoxication of target cells by Bordetella pertussis. Mol Microbiol 53(6):1709–1719

Grimm M, Gsell S, Mittmann C, Nose M, Scholz H, Weil J, Eschenhagen T (1998) Inactivation of (Gialpha) proteins increases arrhythmogenic effects of beta-adrenergic stimulation in the heart. J Mol Cell Cardiol 30(10):1917–1928. https://doi.org/10.1006/jmcc.1998.0769

Gross MK, Au DC, Smith AL, Storm DR (1992) Targeted mutations that ablate either the adenylate cyclase or hemolysin function of the bifunctional cyaA toxin of Bordetella pertussis abolish virulence. Proc Natl Acad Sci U S A 89(11):4898–4902

Gueirard P, Druilhe A, Pretolani M, Guiso N (1998) Role of adenylate cyclase-hemolysin in alveolar macrophage apoptosis during Bordetella pertussis infection in vivo. Infect Immun 66(4):1718–1725

Guermonprez P, Khelef N, Blouin E, Rieu P, Ricciardi-Castagnoli P, Guiso N, Ladant D, Leclerc C (2001) The adenylate cyclase toxin of Bordetella pertussis binds to target cells via the alpha(M)beta(2) integrin (CD11b/CD18). J Exp Med 193(9):1035–1044

Guo Q, Shen Y, Lee YS, Gibbs CS, Mrksich M, Tang WJ (2005) Structural basis for the interaction of Bordetella pertussis adenylyl cyclase toxin with calmodulin. EMBO J 24(18):3190–3201

Hackett M, Guo L, Shabanowitz J, Hunt DF, Hewlett EL (1994) Internal lysine palmitoylation in adenylate cyclase toxin from Bordetella pertussis. Science 266 (5184):433–435

Hackett M, Walker CB, Guo L, Gray MC, Van Cuyk S, Ullmann A, Shabanowitz J, Hunt DF, Hewlett EL, Sebo P (1995) Hemolytic, but not cell-invasive activity, of adenylate cyclase toxin is selectively affected by differential fatty-acylation in Escherichia coli. J Biol Chem 270(35):20250–20253

Hanski E, Farfel Z (1985) Bordetella pertussis invasive adenylate cyclase. Partial resolution and properties of its cellular penetration. J Biol Chem 260(9):5526–5532

Harvill ET, Cotter PA, Yuk MH, Miller JF (1999) Probing the function of Bordetella bronchiseptica adenylate cyclase toxin by manipulating host immunity. Infect Immun 67(3):1493–1500

Hazes B, Boodhoo A, Cockle SA, Read RJ (1996) Crystal structure of the pertussis toxin-ATP complex: a molecular sensor. J Mol Biol 258(4):661–671

Hebeler BH, Young FE (1976) Chemical composition and turnover of peptidoglycan in Neisseria gonorrhoeae. J Bacteriol 126(3):1180–1185

Hewitt M, Canning BJ (2010) Coughing precipitated by Bordetella pertussis infection. Lung 188(Suppl 1): S73–S79

Hewlett EL, Urban MA, Manclark CR, Wolff J (1976) Extracytoplasmic adenylate cyclase of Bordetella pertussis. Proc Natl Acad Sci U S A 73(6):1926–1930

Hewlett EL, Manclark CR, Wolff J (1977) Adenyl cyclase in Bordetella pertussis vaccines. J Infect Dis 136 (Suppl):S216–S219

Hewlett EL, Donato GM, Gray MC (2006) Macrophage cytotoxicity produced by adenylate cyclase toxin from Bordetella pertussis: more than just making cyclic AMP! Mol Microbiol 59(2):447–459

Hinds PW 2nd, Yin C, Salvato MS, Pauza CD (1996) Pertussis toxin induces lymphocytosis in rhesus macaques. J Med Primatol 25(6):375–381

Hodge G, Hodge S, Markus C, Lawrence A, Han P (2003) A marked decrease in L-selectin expression by leucocytes in infants with Bordetella pertussis infection: leucocytosis explained? Respirology 8 (2):157–162

Hudnall SD, Molina CP (2000) Marked increase in L-selectin-negative T cells in neonatal pertussis. The lymphocytosis explained? Am J Clin Pathol 114 (1):35–40

Jacobs C, Huang LJ, Bartowsky E, Normark S, Park JT (1994) Bacterial cell wall recycling provides cytosolic muropeptides as effectors for beta-lactamase induction. EMBO J 13(19):4684–4694

Jo YY, Lee JY, Park CK (2016) Resolvin E1 inhibits substance P-Induced potentiation of TRPV1 in primary sensory neurons. Mediat Inflamm 2016:5259321

Kamanova J, Kofronova O, Masin J, Genth H, Vojtova J, Linhartova I, Benada O, Just I, Sebo P (2008) Adenylate cyclase toxin subverts phagocyte function by RhoA inhibition and unproductive ruffling. J Immunol 181(8):5587–5597

Kapil P, Papin JF, Wolf RF, Zimmerman LI, Wagner LD, Merkel TJ (2018) Maternal vaccination with a monocomponent pertussis toxoid vaccine is sufficient to protect infants in a baboon model of whooping cough. J Infect Dis 217(8):1231–1236

Katada T (2012) The inhibitory G protein G(i) identified as pertussis toxin-catalyzed ADP-ribosylation. Biol Pharm Bull 35(12):2103–2111

Khelef N, Sakamoto H, Guiso N (1992) Both adenylate cyclase and hemolytic activities are required by Bordetella pertussis to initiate infection. Microb Pathog 12(3):227–235

Khelef N, Zychlinsky A, Guiso N (1993) Bordetella pertussis induces apoptosis in macrophages: role of adenylate cyclase-hemolysin. Infect Immun 61 (10):4064–4071

Khelef N, Bachelet CM, Vargaftig BB, Guiso N (1994) Characterization of murine lung inflammation after infection with parental Bordetella pertussis and mutants deficient in adhesins or toxins. Infect Immun 62(7):2893–2900

Kirimanjeswara GS, Agosto LM, Kennett MJ, Bjornstad ON, Harvill ET (2005) Pertussis toxin inhibits neutrophil recruitment to delay antibody-mediated clearance of Bordetella pertussis. J Clin Invest 115 (12):3594–3601

Kotani S, Tsujimoto M, Koga T, Nagao S, Tanaka A, Kawata S (1986) Chemical structure and biological activity relationship of bacterial cell walls and muramyl peptides. Fed Proc 45(11):2534–2540

Krishnamoorthy S, Recchiuti A, Chiang N, Yacoubian S, Lee CH, Yang R, Petasis NA, Serhan CN (2010) Resolvin D1 binds human phagocytes with evidence for proresolving receptors. Proc Natl Acad Sci U S A 107(4):1660–1665

Krueger JM, Karnovsky ML (1987) Sleep and the immune response. Ann N Y Acad Sci 496:510–516

Krueger JM, Pappenheimer JR, Karnovsky ML (1982) Sleep-promoting effects of muramyl peptides. Proc Natl Acad Sci U S A 79(19):6102–6106

Levy BD, Serhan CN (2014) Resolution of acute inflammation in the lung. Annu Rev Physiol 76:467–492

Lim JH, Kim MS, Kim HE, Yano T, Oshima Y, Aggarwal K, Goldman WE, Silverman N, Kurata S, Oh BH (2006) Structural basis for preferential recognition of diaminopimelic acid-type peptidoglycan by a subset of peptidoglycan recognition proteins. J Biol Chem 281(12):8286–8295

Maddox JF, Hachicha M, Takano T, Petasis NA, Fokin VV, Serhan CN (1997) Lipoxin A4 stable analogs are potent mimetics that stimulate human monocytes and THP-1 cells via a G-protein-linked lipoxin A4 receptor. J Biol Chem 272(11):6972–6978

Magalhaes JG, Philpott DJ, Nahori MA, Jehanno M, Fritz J, Le Bourhis L, Viala J, Hugot JP, Giovannini M, Bertin J, Lepoivre M, Mengin-Lecreulx D, Sansonetti PJ, Girardin SE (2005) Murine Nod1 but not its human orthologue mediates innate immune detection of tracheal cytotoxin. EMBO Rep 6(12):1201–1207

Maher SA, Dubuis ED, Belvisi MG (2011) G-protein coupled receptors regulating cough. Curr Opin Pharmacol 11(3):248–253

Mangmool S, Kurose H (2011) G(i/o) protein-dependent and -independent actions of pertussis toxin (PTX). Toxins 3(7):884–899

Martin SA, Karnovsky ML, Krueger JM, Pappenheimer JR, Biemann K (1984) Peptidoglycans as promoters of slow-wave sleep. I. Structure of the sleep-promoting factor isolated from human urine. J Biol Chem 259 (20):12652–12658

Martin C, Requero MA, Masin J, Konopasek I, Goni FM, Sebo P, Ostolaza H (2004) Membrane restructuring by Bordetella pertussis adenylate cyclase toxin, a member of the RTX toxin family. J Bacteriol 186 (12):3760–3765

Masin J, Osicka R, Bumba L, Sebo P (2015) Bordetella adenylate cyclase toxin: a unique combination of a pore-forming moiety with a cell-invading adenylate cyclase enzyme. Pathog Dis 73(8):ftv075

Masin J, Osicka R, Bumba L, Sebo P (2018) Phospholipase A activity of adenylate cyclase toxin? Proc Natl Acad Sci U S A 115(11):E2489–E2490

Mauck J, Chan L, Glaser L (1971) Turnover of the cell wall of Gram-positive bacteria. J Biol Chem 246 (6):1820–1827

Melly MA, McGee ZA, Rosenthal RS (1984) Ability of monomeric peptidoglycan fragments from Neisseria gonorrhoeae to damage human fallopian-tube mucosa. J Infect Dis 149(3):378–386

Mielcarek N, Riveau G, Remoue F, Antoine R, Capron A, Locht C (1998) Homologous and heterologous protection after single intranasal administration of live attenuated recombinant Bordetella pertussis. Nat Biotechnol 16(5):454–457

Mielcarek N, Debrie AS, Raze D, Bertout J, Rouanet C, Younes AB, Creusy C, Engle J, Goldman WE, Locht C (2006) Live attenuated B. pertussis as a single-dose nasal vaccine against whooping cough. PLoS Pathog 2(7):e65

Mobberley-Schuman PS, Connelly B, Weiss AA (2003) Phagocytosis of Bordetella pertussis incubated with convalescent serum. J Infect Dis 187(10):1646–1653

Montgomery MK, McFall-Ngai M (1994) Bacterial symbionts induce host organ morphogenesis during early postembryonic development of the squid Euprymna scolopes. Development 120(7):1719–1729

Morse SI, Morse JH (1976) Isolation and properties of the leukocytosis- and lymphocytosis-promoting factor of Bordetella pertussis. J Exp Med 143(6):1483–1502

Munoz JJ, Arai H, Bergman RK, Sadowski PL (1981) Biological activities of crystalline pertussigen from Bordetella pertussis. Infect Immun 33(3):820–826

Munoz JJ, Bernard CC, Mackay IR (1984) Elicitation of experimental allergic encephalomyelitis (EAE) in mice with the aid of pertussigen. Cell Immunol 83 (1):92–100

Nguyen AW, Wagner EK, Laber JR, Goodfield LL, Smallridge WE, Harvill ET, Papin JF, Wolf RF, Padlan EA, Bristol A, Kaleko M, Maynard JA (2015) A cocktail of humanized anti-pertussis toxin antibodies limits disease in murine and baboon models of whooping cough. Sci Transl Med 7(316):316ra195

Njamkepo E, Pinot F, Francois D, Guiso N, Polla BS, Bachelet M (2000) Adaptive responses of human monocytes infected by Bordetella pertussis: the role of adenylate cyclase hemolysin. J Cell Physiol 183 (1):91–99

Nogimori K, Ito K, Tamura M, Satoh S, Ishii S, Ui M (1984) Chemical modification of islet-activating protein, pertussis toxin. Essential role of free amino groups in its lymphocytosis-promoting activity. Biochim Biophys Acta 801(2):220–231

Orr B, Douce G, Baillie S, Parton R, Coote J (2007) Adjuvant effects of adenylate cyclase toxin of Bordetella pertussis after intranasal immunisation of mice. Vaccine 25(1):64–71

Osickova A, Osicka R, Maier E, Benz R, Sebo P (1999) An amphipathic alpha-helix including glutamates 509 and 516 is crucial for membrane translocation of adenylate cyclase toxin and modulates formation and cation selectivity of its membrane channels. J Biol Chem 274(53):37644–37650

Osickova A, Masin J, Fayolle C, Krusek J, Basler M, Pospisilova E, Leclerc C, Osicka R, Sebo P (2010) Adenylate cyclase toxin translocates across target cell membrane without forming a pore. Mol Microbiol 75 (6):1550–1562

Oskouizadeh K, Selk-Ghafari M, Zahraei-Salehi T, Dezfolian O (2011) Isolation of Bordetella bronchiseptica in a dog with tracheal collapse. Comp Clin Pathol 20(5):153–158

Ostolaza H (2018) Reply to Masin et al: to be or not to be a phospholipase A. Proc Natl Acad Sci U S A 115(11): E2491

Ostolaza H, Martin C, Gonzalez-Bullon D, Uribe KB, Etxaniz A (2017) Understanding the mechanism of translocation of adenylate cyclase toxin across biological membranes. Toxins (Basel) 9(10)

Paddock CD, Sanden GN, Cherry JD, Gal AA, Langston C, Tatti KM, Wu KH, Goldsmith CS, Greer PW, Montague JL, Eliason MT, Holman RC, Guarner J, Shieh WJ, Zaki SR (2008) Pathology and pathogenesis of fatal Bordetella pertussis infection in infants. Clin Infect Dis 47(3):328–338

Paik D, Monahan A, Caffrey DR, Elling R, Goldman WE, Silverman N (2017) SLC46 family transporters facilitate cytosolic innate immune recognition of monomeric peptidoglycans. J Immunol 199(1):263–270

Pande AH, Moe D, Jamnadas M, Tatulian SA, Teter K (2006) The pertussis toxin S1 subunit is a thermally unstable protein susceptible to degradation by the 20S proteasome. Biochemistry 45(46):13734–13740

Parton R, Hall E, Wardlaw AC (1994) Responses to Bordetella pertussis mutant strains and to vaccination in the coughing rat model of pertussis. J Med Microbiol 40(5):307–312

Pearson RD, Symes P, Conboy M, Weiss AA, Hewlett EL (1987) Inhibition of monocyte oxidative responses by Bordetella pertussis adenylate cyclase toxin. J Immunol 139(8):2749–2754

Pham TH, Okada T, Matloubian M, Lo CG, Cyster JG (2008) S1P1 receptor signaling overrides retention mediated by G alpha i-coupled receptors to promote T cell egress. Immunity 28(1):122–133

Pierce C, Klein N, Peters M (2000) Is leukocytosis a predictor of mortality in severe pertussis infection? Intensive Care Med 26(10):1512–1514

Plaut RD, Carbonetti NH (2008) Retrograde transport of pertussis toxin in the mammalian cell. Cell Microbiol 10(5):1130–1139

Plaut RD, Scanlon KM, Taylor M, Teter K, Carbonetti NH (2016) Intracellular disassembly and activity of pertussis toxin require interaction with ATP. Pathog Dis 74(6):ftw065

Rogel A, Meller R, Hanski E (1991) Adenylate cyclase toxin from Bordetella pertussis. The relationship between induction of cAMP and hemolysis. J Biol Chem 266(5):3154–3161

Rose T, Sebo P, Bellalou J, Ladant D (1995) Interaction of calcium with Bordetella pertussis adenylate cyclase toxin. Characterization of multiple calcium-binding sites and calcium-induced conformational changes. J Biol Chem 270(44):26370–26376

Rosenthal RS, Nogami W, Cookson BT, Goldman WE, Folkening WJ (1987) Major fragment of soluble peptidoglycan released from growing Bordetella pertussis is tracheal cytotoxin. Infect Immun 55(9):2117–2120

Ross PJ, Lavelle EC, Mills KH, Boyd AP (2004) Adenylate cyclase toxin from Bordetella pertussis synergizes with lipopolysaccharide to promote innate interleukin-10 production and enhances the induction of Th2 and regulatory T cells. Infect Immun 72(3):1568–1579

Ross PJ, Sutton CE, Higgins S, Allen AC, Walsh K, Misiak A, Lavelle EC, McLoughlin RM, Mills KH (2013) Relative contribution of Th1 and Th17 cells in adaptive immunity to Bordetella pertussis: towards the rational design of an improved acellular pertussis vaccine. PLoS Pathog 9(4):e1003264

Rowlands HE, Goldman AP, Harrington K, Karimova A, Brierley J, Cross N, Skellett S, Peters MJ (2010) Impact of rapid leukodepletion on the outcome of severe clinical pertussis in young infants. Pediatrics 126(4):e816–e827

Rubin K, Glazer S (2016) The potential role of subclinical Bordetella Pertussis colonization in the etiology of multiple sclerosis. Immunobiology 221(4):512–515

Scanlon KM, Gau Y, Zhu J, Skerry C, Wall SM, Soleimani M, Carbonetti NH (2014) Epithelial anion transporter pendrin contributes to inflammatory lung pathology in mouse models of Bordetella pertussis infection. Infect Immun 82(10):4212–4221

Scanlon KM, Snyder YG, Skerry C, Carbonetti NH (2017) Fatal pertussis in the neonatal mouse model is associated with pertussis toxin-mediated pathology beyond the airways. Infect Immun 85(11)

Schenkel AR, Pauza CD (1999) Pertussis toxin treatment in vivo reduces surface expression of the adhesion integrin leukocyte function antigen-1 (LFA-1). Cell Adhes Commun 7(3):183–193

Schneider OD, Weiss AA, Miller WE (2009) Pertussis toxin signals through the TCR to initiate cross-desensitization of the chemokine receptor CXCR4. J Immunol 182(9):5730–5739

Sebo P, Ladant D (1993) Repeat sequences in the Bordetella pertussis adenylate cyclase toxin can be recognized as alternative carboxy-proximal secretion signals by the Escherichia coli alpha-haemolysin translocator. Mol Microbiol 9(5):999–1009

Skerry CM, Mahon BP (2011) A live, attenuated Bordetella pertussis vaccine provides long-term protection against virulent challenge in a murine model. Clin Vaccine Immunol 18(2):187–193

Skerry CM, Cassidy JP, English K, Feunou-Feunou P, Locht C, Mahon BP (2009) A live attenuated Bordetella pertussis candidate vaccine does not cause disseminating infection in gamma interferon receptor knockout mice. Clin Vaccine Immunol 16(9):1344–1351

Skerry C, Goldman WE, Carbonetti NH (2019) Peptidoglycan recognition protein 4 suppresses early inflammatory responses to Bordetella pertussis and contributes to Sphingosine-1-Phosphate receptor agonist-mediated disease attenuation. Infect Immun 87(2)

Skopova K, Tomalova B, Kanchev I, Rossmann P, Svedova M, Adkins I, Bibova I, Tomala J, Masin J, Guiso N, Osicka R, Sedlacek R, Kovar M, Sebo P (2017) Cyclic AMP-elevating capacity of adenylate cyclase toxin-hemolysin is sufficient for lung infection but not for full virulence of Bordetella pertussis. Infect Immun 85(6):pii: e00937-16

Spensieri F, Fedele G, Fazio C, Nasso M, Stefanelli P, Mastrantonio P, Ausiello CM (2006) Bordetella pertussis inhibition of interleukin-12 (IL-12) p70 in human monocyte-derived dendritic cells blocks IL-12 p35 through adenylate cyclase toxin-dependent cyclic AMP induction. Infect Immun 74(5):2831–2838

Surridge J, Segedin ER, Grant CC (2007) Pertussis requiring intensive care. Arch Dis Child 92(11):970–975

Svedova M, Masin J, Fiser R, Cerny O, Tomala J, Freudenberg M, Tuckova L, Kovar M, Dadaglio G, Adkins I, Sebo P (2016) Pore-formation by adenylate cyclase toxoid activates dendritic cells to prime CD8+ and CD4+ T cells. Immunol Cell Biol 94(4):322–333

Szabo G, Gray MC, Hewlett EL (1994) Adenylate cyclase toxin from Bordetella pertussis produces ion conductance across artificial lipid bilayers in a calcium- and polarity-dependent manner. J Biol Chem 269(36):22496–22499

Thierry-Carstensen B, Dalby T, Stevner MA, Robbins JB, Schneerson R, Trollfors B (2013) Experience with monocomponent acellular pertussis combination vaccines for infants, children, adolescents and adults--a review of safety, immunogenicity, efficacy and effectiveness studies and 15 years of field experience. Vaccine 31(45):5178–5191

Thorstensson R, Trollfors B, Al-Tawil N, Jahnmatz M, Bergstrom J, Ljungman M, Torner A, Wehlin L, Van Broekhoven A, Bosman F, Debrie AS, Mielcarek N, Locht C (2014) A phase I clinical study of a live attenuated Bordetella pertussis vaccine--BPZE1; a single centre, double-blind, placebo-controlled, dose-escalating study of BPZE1 given intranasally to healthy adult male volunteers. PLoS One 9(1):e83449

Uehara T, Park JT (2008) Peptidoglycan recycling. EcoSal Plus 3(1)

Uribe KB, Etxebarria A, Martin C, Ostolaza H (2013) Calpain-mediated processing of adenylate cyclase toxin generates a cytosolic soluble catalytically active n-terminal domain. PLoS One 8(6):e67648

Valdez HA, Oviedo JM, Gorgojo JP, Lamberti Y, Rodriguez ME (2016) Bordetella pertussis modulates human macrophage defense gene expression. Pathog Dis 74(6):ftw073

Vojtova J, Kamanova J, Sebo P (2006) Bordetella adenylate cyclase toxin: a swift saboteur of host defense. Curr Opin Microbiol 9(1):69–75

Wainford RD, Kurtz K, Kapusta DR (2008) Central G-alpha subunit protein-mediated control of cardiovascular function, urine output, and vasopressin secretion in conscious Sprague-Dawley rats. Am J Physiol Regul Integr Comp Physiol 295(2):R535–R542

Wang X, Gray MC, Hewlett EL, Maynard JA (2015) The Bordetella adenylate cyclase repeat-in-toxin (RTX) domain is immunodominant and elicits neutralizing antibodies. J Biol Chem 290(6):3576–3591

Warfel JM, Merkel TJ (2014) The baboon model of pertussis: effective use and lessons for pertussis vaccines. Expert Rev Vaccines 13(10):1241–1252

Warfel JM, Beren J, Kelly VK, Lee G, Merkel TJ (2012) Nonhuman primate model of pertussis. Infect Immun 80(4):1530–1536

Warnock RA, Askari S, Butcher EC, von Andrian UH (1998) Molecular mechanisms of lymphocyte homing to peripheral lymph nodes. J Exp Med 187(2):205–216

Weingart CL, Weiss AA (2000) Bordetella pertussis virulence factors affect phagocytosis by human neutrophils. Infect Immun 68(3):1735–1739

Weingart CL, Mobberley-Schuman PS, Hewlett EL, Gray MC, Weiss AA (2000) Neutralizing antibodies to adenylate cyclase toxin promote phagocytosis of Bordetella pertussis by human neutrophils. Infect Immun 68(12):7152–7155

Weiss AA, Goodwin MS (1989) Lethal infection by Bordetella pertussis mutants in the infant mouse model. Infect Immun 57(12):3757–3764

Weiss AA, Hewlett EL, Myers GA, Falkow S (1983) Tn5-induced mutations affecting virulence factors of Bordetella pertussis. Infect Immun 42(1):33–41

Winter K, Zipprich J, Harriman K, Murray EL, Gornbein J, Hammer SJ, Yeganeh N, Adachi K, Cherry JD (2015) Risk factors associated with infant deaths from pertussis: a case-control study. Clin Infect Dis 61(7):1099–1106

Witvliet MH, Burns DL, Brennan MJ, Poolman JT, Manclark CR (1989) Binding of pertussis toxin to eucaryotic cells and glycoproteins. Infect Immun 57 (11):3324–3330

Wolff J, Cook GH, Goldhammer AR, Berkowitz SA (1980) Calmodulin activates prokaryotic adenylate cyclase. Proc Natl Acad Sci U S A 77(7):3841–3844

Wong WS, Rosoff PM (1996) Pharmacology of pertussis toxin B-oligomer. Can J Physiol Pharmacol 74 (5):559–564

Worthington ZE, Carbonetti NH (2007) Evading the proteasome: absence of lysine residues contributes to pertussis toxin activity by evasion of proteasome degradation. Infect Immun 75(6):2946–2953

Yajima M, Hosoda K, Kanbayashi Y, Nakamura T, Nogimori K, Mizushima Y, Nakase Y, Ui M (1978) Islets-activating protein (IAP) in Bordetella pertussis that potentiates insulin secretory responses of rats. Purification and characterization. J Biochem 83 (1):295–303

Zaretzky FR, Gray MC, Hewlett EL (2002) Mechanism of association of adenylate cyclase toxin with the surface of Bordetella pertussis: a role for toxin-filamentous haemagglutinin interaction. Mol Microbiol 45 (6):1589–1598

Zhao CB, Coons SW, Cui M, Shi FD, Vollmer TL, Ma CY, Kuniyoshi SM, Shi J (2008) A new EAE model of brain demyelination induced by intracerebroventricular pertussis toxin. Biochem Biophys Res Commun 370 (1):16–21

Zheng M, Zhu W, Han Q, Xiao RP (2005) Emerging concepts and therapeutic implications of beta-adrenergic receptor subtype signaling. Pharmacol Ther 108(3):257–268

Zocchi MR, Contini P, Alfano M, Poggi A (2005) Pertussis toxin (PTX) B subunit and the nontoxic PTX mutant PT9K/129G inhibit Tat-induced TGF-beta production by NK cells and TGF-beta-mediated NK cell apoptosis. J Immunol 174(10):6054–6061

Adv Exp Med Biol - Advances in Microbiology, Infectious Diseases and Public Health (2019) 1183: 53–80
https://doi.org/10.1007/5584_2019_404
© Springer Nature Switzerland AG 2019
Published online: 21 August 2019

Functional Programming of Innate Immune Cells in Response to *Bordetella pertussis* Infection and Vaccination

Joshua Gillard, Evi van Schuppen,
and Dimitri A. Diavatopoulos

Abstract

Despite widespread vaccination, *B. pertussis* remains one of the least controlled vaccine-preventable diseases. Although it is well known that acellular and whole cell pertussis vaccines induce distinct immune functionalities in memory cells, much less is known about the role of innate immunity in this process. In this review, we provide an overview of the known differences and similarities in innate receptors, innate immune cells and inflammatory signalling pathways induced by the pertussis vaccines either licensed or in development and compare this to primary infection with *B. pertussis*. Despite the crucial role of innate immunity in driving memory responses to *B. pertussis,* it is clear that a significant knowledge gap remains in our understanding of the early innate immune response to vaccination and infection. Such knowledge is essential to develop the next generation of pertussis vaccines with improved host defense against *B. pertussis*.

Authors Joshua Gillard and Evi van Schuppen have equally contributed to this chapter.

J. Gillard, E. van Schuppen, and D. A. Diavatopoulos (✉)
Section Pediatric Infectious Diseases, Laboratory of Medical Immunology, Radboud Institute for Molecular Life Sciences, Radboudumc, Nijmegen, The Netherlands

Radboud Center for Infectious Diseases, Radboudumc, Nijmegen, The Netherlands
e-mail: Dimitri.Diavatopoulos@radboudumc.nl

Keywords

Immune programming · Innate immunity · Pertussis vaccination

1 Introduction

After almost 70 years of vaccination, *Bordetella pertussis* remains (one of) the least controlled vaccine-preventable diseases. Although the differences in surveillance systems and diagnostic methods between countries make it difficult to formally compare pertussis epidemiology between countries, it is clear that *B. pertussis* disease incidence has increased in several countries in recent decades (Tan et al. 2015; Domenech de Celles et al. 2016; Pillsbury et al. 2014). Whole cell and acellular pertussis vaccines are both highly efficacious against pertussis disease, at least in the short term (Greco et al. 1996; Gustafsson et al. 1996). However, there are indications that these vaccines differ in their ability to clear asymptomatic infection (Warfel et al. 2014). Furthermore, recent studies in the baboon challenge model demonstrated that primary infection may lead to sterilizing immunity against subsequent challenge (Warfel et al. 2014). These findings demonstrate that immunity against specific aspects of infection with *B. pertussis* may be achieved via multiple mechanisms.

Innate Immunity Vaccines, including those for *B. pertussis*, have traditionally been developed by isolating the microorganisms, inactivating them with heat or chemicals, and administering some or all of the products as an injection. The first vaccine was the whole cell pertussis vaccine (wP), which was derived from heat-killed *B. pertussis* shortly after its isolation by Jules Bordet in the early 1900s (Bordet and Genou 1906). The acellular pertussis (aP) vaccine was developed much later once protective antigens had been identified, although without a deep understanding of the protective mechanism. Each of these vaccines has come under scrutiny for different reasons, some of which are attributable to suboptimal stimulation of innate immunity. Innate immune signalling plays a key role in the induction of antigen presentation and activation of functional response programs in memory T and B cells. It is therefore critical that we understand the differences and similarities not only between the various pertussis vaccine types, but also in relation to infection. In this review, we describe recent insights into the mechanisms through which innate immunity is programmed by *B. pertussis* and pertussis vaccines, how these innate responses prime adaptive immunity, as well as how the innate immune system can be (re)directed to drive superior protective responses with adjuvants and new vaccine formulations.

Detection of *Bordetella pertussis* during infection, or the vaccine components after vaccination, and the final generation of an effective memory response is primarily mediated through pattern recognition receptors (PRRs) on innate immune cells. These proteins sense microorganisms – and vaccines – through interaction with evolutionarily conserved microbial ligands, which are expressed by bacteria, viruses, parasites, and fungi but not mammals. Thus far, four classes of PRRs have been described: Toll-like receptors (TLRs), nucleotide-binding oligomerization domain-like receptors (NLRs), C-type lectin receptors (CLRs), and retinoic acid-inducible gene-I-like receptors (RLRs). Engagement of a PRR ligand with its cognate receptor triggers an intracellular signalling cascade, ultimately leading to

inflammatory gene transcription, maturation of the stimulated cell, and the production of cytokines and chemokines that regulate both innate and adaptive immune responses (thoroughly reviewed elsewhere, (Brubaker et al. 2015; Kawai and Akira 2009, 2010)). PRR ligands are also expressed by antigen presenting cells, which present processed antigens to CD4$^+$ T cells via the major histocompatibility complex class II molecules. Activation of PRRs by a particular pathogen or vaccine leads to intracellular re-wiring of innate immune cells, thereby driving maturation into functional cell types, and priming adaptive immune responses that lead to pathogen clearance (Schenten and Medzhitov 2011). Thus, immune memory cells are 'functionally programmed' by the unique combination of PRRs and inflammatory pathways that are activated during infection. Natural infection induces long-lasting protection, which can likely be attributed to the induction of the most 'relevant' signalling pathways for clearance and the location of immune induction. Vaccination will ideally mimic this process but may not be identical, depending on the vaccine and administration route that is used.

Innate Control of Adaptive Immunity An important consequence of an effective innate immune response is the stimulation of enduring adaptive immunity and polarization of T helper (Th) cells. Several classes of T helper cells have been described, including Th1, Th2, Th17, Th9, and many others. It has also been demonstrated that infection with microbes induces different subpopulations within each class of T helper cells (reviewed in (Sallusto 2016)). The cellular circuits and intracellular signalling pathways that lead to the induction of these subpopulations is an area of active research. Here, we will briefly review how Th1, Th2, and Th17 cells are induced by innate immunity, and their role during pertussis infection. Naive T cells are driven to differentiate into Th1 cells by early exposure to IFNγ and IL-12 at the time of T cell priming. Natural killer (NK) cells are the earliest source of IFNγ production, while Dendritic cells (DCs) are the source of

IL-12 during priming (O'Garra 1998). IFNγ signals via the STAT1 pathway and induces the expression of T-bet, a transcriptional regulator essential for promoting IFNγ expression and at the same time downregulating the expression of IL-4, which dampens the differentiation of naive T cells into Th2 cells (Takeda et al. 2003). IL-12 produced by DCs binds to its receptor and signals via the STAT4 pathway, enhancing IFNγ production by the differentiated Th1 cells (Athie-Morales et al. 2004). There is evidence that IFNγ and Th1 immunity is essential for clearance of *B. pertussis* from the respiratory tract. Several mouse studies examined T cell responses in the lungs during infection with *B. pertussis*, demonstrating a role for IL-2 and IFNγ secreting antigen-specific CD4$^+$ T cells (Mills et al. 1993; Barnard et al. 1996). Studies using cytokine- and receptor-deficient mice found that IFNγ-depleted mice exhibited a significant attenuation in bacterial clearance, further supporting the role of IFNγ in protection (Barbic et al. 1997). Efforts to investigate the contribution of T cells in the protection against *B. pertussis* in humans have confirmed that pertussis-specific Th1 responses are important (Ryan et al. 1997; Ausiello et al. 1998; Fedele et al. 2013). Naive T cells are polarized towards Th17 cells via a different signalling pathway, and are induced by wPs and natural infection. Th17 differentiation can be initiated by TGFβ, IL-6, IL-21, and IL-23, which are produced by DCs. Similar to T-bet in Th1 polarization, RORγt and RORα are highly induced in Th17 cells and play an important role in this lineage differentiation via the STAT3 pathway (Huang et al. 2012). IL-17 producing CD4$^+$ T cells were also linked to the control of *B. pertussis* infection. Antigen-specific Th17 cells were induced upon *B. pertussis* infection in mice after infection (Feunou et al. 2010; Ross et al. 2013; Scanlon et al. 2014) and IL-17-deficient mice were shown to be attenuated in bacterial clearance (Ross et al. 2013). Further studies showed that γδ T cells in particular play a key role in clearance by providing an early source of IL-17. These cells are pathogen-specific and tissue-resident and expand in the lungs following re-infection of

mice (Misiak et al. 2017b). The same group also showed that antigen-specific tissue resident memory T cells (Trm cells) play a critical role in mediating the rapid clearance seen following re-challenge of *B. pertussis*-convalescent animals (Wilk et al. 2017). The recently established baboon pertussis challenge model provides a unique opportunity to evaluate the histopathology of severe pertussis infection. This model has been shown to recapitulate the characteristic clinical signs of pertussis observed in humans, including mucosal damage, influx of immune cells and paroxysmal coughing. In line with the studies in mice, IL-17- and IFNγ-secreting CD4$^+$ T cells were also detected in convalescent, but not naive baboons, which remained detectable 2 years after challenge. These *B. pertussis*-specific Th1 and Th17 cells are thought to contribute to the immunity conferred by natural infection (Warfel and Merkel 2013).

Th2 responses are generated when naive T cells are exposed to IL-4 or IL-2 during T cell priming. IL-4 signalling occurs via the STAT6 pathway to express GATA3, which is the master regulator of Th2 differentiation. IL-2 signals via the STAT5 pathway and both enhances IL-4 production by differentiated Th2 cells as well as dampening T-bet for Th1 differentiation (Pulendran et al. 2010). Regulatory T cells (Treg cells), also known as suppressor cells, are another subset of T cells that modulate the immune response. In the lungs of *B. pertussis* infected mice, a high number of Treg cells were found to produce IL-10 (Coleman et al. 2012). These regulatory T cells allow the bacteria to survive inside the host by dampening the immune response, but also provide a benefit to the host by minimizing the damage induced by an overactive immune system. Both IL-4-producing CD4$^+$ T and Treg cells were also found to be dispensable for bacterial clearance, since deletion of the IL-4 gene did not attenuate clearance in protective immunity induced by aP in mice (Ross et al. 2013).

CD4$^+$ T cells are essential for promoting the maturation and differentiation of antigen-specific B cells and the production of antibodies, the last of which are critical for preventing infection. B cells also play a role in protection against *B. pertussis*, since repeated immunization of B cell knock-out mice resulted in partial protection, and complete protection could be reconstituted by transfer of pertussis-specific B cells (Leef et al. 2000). A study investigating pertussis patients showed that age and 'closeness' to the last pertussis infection have an effect on the levels of memory B cells specific for pertussis. Waning of cellular immunity occurred within 9 months, and the highest levels of memory B cells were found in adults and elderly compared to infants and children (van Twillert et al. 2014). Although memory B cells may not be able to directly *prevent* infection, they can rapidly proliferate and differentiate into antibody producing cells after re-exposure to pertussis. As such, the presence of memory B cells may be critical in controlling the infection at an early stage and preventing the development of disease symptoms (Ahmed and Gray 1996).

PRR Ligands Classic wP vaccines contain PRR ligands similar to those released during infection, thus inducing similar functional immune memory responses. While many immunogenic PRR ligands and antigens have been described for *B. pertussis*, few, if any classic PRR ligands exist in modern aP formulations. Here we will describe the components of *B. pertussis* and *B. pertussis* vaccines that have been shown to demonstrate some PRR stimulation capacity.

Lipopolysaccharide (LPS) is typically bound by soluble CD14, which is then recognized by TLR4 on the surface of immune cells. This initiates an intracellular signalling pathway that ends in the secretion of IL-6 and type-I interferons (Kawai and Akira 2010). Unlike most Gram-negative bacteria, *B. pertussis* produces a structurally distinct LPS (lipooligosaccharide, LOS) characterised by the replacement of the typical polysaccharide O-side chain with a complex branching structure with unusual sugars (Caroff et al. 2002). There is accumulating evidence that this change impacts on the immunogenicity of the molecule. Reports indicate that *B. pertussis* LOS activates TLR4 via a CD14-independent mechanism, and that it is less capable than *E. coli* LPS in inducing inflammatory cytokines IL-12, IL-6 and IL-23 in human monocyte-derived DCs. Consequently, DCs stimulated with *B. pertussis* LOS, compared to *E. coli* LPS, induce Th2/Th17 T cell polarization instead of Th1 responses (Fedele et al. 2007, 2008). There is also some evidence that LOS induces IL-10 production in murine DCs and leading to the generation of immunosuppressive Tr1 cells *in vivo* (Higgins et al. 2003). During infection, *B. pertussis* LOS has been shown to play an important role in evading other host defence mechanisms, such as the complement system (reviewed in (Jongerius et al. 2015)). *B. pertussis* LOS enables the evasion of complement-mediated killing and enhanced colonization in the mouse model (Ganguly et al. 2014). Similarly, surfactant proteins (SPs) are lipid-binding proteins in the lower respiratory tract that possess antimicrobial activity by binding, damaging, and opsonising microbial membranes. While SP-A is effective against membranes containing LPS, LOS enables *B. pertussis* to escape recognition and clearance (Schaeffer et al. 2004).

Pertussis toxin (PT) is perhaps the most well-known of all the *B. pertussis* virulence factors. Originally identified as lymphocyte-promoting factor (LPF), its name was later changed due to the observation that it is one of the most important toxins for inducing pathogenesis (Cherry 2015). One of the first acellular pertussis vaccines that was developed and tested in humans was a monocomponent vaccine consisting only of inactivated PT adjuvanted with alum (Trollfors et al. 1995). It was later determined that 3- and 5- component aP formulations are more effective, though today all modern aPs contain at least PT as a vaccine antigen (Decker and Edwards 2000).

Active PT displays a range of potent immunological activities, including stimulation of lymphocyte proliferation (Berstad et al. 2000) and immunoglobulin synthesis (Samore and Siber 1996), as well as the induction of T cell responses (Ryan et al. 1998a; Ronchi et al. 2016) and experimental autoimmune diseases (Ronchi et al. 2016;

Dumas et al. 2014). Interestingly, some of these characteristics can be linked to its capacity to stimulate innate immune cells via PRRs. PT has been shown to stimulate TLR2/4 on DC, leading to the activation of NFκB and the production of inflammatory cytokines IL-6, TNFα (Wang et al. 2006; Nasso et al. 2009; Fedele et al. 2010), and IL-12, thereby enhancing Th1 immunity (Ryan et al. 1998a; Tonon et al. 2002; Ausiello et al. 2002). PT can also induce TLR4-dependent pro-IL1β, and activate the inflammasome through a separate mechanism, enabling secretion of active IL-1β in myeloid cells and driving Th17 T cell responses *in vivo* (Ronchi et al. 2016; Dumas et al. 2014). Treatment with aldehyde or heat (common detoxification methods used to make pertussis vaccines) destroys not only many epitopes but also the adjuvanticity of PT, suggesting that either its enzymatic properties or molecular conformation is important for its capacity to stimulate innate immune responses (Ryan et al. 1998a; Phongsisay et al. 2017). Genetically detoxified PT, which remains structurally unchanged but has lost its enzymatic activity, remains capable of inducing TLR2/4-mediated Th1 & Th17 immune responses, albeit via a mechanism dependent on IL-10 secretion (Nasso et al. 2009). Stimulation of IL-10 production, which is unique to genetically detoxified PT, may potentially reduce adjuvanticity and also eliminate the capacity to induce experimental autoimmune disease (Zhou et al. 2014).

Several other *B. pertussis* toxins have also been shown to demonstrate stimulation of the intracellular NLRs. These are currently only present in wPs and actively produced during infection, though they represent potential targets for aP development. Adenylate cyclase toxin (ACT) is one such protein that both stimulates innate immunity and induces protective antibody responses (Hormozi et al. 1999). ACT is composed of two independent domains: while the C-terminal cytolysin (RTX) domain binds to the surface of innate immune cells via the integrin receptors CD11b/CD18 (Guermonprez et al. 2001; Benz et al. 1994), the enzymatic adenylate cyclase moiety is delivered into the cell and disrupts cell signalling by converting ATP into cAMP, thereby disrupting cell signalling (Vojtova et al. 2006). As a consequence, DC maturation and TLR4- mediated IL-12 production is inhibited, instead favouring IL-10 secretion and ultimately the repolarization of T cell responses away from Th1 and towards the development of immunosuppressive T regulatory cells (Ross et al. 2004; Spensieri et al. 2006; Wang et al. 2015). Further investigation has revealed that RTX domain alone has some adjuvant properties by signalling through NLRP3, stimulating the production of IL-1β and protective Th17 immunity (Dunne et al. 2010). Aside from its adjuvant properties, RTX has also been shown to raise antibodies to ACT suggesting that current acellular vaccines might benefit from the addition of RTX (Hickey et al. 2008; Boehm et al. 2018).

Tracheal cytotoxin (TCT) is essentially a by-product of *B. pertussis* peptidoglycan synthesis, which was initially shown to have some cytotoxic effects on epithelial cells (Goldman et al. 1982). The NLR NOD1 was initially identified as the mammalian sensor of TCT in both epithelial cells and immune cells (Magalhaes et al. 2005), inducing inflammatory cytokine responses both *in vitro* and *in vivo*. One remaining question pertained to how TCT gains access to the cytoplasm where it can stimulate NOD1. Recently, a novel transporter has been identified in the fruit fly, mouse, and human cells that facilitates cytosolic recognition of TCT in epithelial cells (Paik et al. 2017). One of the human homologs, SLC56A2s, was shown to localize to acidic organelles such as late-endosomes or endolysosomes after TCT stimulation. Delivery of TCT via SLC56A2s led to NOD1 dependent activation of NFkβ. SLC56A2s also appears to be moderately expressed in CD14[+] monocytes, suggesting a new mechanism through which innate immunity can detect and respond to pathogens like *B. pertussis*. Additionally, it has also been observed that many circulating strains of *B. pertussis* can elicit NOD2 signalling in reporter cell lines. However, it is not yet clear which specific *B. pertussis* component may be driving this response (Hovingh et al. 2017). The importance of NLRs in mediating vaccine-induced immunity is highlighted by the observation that Freund's adjuvant, which is considered to be one of the most potent adjuvants, signals through NOD2. NOD2 detects cytosolic bacteria by sensing peptidoglycan (PGN) via muramyl dipeptide (Behr and Divangahi 2015). Since NOD2 activation is particularly associated with the induction of Th17 immunity (van Beelen et al. 2007), which is highly relevant for protection against *B. pertussis*, it represents a target for further vaccine development.

Filamentous hemagglutinin (FHA), fimbriae (FIM), and pertactin (PRN) are *B. pertussis* virulence factors that are important for adhesion to and colonization of the respiratory epithelium. Although it appears that they primarily orchestrate an anti-inflammatory or immunomodulatory activity on innate immunity, they are nonetheless capable of mounting protective antibody responses. As primary components in aPs, but also in wP formulations, it is important to describe the influence these antigens have on inflammation and the induction of adaptive immunity. FHA is a large 220kDA protein that is both secreted and surface-associated (Renauld-Mongenie et al. 1996), and which enables binding to a variety of extracellular structures on epithelial cells and macrophages (Locht et al. 1993). The first studies in mice indicated that FHA is required by *B. pertussis* to evade host immunity by suppressing IL-12 and stimulating IL-10 production, thereby avoiding protective Th1 immunity and favouring the generation of tolerogenic Tr1 cells (McGuirk and Mills 2000; McGuirk et al. 2002). Perhaps the most striking evidence for the immunosuppressive role of FHA is the finding that experimental colitis is completely prevented following parenteral injection (Braat et al. 2007), once again mediated by IL-10 and the suppression of adaptive immunity. Later studies used similar methods to determine the role of FHA in a related species *Bordetella bronchiseptica*. The earliest report is concordant from the experiments with *B. pertussis*, suggesting that FHA is critical for adherence

and possesses immunomodulatory characteristics (Inatsuka et al. 2005). Insight into the potential mechanism came with the observation that FHA inhibits the development of IL-17 responses and neutrophil recruitment and activation, leading to impaired bacterial clearance (Henderson et al. 2012). Until this moment, human studies have been exclusively performed *in vitro* and highlight an opposite role for FHA. In this context, stimulation of peripheral blood mononuclear cells, monocytes, and macrophages have all indicated that FHA is highly pro-inflammatory both in terms of cytokine secretion and gene transcription (Dieterich and Relman 2011; Abramson et al. 2001, 2008). These differences in responses might be explained by host specific-responses to FHA. However, one investigation suggested that highly pure FHA that is depleted of co-purified LPS fails to elicit IL-10 or pro-inflammatory cytokine secretion (Villarino Romero et al. 2016), and another suggests that FHA interacts with TLR2, but not TLR4 (which is a well-described receptor for LPS) (Asgarian-Omran et al. 2015).

B. pertussis FIM are composed of either FIM2 or FIM3 depending on the serotype, and are often present together in some of the acellular pertussis vaccine formulations (Ashworth et al. 1982). wP formulations often contain multiple strains to cover the various FIM types. Besides its role in attachment to epithelial cells, FIM also plays an important role in stimulating antibody responses during experimental infection, not only against FIM itself but also against other *B. pertussis* antigens. One study found that mice infected with mutant FIM strains of *B. bronchiseptica* produced significantly lower overall antibody responses both early and late after infection (Mattoo et al. 2000). Similarly, supplementation of Fim2/3 to FIM-containing acellular pertussis vaccine formulations appears to increase their protective capacity *in vivo*, but without affecting the secretion of innate inflammatory cytokines *in vitro* (Queenan et al. 2019). A clearer role for the immunosuppressive effects on innate immunity of FIM have been identified. The lungs of mice infected with FIM-deficient strains of

B. pertussis show markedly increased IL-1β and inflammatory chemokine MCP-1 after 1- and 3-days post-infection, corresponding to lower bacterial burden at later time points (Scheller et al. 2015). Altogether, it appears that while FIM is important for humoral immunity, it remains to be seen if it triggers innate immune receptors and whether FIM mediates some effect on T cell polarization.

Although it is a protective antigen, the role of PRN as a virulence factor is controversial. Initial *in vitro* studies have indicated that it promotes adhesion to epithelial cells, but experiments in the mouse model have demonstrated that it is not essential for colonization (Leininger et al. 1991; Roberts et al. 1991). Until recently, it was also unknown whether this protein had any innate immune activating capacity. It was demonstrated that mutant PRN-deficient strains stimulated higher neutrophil responses soon after infection, suggesting that PRN could inhibit phagocytosis (Inatsuka et al. 2010). Infection of human DC in a separate study similarly reported that *B. pertussis* lacking PRN displayed increased invasion, although no changes in differences in DC maturation were observed (Stefanelli et al. 2009). One shortcoming of these previous studies is that the comparison between PRN-deficient and PRN-expressing *B. pertussis* was made using strains with different genetic backgrounds. Recently, Hovingh et al. (Hovingh et al. 2018) overcame this limitation and identified the specific immunomodulatory role of PRN *in vitro* and *in vivo* by comparing multiple isogenic strain pairs, presumably differing only in PRN expression. They found that both human and mouse DCs secrete increased levels of pro-inflammatory cytokines TNFa, IL-6, IL-8, and growth factor G-CSF in response to stimulation with PRN-deficient strains. Further investigation revealed a transcriptomic signature in these cells characterized by increased expression of important pro-inflammatory transcription factors such as NFκB. In mice, they reported a trend towards increased inflammatory cytokines in sera 3- and 7-days post-infection, as well as significantly increased gene expression for cell death pathways in the lungs. Taken together, it appears that PRN

helps *B. pertussis* evade host innate immunity by inhibiting phagocytosis and activation.

Almost two decades ago, a new process called "reverse vaccinology" was described (reviewed in (Rappuoli 2000; Rappuoli et al. 2016) whereby genomic, proteomic and bioinformatic technologies enable the discovery of novel vaccine components. This approach has been used in recent years to identify promising new antigens and adjuvants for *B. pertussis*. One of the first studies to apply this molecular information to *B. pertussis* discovered several putatively immunogenic lipoproteins (Dunne et al. 2015). Ultimately, two of the candidates displayed significant TLR2 stimulation and elicited pro-inflammatory cytokine secretion *in vitro*, including the Th1 polarizing cytokine IL-12. A representative synthetic lipopeptide (LP1569) was then generated and shown to possess similar adjuvanticity in mouse and human DCs. In another instance, an unconventional immunoproteomics approach was employed to describe the dominant MHC-II-presented CD4[+] T cell epitopes (Stenger et al. 2014). Curiously, none were derived from the known virulence factors, but rather from cytosolic and envelope proteins. Synthetic peptides that were based on these proteins were generated and screened for their capability to elicit T cell responses. All of them induced T cell proliferation and Th1/Th2 polarized responses, thus establishing a role for proteins other than the known virulence factors in immunity to *B. pertussis*. In light of the central role for DCs in triggering adaptive responses, an interesting question arising out of these findings is whether the identified proteins also possess some capacity to stimulate innate immune responses via PRRs. There are many *B. pertussis* factors that have some effect on the subsequent immune response, either directly by triggering PRRs or indirectly by modulating the subsequent inflammatory response, for example, through the recruitment of cells, or production of cytokines. Thus far, there has been no concerted effort to identify all the PRRs that are activated by *B. pertussis* or by pertussis vaccines. Thus, it stands to reason that there are very likely many other PRR ligands besides the ones described above that play a role that are currently unknown. TLR9 induction by *B. pertussis* is one example. TLR9 is activated by bacterial DNA, which contains foreign CpG motifs (Krieg 2002). Since *B. pertussis* can survive and even replicate inside phagocytic cells, this intracellular TLR pathway may be selectively activated during infection but not during vaccination. It bears emphasizing that the above list is far from complete and may be highly biased, especially towards LOS/LPS and TLR4. Finally, another important feature of innate immune recognition of foreign material is the information processing that occurs during phagocytosis. What this implies is that signals derived from soluble antigens versus whole bacterial particles are fundamentally different (reviewed in (Underhill and Goodridge 2012)). The latter will induce different co-signalling and inflammatory responses upon recognition, as they provide relevant context to the immune system about the clearance mechanisms that need to be activated.

wP vs aP Vaccines Owing to its repertoire of PRR ligands, including LOS and residual toxins, the memory response to wP is likely similar to that of natural infection, stimulating broad and strong innate responses. The potency of wP vaccines that were used prior to the introduction of the aP vaccines has been reported to vary, which may be related to differences in the vaccine production process or the strains used in different countries. The structure of LOS is different between *B. pertussis* strains, which impacts the endotoxicity (Marr et al. 2010). Thus, an important component in driving the innate immune response to wPs is the LOS that binds to TLR4 (Dunne et al. 2015). The consequence on innate immunity is that wPs stimulate DCs and monocytes to produce large quantities of pro-inflammatory cytokines such as IL-1β, IL-6, IL-12, as well as markers for activation and maturation (Hoonakker et al. 2015). Interestingly, wPs are also capable of stimulating NK cell responses. During natural infection, IFNγ from NK cells controls the infection and facilitates Th1 responses (Byrne et al. 2004). Similarly, stimulation with wP triggers progressive

maturation of NK cells, including shifts in IFNγ and cytotoxicity. These changes were shown to be highly associated with the *B. pertussis* vaccine component and not the tetanus or diphtheria toxoids, which are commonly administered together in wP formulations (White et al. 2014). Ultimately, cellular immunity induced by wPs is mediated by Th1 and Th17 T cell responses, leading to neutrophil recruitment and killing of *B. pertussis* (Ross et al. 2013). Unfortunately, the same factors that steer this adaptive immune response have led to its discontinuation. The main issues with wPs were the associated side-effects relating to local and systemic inflammation, which in the mouse model is driven by IL-1β (Donnelly et al. 2001). As a consequence, despite the high efficacy of wP vaccination, reactogenicity was severe enough to instigate a decrease in vaccine coverage in many countries (Clark 2014; Plotkin 2014). Since LOS signalling through TLR4 is generally recognised as the main mediator of both wP effectiveness *and* reactogenicity (Donnelly et al. 2001), there have been efforts to detoxify or replace this constituent. A phase I clinical trial reported that infants vaccinated with wP that has reduced LOS content (wP_low) showed similarly robust immunogenicity, though disappointingly no significant differences between the wP and wP_low groups in terms of reactogenicity (Zorzeto et al. 2009). Another strategy has been to supplement wPs with immunogenic but non-toxic LPS analogs. Monophosphoryl lipid A (MPL) is one such candidate that has been approved for human vaccines. wP vaccines modified in this way were shown to elicit similarly potent IFNγ and neutrophil responses as classic wP, but with significantly less serum IL-6, suggesting that they may be a suitable solution (Geurtsen et al. 2008).

aPs were developed in the 1990s in order to address the safety concerns of wPs, and today they are the principal pertussis paediatric and booster vaccine in most industrialized countries. Several hypotheses have been proposed to explain the failure of modern aP vaccines. First, there is accumulating evidence that circulating strains of *B. pertussis* display substantial variation in many of the protective antigens in aPs, including PT, FIM2/3 (Bart et al. 2014; Tsang et al. 2004; van Loo et al. 2002), as well as the reported loss of PRN in many countries (Zeddeman et al. 2014). Another likely explanation is founded on the observation that aPs fail to induce optimal cellular and humoral immune responses, leading to failure to generate enduring immunological memory and failure to protect against *B. pertussis* infection (Diavatopoulos and Edwards 2017; Ross et al. 2013). Importantly, the adaptive immune response profile of aP is different from natural infection and wP vaccination in many models, including mouse, human, and baboon. The pattern in immune response is generally similar: while natural infection and wP vaccination elicit strong Th1/Th17 T cell polarization, aPs elicit potent Th2 immunity, as well as a moderate Th1 (humans & baboon, (Ryan et al. 1998b; Warfel et al. 2014)), or Th17 (mouse, (Ross et al. 2013)) response. Thus far, no clear protective role of the Th2 component has been identified in any model with regards to clearance of *B. pertussis*. This sub-optimal polarization is also reflected in the quality of the humoral response, as significant differences in antibody class-switching have been observed in both mice (Raeven et al. 2015) and humans. Antibody classes and isotypes were compared in children boosted with aP, but who were either primed with aP or wP in infancy. Both groups showed strong IgG1 responses, which is able to fix complement, opsonise, and is associated with Th1 immunity. However, only the aP-primed children showed significant induction of Th2-associated IgG4, which not only lacks the functionality of IgG1, but also highly correlates with IgE levels and restricts the inflammatory response through binding of the inhibitory FcgRIIB receptor (Hendrikx et al. 2011; James and Till 2016). Not surprisingly, a mechanism to explain these observations can be found in examining the capacity of aPs to stimulate innate immunity. Much insight into the role of antigen presenting cells (APCs) in programming vaccine-induced immunity has been gained from analysing normal (NVR) and low-vaccine responder infants (LVR), who fail to raise protective antibody titres to

many vaccine antigens and have reduced T cell function (Pichichero et al. 2016). Although there were no differences in terms of frequencies of monocytes or conventional DCs in blood, LVR infants display lower MHC-II expression on resting APCs. Consistent with this, peripheral blood mononuclear cells show reduced IRF7 transcriptional upregulation, as well as reduced IL-12, IL-1β, and IFNα secretion to stimulation with TLR7/8 agonist R848, suggesting diminished type I IFN signalling and suboptimal APC molecular programming (Pichichero et al. 2016; Surendran et al. 2017; Surendran et al. 2016) post-vaccination. Taken together, these findings highlight the role of these pathways in eliciting robust antibody and T cell responses after vaccination. There is accumulating evidence that the composition of aPs is not ideal, or may even be detrimental to stimulating innate immunity in this manner. aPs do not contain any classical PRR ligands, and in fact, some of the vaccine antigens, such as PRN, FIM, and FHA, possess dampening properties (Hovingh et al. 2018; Inatsuka et al. 2005; Scheller et al. 2015). Instead, a chief driver of the innate immune response to aP vaccines is the adjuvant alum, which produces strong Th2 immunity (McKee et al. 2009). Despite being one of the oldest and most widely used adjuvants, its mechanism of action remains elusive (Coffman et al. 2010; Mbow et al. 2010). For years, it was thought that alum salts only served as passive depots for vaccine antigens, thereby enhancing their uptake. Interest in the immunostimulatory activity of alum peaked with the discovery that cell damage induced by alum at the site of injection caused inflammasome activation through the induction of uric acid (Eisenbarth et al. 2008; Hornung et al. 2008). However, activation of the inflammasome was shown not to be strictly essential for the adjuvanticity of alum (McKee et al. 2009; Spreafico et al. 2010). One recent study compared the aP head-to-head with standard alum in order to discern the specific effect of aP antigens to stimulate monocytes. Overall, they found that while alum alone was capable of producing a mild type-I IFN response, aP generated an opposite reaction and also stimulated the production of the anti-inflammatory cytokine IL-10 (Kooijman et al. 2018). Cumulatively, the lesson from studies on the immunogenicity of aPs and wPs seems to be that a more suitable adjuvant is necessary in order to enhance the effectiveness of aP formulations. To that end, much effort has been made to find an appropriate candidate that is non-toxic, capable of triggering innate responses, and elicits an optimal adaptive immune profile. Owing to their central role in initiating immune responses, almost all PRRs can be considered targets, and many of them have been tested as either a supplement to aP vaccines or as a replacement for alum. Initial attempts involved complementing aPs with constituents analogous to those normally expressed by *B. pertussis*. As previously discussed, the LOS that is present in wPs is an important TLR4 ligand and is favourable for its Th1-promoting capacity. As with wPs, several investigations have been conducted to explore the effect of LPS analogs on aP potency. Supplementation in such a way improved the vaccine by reducing colonization in the lungs post-vaccination and skewed T cell immunity from a Th2 to a Th1/Th17 phenotype (Brummelman et al. 2015; Geurtsen et al. 2007). Similarly, classic aP antigens combined with LP1569, a recently identified TLR2 ligand from *B. pertussis,* induced significantly more Th1-associated humoral and cellular responses than the same antigens supplemented with alum (Dunne et al. 2015). ACT is often referred to as a "multi-talented" toxin due to its flexibility for use in biological applications (Ladant and Ullmann 1999). Besides its established adjuvanticity in augmenting aPs by stimulating the inflammasome and IL-1β production (Dunne et al. 2010; Macdonald-Fyall et al. 2004), ACT can also be used as an antigen delivery vector for other therapeutic applications. Since the RTX domain effectively binds CD11b on DCs and can tolerate the insertion of exogenous peptides, recombinant detoxified ACT containing antigens of interest has been effectively targeted to DCs, where it can stimulate TLR4 and upregulate pro-inflammatory cytokines, antigen presentation, and costimulatory molecules

(Dadaglio et al. 2014). Other approaches for improving aPs have explored the use of chemical or synthetic agonists that are known to strongly stimulate innate immune cells.

Advances in Vaccine Development Bacterial lysates have been recognized since the late 1800s for their capacity to limit tumour growth and historically were used to treat cancer (Coley 1991). Among the components, bacterial DNA was identified as possessing significant antitumor activity and has since been shown to elicit Th1 adaptive immune responses (Relyveld et al. 1998). The immunostimulatory properties of bacterial genomic DNA were shown to be mediated by the detection of unmethylated CpG motifs by TLR9 (Krieg et al. 1995). Monocytes, macrophages, and DCs express TLR9 in endosomes where DNA is released from phagocytosed bacteria, and activation of TLR9 leads to IL-12 production, MHC-II expression, and maturation of the stimulated cell (Krieg 2002). This discovery has led to the development and optimization of synthetic CpG oligodeoxynucelotides (ODNs) that specifically target TLR9, one of which has already been licensed for use as an adjuvant in hepatitis B vaccines (Campbell 2017). The other intracellular TLRs, such as TLR7 and TLR8, are also attractive targets for adjuvant development since they also possess Th1 polarizing potential. TLR7 normally binds virus-derived single stranded RNA, and promisingly, a TLR7 agonist adsorbed to alum and administered with the usual aP vaccine antigens was recently shown to provide equal or greater protection than the wP vaccine (Misiak et al. 2017a). STING (STimulator of INterferon Genes) is yet another intracellular sensor of microbial DNA that has been linked to antiviral responses and innate immunity to many different intracellular pathogens. It recognises unique bacterial nucleic acids called cyclic dinucleotides and mediates transcriptional activation of type I interferon responses (reviewed in (Burdette and Vance 2013)). C-di-GMP is one such cyclic dinucleotide that has been shown to significantly improve protective immunity in murine models of experimental infection with several species of pathogenic bacteria (Brouillette et al. 2005; Yan et al. 2009; Karaolis et al. 2007). One recent study investigated whether combining this adjuvant with the (previously described) TLR2 ligand LP1569 in the typical aP formulation might help enhance otherwise variable or absent Th17 immune responses *in vivo* and *in vitro*. Intranasal administration of this experimental vaccine was capable of inducing long-lived, IL-17 secreting tissue-resident CD4$^+$ T cells, which was correlated with enhanced protection in nasal tissue up to 10 months after vaccination. In line with this, *in vitro* stimulation of DCs was associated with the secretion of Th1- and Th17 -polarizing cytokines (IFNβ, IL-12, IL-1β, IL-23, and IL-6) (Allen et al. 2018). Taken together, these promising results demonstrate that individual adjuvants may potentiate each other's activity, ultimately leading to robust innate and adaptive immunity, as well as long-lasting protection from infection.

Given the apparent importance of simultaneous PRR ligation for vaccine induced immunity and effectiveness, this aspect has not been fully explored for many modern aPs, which sometimes contain many other components in addition to the classic pertussis, diphtheria, and tetanus antigens. For example, licensed paediatric acellular pertussis vaccines exist that contain inactivated poliovirus (IPV) and *Haemophilus influenzae* type B (HiB). Until this moment, the capacity of these components to adjuvant the response to the other vaccine antigens has been poorly described. One study investigated a wide range of licensed prophylactic vaccines for their capacity to drive the maturation of monocyte-derived DCs, with the intention to evaluate the cells' suitability for clinical use as a DC-based cancer vaccine (Schreibelt et al. 2010). They found that the conjugate HiB vaccine does in fact possess some TLR4 stimulation capacity. In line with this, they also found that a hexavalent pertussis combination vaccine containing HiB and IPV components also possesses some TLR4 stimulation activity, which potentially may augment the immunogenicity of the vaccine. We speculate that the adaptive immune profile elicited by combination vaccines may be different compared to those measured from classic aPs. One shortfall of the aforementioned study is that only TLR2, TLR4, and TLR5 pathways were tested for each vaccine. Thus, it is still unclear to what effect the other PRRs may be involved in shaping the vaccine innate immune response.

As has been discussed, there has been considerable effort to revise current wPs and aPs through modulating their capacity to stimulate innate immunity via PRRs. However, there is significant evidence suggesting that entirely replacing these vaccines may be an attractive option. Thus far, two different approaches have gained traction in the past years: outer membrane vesicles (OMVs) derived from *B. pertussis*, and intranasal administration of live attenuated *B. pertussis*. OMVs are naturally shed from Gram-negative bacteria during growth (Beveridge 1999), and contain naturally incorporated bacterial surface antigens as well as periplasm components (Holst et al. 2009). The particles derived from *B. pertussis* contain over 200 protein antigens, which is similar to wPs, and include the antigens typically found in aPs (Raeven et al. 2015). These OMVs were also found to contain adhesins that represent important virulence factors for colonization, many of which also displayed immunogenic properties (Gasperini et al. 2018). In a murine intranasal vaccination model, OMVs were shown to be much more protective than aPs (Gaillard et al. 2014). This was associated with significant Th1/Th17 adaptive immune responses, as well as a higher proportion of Th1-associated antibodies (Bottero et al. 2016; Gaillard et al. 2014). OMVs also contain at least 50 times less LOS than wPs, indicating that they may have a superior safety profile (Hozbor 2017). In line with this, OMVs were shown to elicit equivalent protection as compared to the wP vaccine, but with significantly less systemic pro-inflammatory cytokine production. Notably, OMVs produced higher neutrophil responses and lung type- I and -II IFN transcriptomic responses than wP (Raeven et al. 2016). Mucosal immunization seems to be important for its protective effect, however, as subcutaneous injection was shown to be far less effective (Raeven et al. 2018).

Trained Immunity The notion that immunological memory is a unique property of the adaptive immune system has been recently challenged. Evidence for innate immune memory, or "trained immunity" (Netea et al. 2016), as it has been called, stems from the initial observation that the innate immune system of plants is capable of responding more potently upon reinfection (Kachroo and Robin 2013). The mechanism of this enhanced resistance has been attributed to epigenetic reprogramming of the innate immune cells, which persists over time and enables the cells to retain long-term changes to their functional programs (Kachroo and Robin 2013). This property of innate immunity was later also found in a range of invertebrates (Kurtz and Franz 2003) including mosquitoes, where it mediates protection against *Plasmodium* (Rodrigues et al. 2010). Looking back, there have been clues that the same phenomenon exists in vertebrates like mice and humans. For example, exposure to β-glucan, a polysaccharide component that occurs naturally in the cell walls of fungi and which signals through the CLR Dectin-1 (Quintin et al. 2012), is capable of protecting mice against heterologous infection with *Staphylococcus aureus* (Di Luzio and Williams 1978). Similarly, other microbial ligands for PRRs have been shown to induce enduring, non-specific protective effects in the murine model, including peptidoglycan components (Krahenbuhl et al. 1981), LPS (Foster et al. 2007), CpG (Ribes et al. 2014), and flagellin (Munoz et al. 2010). Since the 1940s, epidemiological studies have shown that vaccination with the tuberculosis (TB) vaccine bacillus Calmette-Guerin (BCG) reduces childhood mortality in populations independently of its effect on TB (Levine and Sackett 1946; Ferguson and Simes 1949; Garly et al. 2003). In humans and in mice, this newfound principle of innate immunity has been shown to be mediated by epigenetic rewiring of monocytes, thereby modulating gene expression including those for inflammation

(Kleinnijenhuis et al. 2012). Considering the relatively short lifespan of monocytes, one question rising out of these observations is how imprinting is maintained through time such that sustained protection is possible. Both innate and adaptive immune cells originate in the bone marrow from long-lived haematopoietic stem cells (HSCs), which are known to respond to haematopoietic stress and inflammation (Takizawa et al. 2011; Manz and Boettcher 2014). In line with this, several studies have shown that HSCs are capable of sensing pathogens via PRR and cytokine/chemokine receptor expression, which induces distinct transcriptional programs that determine which immune cell lineages are released into circulation (King and Goodell 2011). Recently, Kaufmann et al. elegantly demonstrated that BCG vaccination in mice not only programs HSCs to favour myelopoiesis, but also to generate monocytes and macrophages with an epigenetic signature that grants long-term innate immune protection from subsequent *Mycobacterium tuberculosis* infection (Kaufmann et al. 2018). In addition to BCG, several other pathogens can exert protective effects through trained immunity, such as *Candida albicans* (Quintin et al. 2012), herpesviruses (Barton et al. 2007), and helminth parasites (Chen et al. 2014), which were also shown to be mediated by epigenetic programming of monocytes or macrophages. Specific evidence for bone marrow HSC reprogramming has only been shown for BCG vaccination, and recently has also been implicated in vaccine-induced immunity to pertussis (Varney et al. 2018). The influence of pertussis vaccines and vaccine components on HSCs is of key interest for vaccine development since it represents an entirely new and exciting mechanism for adjuvanticity. To address this knowledge gap, Varney and colleagues examined the capacity of aP and wP to alter the characteristics of HSCs and the downstream peripheral immune response. In this context, it was determined that wP, but not aP, exerted a profound effect on innate and adaptive immune cell development. Vaccination with wP led to the expansion HSCs and of the myeloid progenitor population, thereby establishing a state of increased innate immune responsiveness. This was supported by the finding that wP vaccination also led to an increase in size, as well as difference in composition of peripheral lymphoid organs. These stores of immature innate immune cells were rapidly released upon subsequent infection. Development of adaptive immune cells in the bone marrow was also modulated by wP vaccination, as common lymphoid progenitor proportions also increased, though cells expressing B220, a pan-marker for B cells in mice, decreased. Upon subsequent infection, wP-vaccinated mice displayed a striking increase in maturing B cell populations. Transcriptomic analysis of HSCs of wP-primed mice revealed significant IFN-induced gene expression, which agrees with the observations made upon profiling of BCG – "educated" HSCs (Kaufmann et al. 2018). Currently, the belief is that a common mechanism may exist for programming of HSCs that ultimately relies on PRR-induced soluble IFN in the blood, which is essential for driving proliferative capacity and lineage polarization (Kaufmann et al. 2018; Varney et al. 2018). Recently, an experimental pertussis vaccine has been shown to display some of these qualities. BPZE1, which is a genetic mutant strain of *B. pertussis* that that does not produce TCT and dermonecrotic toxin and has been modified to express genetically inactivated PT, has shown promise as a live attenuated vaccine (reviewed in (Cauchi and Locht 2018)). It has been demonstrated to be protective in mice (Mielcarek et al. 2006) and baboons (Locht et al. 2017), as well as safe for use in humans (Thorstensson et al. 2014), where it also elicited high specific antibody responses. Similar to natural infection, BPZE1 elicited potent Th1 and Th17 immunity in pre-clinical models following primary vaccination (Solans et al. 2018; Feunou et al. 2010). BPZE1 vaccination also appears to elicit an immune dampening and protective effect against heterologous diseases, such as influenza (Li et al. 2010) and respiratory syncytial virus (RSV) (Schnoeller et al. 2014) infection, as well as allergic asthma (Kavanagh et al. 2010). These effects were shown not to be mediated by adaptive

immunity, potentially pointing to a role for trained immunity. Conversely, BPZE1 treatment results in predominantly reduced inflammatory responses upon secondary stimulation, whereas training by BCG vaccination predominately induces higher inflammatory responses. *In vitro* studies have demonstrated that BPZE1 can impact on human monocyte derived DC maturation and cytokine secretion, enabling them to not only produce IL-12, but also influence the antiviral response when the cells were also infected with RSV (Schiavoni et al. 2014). While both wild type *B. pertussis* and BZPE1 are capable of activating DCs in a similar manner, only BZPE1-primed DCs retained the capacity to migrate in response to a chemokine gradient. This unusual finding was found to be linked to the lack of PT expression, which is known to exert an inhibitory effect on chemotaxis (Fedele et al. 2011). Since these findings do not constitute direct evidence for the transcriptional alteration of HSCs by BPZE1, it remains to be seen if the non-specific, innate immune protective effects of BPZE1 vaccination are mediated through a mechanism similar to BCG or wP vaccination.

Innate Immune Cells and Infection To provide context to the vaccine responses it is worthwhile to review the mechanisms by which immunity to natural infection is induced. Upon primary infection, *B. pertussis* is recognised by the innate immune system by both epithelial cells and resident antigen-presenting cells through various PRRs (Fedele et al. 2013). This recognition induces a signalling cascade that triggers the release of chemokines and cytokines. These soluble mediators promote inflammation and recruit immune cells to the site of infection (Raeven et al. 2014). The subsequent activation of both local and recruited innate immune cells then induces antimicrobial programs that help contain *B. pertussis* replication in the respiratory tract and limit damage to the host (de Gouw et al. 2011).

Most animal studies to date have focused on the primary response to infection. However, most *B. pertussis* infections in humans do not occur in immunologically naive individuals. There is a significant knowledge gap in our understanding of how pre-existing immunity – either induced by wP or aP primary vaccination – modulates the innate immune response to secondary exposure by infection and if and how this affects the recall response and the development of *de novo* immunological memory. For instance, the presence of vaccination-induced antibodies or antigen-specific B cells can modulate the ensuing immune response by either promoting or by dampening inflammation, depending on the type of antibody induced by vaccination (Kampmann and Jones 2015; Netea and van Crevel 2014).

To facilitate survival, replication, and transmission, *B. pertussis* has evolved a number of strategies to modulate or prevent various aspects of the immune response, i.e. *B. pertussis* can modulate bacterial recognition, inhibit the recruitment and phagocytic capacity of immune cells and suppress humoral and cellular responses. Since these interactions have extensively been reviewed elsewhere (de Gouw et al. 2011), here we will discuss how the innate immune response of the host is induced and (re)programmed after infection with *B. pertussis*. Following the initial release of chemokines, recruitment of immune cells to the respiratory tract is an essential early response after the initial recognition of *B. pertussis*. This process has almost exclusively been investigated in the mouse model with a strong focus on the lower respiratory tract. Vandebriel et al. described the recruitment of granulocytes, alveolar macrophages, and lymphocytes to the lungs of infected mice (Vandebriel et al. 2003). This is in line with other studies in mice that observed infiltration of DCs and macrophages first, followed by neutrophils and NK cells second, and finally T cells (Dunne et al. 2009; McGuirk et al. 1998). CD11c$^+$ CD8α^+ DCs were found to be infiltrated into cervical lymph nodes and then migrated to the lungs 1–7 days after infection of mice. These DCs were found to have an upregulated expression of MHC and co-stimulatory molecules and increased secretion of IFN-γ, IL-4, and IL-10. Depletion experiments showed that CD11c$^+$ CD8α^+ DCs play an important role, especially in protection early during infection (Dunne et al. 2009). In contrast, DCs have also been shown to be capable of inhibiting the protective immune response to *B. pertussis*. For instance, plasmacytoid (p) DCs were shown to contribute to early pathogenesis during *B. pertussis* infection in mice. pDCs secreted IFNα, which had an inhibitory effect on Th17 differentiation and thus infection, which facilitated early pulmonary colonization with *B. pertussis* (Wu et al. 2016). DCs can also secrete IL-10 in response to *B. pertussis*, which directs the induction of Tr1 cells from naive T cells in the respiratory tract of mice during an acute infection with *B. pertussis*. These regulatory cells inhibited the protective responses of Th1 cells and thus lead to evasion of protective immunity (McGuirk et al. 2002).

The second wave of innate immune cells that is recruited to the site of infection are the macrophages. Macrophages were shown to be important for clearance of *B. pertussis*, since their depletion in mice exacerbated the infection (Carbonetti et al. 2007). Macrophages mediate immunity to *B. pertussis* via MyD88 adaptor-like protein (Mal), since mice deficient in this protein display exacerbating infection and enhanced morbidity and mortality. Macrophages have an important role in controlling infection and demonstrate a critical role for Mal as a major innate signalling pathway (Bernard et al. 2015). In humans, analysis of bronchoalveolar lavage specimens found that *B. pertussis* was almost entirely associated with macrophages (Bromberg et al. 1991). Although macrophages may be a key cell type for phagocytosis of *B. pertussis* infection, there is evidence that *B. pertussis* can also survive in alveolar macrophages, as demonstrated from bronchoalveolar lavage samples from mice challenged with *B. pertussis*. In this study, macrophages were found to internalize the bacteria but not kill them, rendering the bacteria resistant to killing via humoral defence mechanisms (Hellwig et al. 1999).

Following the recruitment of DCs and macrophages, neutrophils are the next cell type to arrive at the site of infection. To assess the role of neutrophils in the early immune response to *B. pertussis* infection in mice, neutrophils were depleted from mice prior to infection with *B. pertussis*. Interestingly, depletion of neutrophils did not significantly alter the bacterial load after primary infection with wildtype *B. pertussis*. Since neutrophils may require further activation before they can effectively kill bacteria, the experiment was repeated in convalescent mice who already developed memory cells and were re-challenged with *B. pertussis*. In this case, a significant increase in bacterial load was observed after infection, suggesting that neutrophils are likely not involved in bacterial clearance during primary infection, but play an important role in controlling *B. pertussis* infections in hosts with pre-existing immunity. The increased neutrophil killing capacity may be due to the presence of opsonising antibodies that enhance phagocytosis, or the presence of higher concentrations of IL-17 or IFN-γ (Andreasen and Carbonetti 2009). In line with these findings, another study showed that *B. pertussis* did not kill naive neutropenic mice, suggesting that neutrophils are not critical to the early inflammatory response to *B. pertussis* (Harvill et al. 1999). Recently, it was shown that neutrophils are essential for the induction of human Th17 cells. Neutrophils release elastase upon activation that processes DC-derived CXCL8 into a truncated form that is capable of inducing Th17 cells (Souwer et al. 2018). This process takes time and may explain why studies have found that neutrophils are critical players in controlling the infection in hosts with pre-existing immunity, but not during the early inflammatory response.

Although NK cells are most well-known for their capacity to kill virus-infected cells, they can also exert direct antimicrobial effects on bacteria, including *B. pertussis*. NK cells were found to be recruited in significant numbers to the lungs of mice following respiratory challenge with *B. pertussis* (Byrne et al. 2004). These cells were found to be the primary source of IFNγ production in the lungs during the acute stage of infection, and may help activate recruited phagocytes in controlling *B. pertussis* replication. Bacterial challenge of NK cell-depleted mice resulted in lethal infection, with increased lung bacterial load and spread of the bacteria to the liver. Depletion of NK cells appeared to switch the Th1-dominant response to a more Th2-biased response, which suggest that NK cells play a role in selectively promoting the induction of Th1 cells during *B. pertussis* infection (Byrne et al. 2004).

Infection-Induced Signalling Pathways A well-known recognition receptor of *B. pertussis* is TLR4, which was confirmed by a study in which different inbred strains of mice were screened for susceptibility to *B. pertussis* after intranasal infection (Banus et al. 2006). The most resistant and the most susceptible strain to *B. pertussis* infection were selected for further genetic and phenotypic characterization, and all detectable genetic differences between the strains could be accounted to a functional mutation in the intracellular domain of *Tlr4*. TLR4 activation leads to upregulation of cytokines essential for efficient clearance of bacteria from the lungs, including IL-1β, IFNγ, and TNFα, and recruitment of neutrophils to the lungs (Mann et al. 2004; Moreno et al. 2013; Banus et al. 2006, 2007). In humans, the role of TLR4 in infectious diseases has also been investigated in genetic association studies (Schroder and Schumann 2005). Most studies have focused on single nucleotide polymorphisms (SNPs) within the gene encoding TLR4, Asp299Gly and Thr399Ile. These SNPs are present in approximately 10% of Caucasians, and have been found to be positively correlated with several infectious diseases. Preliminary studies also indicated that susceptibility to infectious diseases could be increased by SNPs in genes encoding other TLRs, such as TLR2 (Schroder and Schumann 2005).

IL-1R signalling is also induced upon respiratory infection with *B. pertussis*. IL-1β-deficiency in mice was associated with more persistent levels of inflammation, characterized by immune cell infiltration. Perhaps not surprisingly, this suggests that IL-1R-mediated signalling is required to efficiently clear *B. pertussis* infection. Of note, the IL-1R signalling pathway was required for subsequent resolution of inflammation, but not the initiation and establishment of inflammation. This was demonstrated by infection of wildtype, IL-1R-deficient, and IL-1β-deficient mice, which showed similar inflammation early after infection but not at day 28, at which time point only the wildtype mice had resolved inflammation and reduced *B. pertussis* load. IL-1β could be processed into its active form in a caspase-1 independent manner in the murine lung (Place et al. 2014). These findings suggest that IL-1R signalling is required, independent of inflammasome activation, to clear *B. pertussis* infection in the mouse model.

More recently, transcriptomic analysis was applied to elucidate the immunological pathways induced during the different phases after primary infection with *B. pertussis* in mice. Differential gene expression analysis suggested enhanced activation of TLR4, TLR9, DNA-dependent activator of IFN-regulatory factors (DAI), and C-type lectin receptor (CLR) signalling pathways, implying recognition of multiple *B. pertussis* ligands (Raeven et al. 2014). Several CLRs showed upregulated gene expression after *B. pertussis* infection, such as *Mgl1* and the mouse DC-SIGN-related proteins *Signr7* and *Signr8* (Powlesland et al. 2006). *Mgl1* recognizes N-acetylgalactosamine (GalNAc), which is present in *B. pertussis* (Vukman et al. 2013). These findings show that microbial ligands produced during *B. pertussis* infection activate multiple PRR pathways during infection.

In addition to these pathogen recognition receptors, several transcription factors were also upregulated in the lungs of mice upon *B. pertussis* infection: *Pu.1*, *Shp-1*, *Nfil3*, *Syk*, *Irf5*, *Irf7*, and *Irf8*. These transcription factors are all regulators of immunological processes and determine the strength and direction of the adaptive immune response (Raeven et al. 2014). The transcription factor SHP1 has been shown to be involved in clearance of *B. pertussis*, since hijacking of this pathway by ACT enables *B. pertussis* to evade killing by neutrophils and macrophages in a complement-dependent oxidative burst and using opsonophagocytic mechanisms (Cerny et al. 2015). ACT is also an inhibitor of SYK activity and blocks CR3 signalling triggered by binding of C3b-opsonised bacteria to the CR3 receptor (Osicka et al. 2015). IRF8 is an important player in the regulation of IL-12, which is crucial for Th1-inducing activity (Spensieri et al. 2006; Trinchieri et al. 2003), but that may also be inhibited by ACT (Spensieri et al. 2006).

2   Conclusion

Although many significant advancements have been made in the past years with respect to understanding the innate immune response to pertussis infection and vaccines, there are still many important knowledge gaps that preclude the identification of innate drivers of correlates of protection and the development of a new generation of vaccines. Most of the vaccine research that exists for pertussis has involved the analysis of individual components in *in vitro* and animal models. An overview of these findings is shown in Fig. 1. However, a comprehensive proteomic or lipidomic cataloguing of microbial components from *B. pertussis* interacting with host factors has not yet been conducted, thus making it difficult to appreciate the full complexity of host-pathogen interactions. In addition, there is little appreciation for how the other vaccine antigens in modern vaccine formulations affect the response to pertussis antigens. Analysing individual components also does not capture the full complexity of the cellular circuits and intracellular signalling that is engaged by simultaneous stimulation of multiple PRRs across multiple innate immune cell types. Fortunately, new tools in computational immunology have recently been developed that are beginning to shed light on these processes (Gaudet and Miller-Jensen 2016). Single-cell technologies have been revolutionizing immunology, and have already uncovered new mechanistic

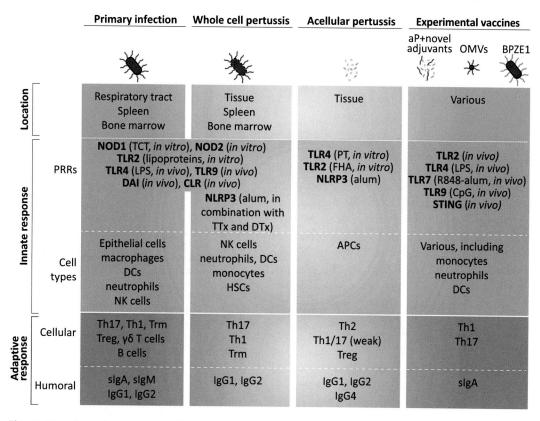

		Primary infection	Whole cell pertussis	Acellular pertussis	Experimental vaccines
					aP+novel adjuvants OMVs BPZE1
Location		Respiratory tract Spleen Bone marrow	Tissue Spleen Bone marrow	Tissue	Various
Innate response	PRRs	**NOD1** (TCT, *in vitro*), **NOD2** (*in vitro*) **TLR2** (lipoproteins, *in vitro*) **TLR4** (LPS, *in vivo*), **TLR9** (*in vivo*) **DAI** (*in vivo*), **CLR** (*in vivo*)	**NLRP3** (alum, in combination with TTx and DTx)	**TLR4** (PT, *in vitro*) **TLR2** (FHA, *in vitro*) **NLRP3** (alum)	**TLR2** (*in vivo*) **TLR4** (LPS, *in vivo*) **TLR7** (R848-alum, *in vivo*) **TLR9** (CpG, *in vivo*) **STING** (*in vivo*)
	Cell types	Epithelial cells macrophages DCs neutrophils NK cells	NK cells neutrophils, DCs monocytes HSCs	APCs	Various, including monocytes neutrophils DCs
Adaptive response	Cellular	Th17, Th1, Trm Treg, γδ T cells B cells	Th17 Th1 Trm	Th2 Th1/17 (weak) Treg	Th1 Th17
	Humoral	sIgA, sIgM IgG1, IgG2	IgG1, IgG2	IgG1, IgG2 IgG4	sIgA

Fig. 1 Overview of the tissues and immune factors induced by primary infection or by licensed and experimental vaccines. Notably, despite the widespread use of acellular pertussis vaccines, we currently do not fully understand which cell types or which subclasses of antigen presenting cells (APCs) are involved in mediating the innate immune response in humans

insights into canonical biology through the precise quantification of dynamics in populations of cells, and through deep phenotyping of individual cells that are processing multiple signals. It remains to be seen how these technologies will sculpt our understanding of vaccine- and infection-induced immunity.

At the moment, most inquiries into improving vaccine responsiveness with new vaccine formulations or adjuvants have been conducted in a model where there is no pre-existing specific immunity. Naturally, however, this is very unlikely to be the case since pertussis vaccine coverage in most countries luckily remains high. In addition, humans are incredibly diverse with respect to genetics, age, nutritional status, microbiome content, the presence of chronic infections or inflammatory disorders, and environmental exposure to name a few. All of these features represent different immunological perturbations that may ultimately sculpt both innate and adaptive immune responses to vaccines. In order to address the natural variation that occurs in human immune responses, a new discipline has emerged called systems vaccinology (Pulendran 2014). By integrating systems biology tools with immunological data from multiple human subjects receiving a common vaccine, systems vaccinology has been demonstrated to be an incredibly powerful tool for identifying molecular signatures early after vaccination that correlate with and predict adaptive immune responses. Such approaches have already uncovered new biological knowledge into how human immune responses are orchestrated, which may have otherwise been impossible with our current canonical understanding of immunity. Systems vaccinology has been successfully applied to influenza and pneumococcal vaccines (Obermoser et al. 2013), measles vaccines (Li et al. 2014), and the yellow fever vaccine (Pulendran 2009), and several others, but as of the time of this writing, not to any of the pertussis vaccines. One concern for this approach is that the most significant relationships are considered to be those connected with immunological endpoints such as antibodies. A potentially better indication of vaccine effectiveness and infection-induced immunity could be the analysis of human immune responses in the context of controlled experimental human infection. Such a model could also shed light on the potential induction of trained immunity by *B. pertussis* or *B. pertussis* vaccines, which represents the frontier of innate immune programming. A method for controlled infection with wild type *B. pertussis* has already been established (de Graaf et al. 2017) within the PERISCOPE consortium (Consortium 2018), and analysis of immune responses in this context is ongoing. In sum, new technology, computational methods applied to immunological data, and study designs with human volunteers stand to completely transform our current understanding of *B. pertussis* infection and *B. pertussis* vaccine-induced immunity.

References

Abramson T, Kedem H, Relman DA (2001) Proinflammatory and proapoptotic activities associated with Bordetella pertussis filamentous hemagglutinin. Infect Immun 69(4):2650–2658. https://doi.org/10.1128/IAI.69.4.2650-2658.2001

Abramson T, Kedem H, Relman DA (2008) Modulation of the NF-kappaB pathway by Bordetella pertussis filamentous hemagglutinin. PLoS One 3(11):e3825. https://doi.org/10.1371/journal.pone.0003825

Ahmed R, Gray D (1996) Immunological memory and protective immunity: understanding their relation. Science 272(5258):54–60

Allen AC, Wilk MM, Misiak A, Borkner L, Murphy D, Mills KHG (2018) Sustained protective immunity against Bordetella pertussis nasal colonization by intranasal immunization with a vaccine-adjuvant combination that induces IL-17-secreting TRM cells. Mucosal Immunol 11(6):1763–1776. https://doi.org/10.1038/s41385-018-0080-x

Andreasen C, Carbonetti NH (2009) Role of neutrophils in response to Bordetella pertussis infection in mice. Infect Immun 77(3):1182–1188. https://doi.org/10.1128/IAI.01150-08

Asgarian-Omran H, Amirzargar AA, Zeerleder S, Mahdavi M, van Mierlo G, Solati S, Jeddi-Tehrani M, Rabbani H, Aarden L, Shokri F (2015) Interaction of Bordetella pertussis filamentous hemagglutinin with human TLR2: identification of the TLR2-binding domain. APMIS 123(2):156–162. https://doi.org/10.1111/apm.12332

Ashworth LA, Irons LI, Dowsett AB (1982) Antigenic relationship between serotype-specific agglutinogen and fimbriae of Bordetella pertussis. Infect Immun 37(3):1278–1281

Athie-Morales V, Smits HH, Cantrell DA, Hilkens CM (2004) Sustained IL-12 signaling is required for Th1 development. J Immunol 172(1):61–69

Ausiello CM, Lande R, la Sala A, Urbani F, Cassone A (1998) Cell-mediated immune response of healthy adults to Bordetella pertussis vaccine antigens. J Infect Dis 178(2):466–470

Ausiello CM, Fedele G, Urbani F, Lande R, Di Carlo B, Cassone A (2002) Native and genetically inactivated pertussis toxins induce human dendritic cell maturation and synergize with lipopolysaccharide in promoting T helper type 1 responses. J Infect Dis 186(3):351–360. https://doi.org/10.1086/341510

Banus HA, Vandebriel RJ, de Ruiter H, Dormans JA, Nagelkerke NJ, Mooi FR, Hoebee B, van Kranen HJ, Kimman TG (2006) Host genetics of Bordetella pertussis infection in mice: significance of toll-like receptor 4 in genetic susceptibility and pathobiology. Infect Immun 74(5):2596–2605. https://doi.org/10.1128/IAI.74.5.2596-2605.2006

Banus S, Pennings J, Vandebriel R, Wester P, Breit T, Mooi F, Hoebee B, Kimman T (2007) Lung response to Bordetella pertussis infection in mice identified by gene-expression profiling. Immunogenetics 59(7):555–564. https://doi.org/10.1007/s00251-007-0227-5

Barbic J, Leef MF, Burns DL, Shahin RD (1997) Role of gamma interferon in natural clearance of Bordetella pertussis infection. Infect Immun 65(12):4904–4908

Barnard A, Mahon BP, Watkins J, Redhead K, Mills KH (1996) Th1/Th2 cell dichotomy in acquired immunity to Bordetella pertussis: variables in the in vivo priming and in vitro cytokine detection techniques affect the classification of T-cell subsets as Th1, Th2 or Th0. Immunology 87(3):372–380

Bart MJ, Harris SR, Advani A, Arakawa Y, Bottero D, Bouchez V, Cassiday PK, Chiang CS, Dalby T, Fry NK, Gaillard ME, van Gent M, Guiso N, Hallander HO, Harvill ET, He Q, van der Heide HG, Heuvelman K, Hozbor DF, Kamachi K, Karataev GI, Lan R, Lutynska A, Maharjan RP, Mertsola J, Miyamura T, Octavia S, Preston A, Quail MA,

Sintchenko V, Stefanelli P, Tondella ML, Tsang RS, Xu Y, Yao SM, Zhang S, Parkhill J, Mooi FR (2014) Global population structure and evolution of Bordetella pertussis and their relationship with vaccination. MBio 5(2):e01074. https://doi.org/10.1128/mBio.01074-14

Barton ES, White DW, Cathelyn JS, Brett-McClellan KA, Engle M, Diamond MS, Miller VL, Virgin HW (2007) Herpesvirus latency confers symbiotic protection from bacterial infection. Nature 447(7142):326–329. https://doi.org/10.1038/nature05762

Behr MA, Divangahi M (2015) Freund's adjuvant, NOD2 and mycobacteria. Curr Opin Microbiol 23:126–132. https://doi.org/10.1016/j.mib.2014.11.015

Benz R, Maier E, Ladant D, Ullmann A, Sebo P (1994) Adenylate cyclase toxin (CyaA) of Bordetella pertussis. Evidence for the formation of small ion-permeable channels and comparison with HlyA of Escherichia coli. J Biol Chem 269(44):27231–27239

Bernard NJ, Finlay CM, Tannahill GM, Cassidy JP, O'Neill LA, Mills KH (2015) A critical role for the TLR signaling adapter Mal in alveolar macrophage-mediated protection against Bordetella pertussis. Mucosal Immunol 8(5):982–992. https://doi.org/10.1038/mi.2014.125

Berstad AK, Oftung F, Korsvold GE, Haugen IL, Froholm LO, Holst J, Haneberg B (2000) Induction of antigen-specific T cell responses in human volunteers after intranasal immunization with a whole-cell pertussis vaccine. Vaccine 18(22):2323–2330

Beveridge TJ (1999) Structures of gram-negative cell walls and their derived membrane vesicles. J Bacteriol 181(16):4725–4733

Boehm DT, Hall JM, Wong TY, DiVenere AM, Sen-Kilic E, Bevere JR, Bradford SD, Blackwood CB, Elkins CM, DeRoos KA, Gray MC, Cooper CG, Varney ME, Maynard JA, Hewlett EL, Barbier M, Damron FH (2018) Evaluation of adenylate cyclase toxoid antigen in acellular pertussis vaccines by using a Bordetella pertussis challenge model in mice. Infect Immun 86(10). https://doi.org/10.1128/IAI.00857-17

Bordet J, Genou O (1906) Le microbe de la coqueluche. Ann Inst Pasteur 20:731–741

Bottero D, Gaillard ME, Zurita E, Moreno G, Martinez DS, Bartel E, Bravo S, Carriquiriborde F, Errea A, Castuma C, Rumbo M, Hozbor D (2016) Characterization of the immune response induced by pertussis OMVs-based vaccine. Vaccine 34(28):3303–3309. https://doi.org/10.1016/j.vaccine.2016.04.079

Braat H, McGuirk P, Ten Kate FJ, Huibregtse I, Dunne PJ, Hommes DW, Van Deventer SJ, Mills KH (2007) Prevention of experimental colitis by parenteral administration of a pathogen-derived immunomodulatory molecule. Gut 56(3):351–357. https://doi.org/10.1136/gut.2006.099861

Bromberg K, Tannis G, Steiner P (1991) Detection of Bordetella pertussis associated with the alveolar macrophages of children with human immunodeficiency virus infection. Infect Immun 59(12):4715–4719

Brouillette E, Hyodo M, Hayakawa Y, Karaolis DK, Malouin F (2005) 3′,5′-cyclic diguanylic acid reduces the virulence of biofilm-forming Staphylococcus aureus strains in a mouse model of mastitis infection. Antimicrob Agents Chemother 49(8):3109–3113. https://doi.org/10.1128/AAC.49.8.3109-3113.2005

Brubaker SW, Bonham KS, Zanoni I, Kagan JC (2015) Innate immune pattern recognition: a cell biological perspective. Annu Rev Immunol 33:257–290. https://doi.org/10.1146/annurev-immunol-032414-112240

Brummelman J, Helm K, Hamstra HJ, van der Ley P, Boog CJ, Han WG, van Els CA (2015) Modulation of the CD4(+) T cell response after acellular pertussis vaccination in the presence of TLR4 ligation. Vaccine 33(12):1483–1491. https://doi.org/10.1016/j.vaccine.2015.01.063

Burdette DL, Vance RE (2013) STING and the innate immune response to nucleic acids in the cytosol. Nat Immunol 14(1):19–26. https://doi.org/10.1038/ni.2491

Byrne P, McGuirk P, Todryk S, Mills KH (2004) Depletion of NK cells results in disseminating lethal infection with Bordetella pertussis associated with a reduction of antigen-specific Th1 and enhancement of Th2, but not Tr1 cells. Eur J Immunol 34(9):2579–2588. https://doi.org/10.1002/eji.200425092

Campbell JD (2017) Development of the CpG adjuvant 1018: a case study. Methods Mol Biol 1494:15–27. https://doi.org/10.1007/978-1-4939-6445-1_2

Carbonetti NH, Artamonova GV, Van Rooijen N, Ayala VI (2007) Pertussis toxin targets airway macrophages to promote Bordetella pertussis infection of the respiratory tract. Infect Immun 75(4):1713–1720. https://doi.org/10.1128/IAI.01578-06

Caroff M, Karibian D, Cavaillon JM, Haeffner-Cavaillon N (2002) Structural and functional analyses of bacterial lipopolysaccharides. Microbes Infect 4(9):915–926

Cauchi S, Locht C (2018) Non-specific effects of live attenuated pertussis vaccine against heterologous infectious and inflammatory diseases. Front Immunol 9:2872. https://doi.org/10.3389/fimmu.2018.02872

Cerny O, Kamanova J, Masin J, Bibova I, Skopova K, Sebo P (2015) Bordetella pertussis adenylate cyclase toxin blocks induction of bactericidal nitric oxide in macrophages through cAMP-dependent activation of the SHP-1 phosphatase. J Immunol 194(10):4901–4913. https://doi.org/10.4049/jimmunol.1402941

Chen F, Wu W, Millman A, Craft JF, Chen E, Patel N, Boucher JL, Urban JF Jr, Kim CC, Gause WC (2014) Neutrophils prime a long-lived effector macrophage phenotype that mediates accelerated helminth expulsion. Nat Immunol 15(10):938–946. https://doi.org/10.1038/ni.2984

Cherry JD (2015) The history of pertussis (Whooping Cough); 1906–2015: facts, myths, and misconceptions. Curr Epidemiol Rep 2(2):120–130. https://doi.org/10.1007/s40471-015-0041-9

Clark TA (2014) Changing pertussis epidemiology: everything old is new again. J Infect Dis 209(7):978–981. https://doi.org/10.1093/infdis/jiu001

Coffman RL, Sher A, Seder RA (2010) Vaccine adjuvants: putting innate immunity to work. Immunity 33 (4):492–503. https://doi.org/10.1016/j.immuni.2010.10.002

Coleman MM, Finlay CM, Moran B, Keane J, Dunne PJ, Mills KH (2012) The immunoregulatory role of CD4 (+) FoxP3(+) CD25(−) regulatory T cells in lungs of mice infected with Bordetella pertussis. FEMS Immunol Med Microbiol 64(3):413–424. https://doi.org/10.1111/j.1574-695X.2011.00927.x

Coley WB (1991) The treatment of malignant tumors by repeated inoculations of erysipelas. With a report of ten original cases. 1893. Clin Orthop Relat Res 262:3–11

Consortium P (2018) PERISCOPE: road towards effective control of pertussis. Lancet Infect Dis. https://doi.org/10.1016/S1473-3099(18)30646-7

Dadaglio G, Fayolle C, Zhang X, Ryffel B, Oberkampf M, Felix T, Hervas-Stubbs S, Osicka R, Sebo P, Ladant D, Leclerc C (2014) Antigen targeting to CD11b+ dendritic cells in association with TLR4/TRIF signaling promotes strong CD8+ T cell responses. J Immunol 193(4):1787–1798. https://doi.org/10.4049/jimmunol.1302974

de Gouw D, Diavatopoulos DA, Bootsma HJ, Hermans PW, Mooi FR (2011) Pertussis: a matter of immune modulation. FEMS Microbiol Rev 35(3):441–474. https://doi.org/10.1111/j.1574-6976.2010.00257.x

de Graaf H, Gbesemete D, Gorringe AR, Diavatopoulos DA, Kester KE, Faust SN, Read RC (2017) Investigating Bordetella pertussis colonisation and immunity: protocol for an inpatient controlled human infection model. BMJ Open 7(10):e018594. https://doi.org/10.1136/bmjopen-2017-018594

Decker MD, Edwards KM (2000) Acellular pertussis vaccines. Pediatr Clin N Am 47(2):309–335

Di Luzio NR, Williams DL (1978) Protective effect of glucan against systemic Staphylococcus aureus septicemia in normal and leukemic mice. Infect Immun 20 (3):804–810

Diavatopoulos DA, Edwards KM (2017) What is wrong with pertussis vaccine immunity? Why immunological memory to pertussis is failing. Cold Spring Harb Perspect Biol 9(12). https://doi.org/10.1101/cshperspect.a029553

Dieterich C, Relman DA (2011) Modulation of the host interferon response and ISGylation pathway by B. pertussis filamentous hemagglutinin. PLoS One 6 (11):e27535. https://doi.org/10.1371/journal.pone.0027535

Domenech de Celles M, Magpantay FM, King AA, Rohani P (2016) The pertussis enigma: reconciling epidemiology, immunology and evolution. Proc Biol Sci 283(1822). https://doi.org/10.1098/rspb.2015.2309

Donnelly S, Loscher CE, Lynch MA, Mills KH (2001) Whole-cell but not acellular pertussis vaccines induce convulsive activity in mice: evidence of a role for toxin-induced interleukin-1beta in a new murine model for analysis of neuronal side effects of vaccination. Infect Immun 69(7):4217–4223. https://doi.org/10.1128/IAI.69.7.4217-4223.2001

Dumas A, Amiable N, de Rivero Vaccari JP, Chae JJ, Keane RW, Lacroix S, Vallieres L (2014) The inflammasome pyrin contributes to pertussis toxin-induced IL-1beta synthesis, neutrophil intravascular crawling and autoimmune encephalomyelitis. PLoS Pathog 10(5):e1004150. https://doi.org/10.1371/journal.ppat.1004150

Dunne PJ, Moran B, Cummins RC, Mills KH (2009) CD11c+CD8alpha+ dendritic cells promote protective immunity to respiratory infection with Bordetella pertussis. J Immunol 183(1):400–410. https://doi.org/10.4049/jimmunol.0900169

Dunne A, Ross PJ, Pospisilova E, Masin J, Meaney A, Sutton CE, Iwakura Y, Tschopp J, Sebo P, Mills KH (2010) Inflammasome activation by adenylate cyclase toxin directs Th17 responses and protection against Bordetella pertussis. J Immunol 185(3):1711–1719. https://doi.org/10.4049/jimmunol.1000105

Dunne A, Mielke LA, Allen AC, Sutton CE, Higgs R, Cunningham CC, Higgins SC, Mills KH (2015) A novel TLR2 agonist from Bordetella pertussis is a potent adjuvant that promotes protective immunity with an acellular pertussis vaccine. Mucosal Immunol 8(3):607–617. https://doi.org/10.1038/mi.2014.93

Eisenbarth SC, Colegio OR, O'Connor W, Sutterwala FS, Flavell RA (2008) Crucial role for the Nalp3 inflammasome in the immunostimulatory properties of aluminium adjuvants. Nature 453 (7198):1122–1126. https://doi.org/10.1038/nature06939

Fedele G, Celestino I, Spensieri F, Frasca L, Nasso M, Watanabe M, Remoli ME, Coccia EM, Altieri F, Ausiello CM (2007) Lipooligosaccharide from Bordetella pertussis induces mature human monocyte-derived dendritic cells and drives a Th2 biased response. Microbes Infect 9(7):855–863. https://doi.org/10.1016/j.micinf.2007.03.002

Fedele G, Nasso M, Spensieri F, Palazzo R, Frasca L, Watanabe M, Ausiello CM (2008) Lipopolysaccharides from Bordetella pertussis and Bordetella parapertussis differently modulate human dendritic cell functions resulting in divergent prevalence of Th17-polarized responses. J Immunol 181(1):208–216

Fedele G, Spensieri F, Palazzo R, Nasso M, Cheung GY, Coote JG, Ausiello CM (2010) Bordetella pertussis commits human dendritic cells to promote a Th1/Th17 response through the activity of adenylate cyclase toxin and MAPK-pathways. PLoS One 5(1):e8734. https://doi.org/10.1371/journal.pone.0008734

Fedele G, Bianco M, Debrie AS, Locht C, Ausiello CM (2011) Attenuated Bordetella pertussis vaccine candidate BPZE1 promotes human dendritic cell CCL21-induced migration and drives a Th1/Th17 response. J Immunol 186(9):5388–5396. https://doi.org/10.4049/jimmunol.1003765

Fedele G, Bianco M, Ausiello CM (2013) The virulence factors of Bordetella pertussis: talented modulators of host immune response. Arch Immunol Ther Exp 61 (6):445–457. https://doi.org/10.1007/s00005-013-0242-1

Ferguson RG, Simes AB (1949) BCG vaccination of Indian infants in Saskatchewan. Tubercle 30(1):5–11

Feunou PF, Bertout J, Locht C (2010) T- and B-cell-mediated protection induced by novel, live attenuated pertussis vaccine in mice. Cross protection against parapertussis. PLoS One 5(4):e10178. https://doi.org/10.1371/journal.pone.0010178

Foster SL, Hargreaves DC, Medzhitov R (2007) Gene-specific control of inflammation by TLR-induced chromatin modifications. Nature 447(7147):972–978. https://doi.org/10.1038/nature05836

Gaillard ME, Bottero D, Errea A, Ormazabal M, Zurita ME, Moreno G, Rumbo M, Castuma C, Bartel E, Flores D, van der Ley P, van der Ark A, Hozbor DF (2014) Acellular pertussis vaccine based on outer membrane vesicles capable of conferring both long-lasting immunity and protection against different strain genotypes. Vaccine 32(8):931–937. https://doi.org/10.1016/j.vaccine.2013.12.048

Ganguly T, Johnson JB, Kock ND, Parks GD, Deora R (2014) The Bordetella pertussis Bps polysaccharide enhances lung colonization by conferring protection from complement-mediated killing. Cell Microbiol 16 (7):1105–1118. https://doi.org/10.1111/cmi.12264

Garly ML, Martins CL, Bale C, Balde MA, Hedegaard KL, Gustafson P, Lisse IM, Whittle HC, Aaby P (2003) BCG scar and positive tuberculin reaction associated with reduced child mortality in West Africa. A non-specific beneficial effect of BCG? Vaccine 21 (21–22):2782–2790

Gasperini G, Biagini M, Arato V, Gianfaldoni C, Vadi A, Norais N, Bensi G, Delany I, Pizza M, Arico B, Leuzzi R (2018) Outer membrane vesicles (OMV)-based and proteomics-driven antigen selection identifies novel factors contributing to Bordetella pertussis adhesion to epithelial cells. Mol Cell Proteomics 17(2):205–215. https://doi.org/10.1074/mcp.RA117.000045

Gaudet S, Miller-Jensen K (2016) Redefining signaling pathways with an expanding single-cell toolbox. Trends Biotechnol 34(6):458–469. https://doi.org/10.1016/j.tibtech.2016.02.009

Geurtsen J, Banus HA, Gremmer ER, Ferguson H, de la Fonteyne-Blankestijn LJ, Vermeulen JP, Dormans JA, Tommassen J, van der Ley P, Mooi FR, Vandebriel RJ (2007) Lipopolysaccharide analogs improve efficacy of acellular pertussis vaccine and reduce type I hypersensitivity in mice. Clin Vaccine Immunol 14 (7):821–829. https://doi.org/10.1128/CVI.00074-07

Geurtsen J, Fransen F, Vandebriel RJ, Gremmer ER, de la Fonteyne-Blankestijn LJ, Kuipers B, Tommassen J, van der Ley P (2008) Supplementation of whole-cell pertussis vaccines with lipopolysaccharide analogs: modification of vaccine-induced immune responses. Vaccine 26(7):899–906. https://doi.org/10.1016/j.vaccine.2007.12.012

Goldman WE, Klapper DG, Baseman JB (1982) Detection, isolation, and analysis of a released Bordetella pertussis product toxic to cultured tracheal cells. Infect Immun 36(2):782–794

Greco D, Salmaso S, Mastrantonio P, Giuliano M, Tozzi AE, Anemona A, Ciofi degli Atti ML, Giammanco A, Panei P, Blackwelder WC, Klein DL, Wassilak SG (1996) A controlled trial of two acellular vaccines and one whole-cell vaccine against pertussis. Progetto Pertosse Working Group. N Engl J Med 334 (6):341–348. https://doi.org/10.1056/NEJM199602083340601

Guermonprez P, Khelef N, Blouin E, Rieu P, Ricciardi-Castagnoli P, Guiso N, Ladant D, Leclerc C (2001) The adenylate cyclase toxin of Bordetella pertussis binds to target cells via the alpha(M)beta(2) integrin (CD11b/CD18). J Exp Med 193(9):1035–1044

Gustafsson L, Hallander HO, Olin P, Reizenstein E, Storsaeter J (1996) A controlled trial of a two-component acellular, a five-component acellular, and a whole-cell pertussis vaccine. N Engl J Med 334 (6):349–355. https://doi.org/10.1056/NEJM199602083340602

Harvill ET, Cotter PA, Miller JF (1999) Pregenomic comparative analysis between bordetella bronchiseptica RB50 and Bordetella pertussis tohama I in murine models of respiratory tract infection. Infect Immun 67 (11):6109–6118

Hellwig SM, Hazenbos WL, van de Winkel JG, Mooi FR (1999) Evidence for an intracellular niche for Bordetella pertussis in broncho-alveolar lavage cells of mice. FEMS Immunol Med Microbiol 26 (3–4):203–207. https://doi.org/10.1111/j.1574-695X.1999.tb01391.x

Henderson MW, Inatsuka CS, Sheets AJ, Williams CL, Benaron DJ, Donato GM, Gray MC, Hewlett EL, Cotter PA (2012) Contribution of Bordetella filamentous hemagglutinin and adenylate cyclase toxin to suppression and evasion of interleukin-17-mediated inflammation. Infect Immun 80(6):2061–2075. https://doi.org/10.1128/IAI.00148-12

Hendrikx LH, Schure RM, Ozturk K, de Rond LG, de Greeff SC, Sanders EA, Berbers GA, Buisman AM (2011) Different IgG-subclass distributions after whole-cell and acellular pertussis infant primary vaccinations in healthy and pertussis infected children. Vaccine 29(40):6874–6880. https://doi.org/10.1016/j.vaccine.2011.07.055

Hickey FB, Brereton CF, Mills KH (2008) Adenylate cyclase toxin of Bordetella pertussis inhibits TLR-induced IRF-1 and IRF-8 activation and IL-12 production and enhances IL-10 through MAPK activation in dendritic cells. J Leukoc Biol 84(1):234–243. https://doi.org/10.1189/jlb.0208113

Higgins SC, Lavelle EC, McCann C, Keogh B, McNeela E, Byrne P, O'Gorman B, Jarnicki A, McGuirk P, Mills KH (2003) Toll-like receptor 4-mediated innate IL-10 activates antigen-specific regulatory T cells and confers resistance to Bordetella pertussis by inhibiting inflammatory pathology. J Immunol 171(6):3119–3127

Holst J, Martin D, Arnold R, Huergo CC, Oster P, O'Hallahan J, Rosenqvist E (2009) Properties and clinical performance of vaccines containing outer membrane vesicles from Neisseria meningitidis. Vaccine 27(Suppl 2):B3–B12. https://doi.org/10.1016/j.vaccine.2009.04.071

Hoonakker ME, Verhagen LM, Hendriksen CF, van Els CA, Vandebriel RJ, Sloots A, Han WG (2015) In vitro innate immune cell based models to assess whole cell Bordetella pertussis vaccine quality: a proof of principle. Biologicals 43(2):100–109. https://doi.org/10.1016/j.biologicals.2014.12.002

Hormozi K, Parton R, Coote J (1999) Adjuvant and protective properties of native and recombinant Bordetella pertussis adenylate cyclase toxin preparations in mice. FEMS Immunol Med Microbiol 23(4):273–282. https://doi.org/10.1111/j.1574-695X.1999.tb01248.x

Hornung V, Bauernfeind F, Halle A, Samstad EO, Kono H, Rock KL, Fitzgerald KA, Latz E (2008) Silica crystals and aluminum salts activate the NALP3 inflammasome through phagosomal destabilization. Nat Immunol 9(8):847–856. https://doi.org/10.1038/ni.1631

Hovingh ES, van Gent M, Hamstra HJ, Demkes M, Mooi FR, Pinelli E (2017) Emerging Bordetella pertussis strains induce enhanced signaling of human pattern recognition receptors TLR2, NOD2 and secretion of IL-10 by dendritic cells. PLoS One 12(1):e0170027. https://doi.org/10.1371/journal.pone.0170027

Hovingh ES, Mariman R, Solans L, Hijdra D, Hamstra HJ, Jongerius I, van Gent M, Mooi F, Locht C, Pinelli E (2018) Bordetella pertussis pertactin knock-out strains reveal immunomodulatory properties of this virulence factor. Emerg Microbes Infect 7(1):39. https://doi.org/10.1038/s41426-018-0039-8

Hozbor DF (2017) Outer membrane vesicles: an attractive candidate for pertussis vaccines. Expert Rev Vaccines 16(3):193–196. https://doi.org/10.1080/14760584.2017.1276832

Huang G, Wang Y, Chi H (2012) Regulation of TH17 cell differentiation by innate immune signals. Cell Mol Immunol 9(4):287–295. https://doi.org/10.1038/cmi.2012.10

Inatsuka CS, Julio SM, Cotter PA (2005) Bordetella filamentous hemagglutinin plays a critical role in immunomodulation, suggesting a mechanism for host specificity. Proc Natl Acad Sci U S A 102 (51):18578–18583. https://doi.org/10.1073/pnas.0507910102

Inatsuka CS, Xu Q, Vujkovic-Cvijin I, Wong S, Stibitz S, Miller JF, Cotter PA (2010) Pertactin is required for Bordetella species to resist neutrophil-mediated clearance. Infect Immun 78(7):2901–2909. https://doi.org/10.1128/IAI.00188-10

James LK, Till SJ (2016) Potential mechanisms for IgG4 inhibition of immediate hypersensitivity reactions.

Curr Allergy Asthma Rep 16(3):23. https://doi.org/10.1007/s11882-016-0600-2

Jongerius I, Schuijt TJ, Mooi FR, Pinelli E (2015) Complement evasion by Bordetella pertussis: implications for improving current vaccines. J Mol Med (Berl) 93 (4):395–402. https://doi.org/10.1007/s00109-015-1259-1

Kachroo A, Robin GP (2013) Systemic signaling during plant defense. Curr Opin Plant Biol 16(4):527–533. https://doi.org/10.1016/j.pbi.2013.06.019

Kampmann B, Jones CE (2015) Factors influencing innate immunity and vaccine responses in infancy. Philos Trans R Soc Lond Ser B Biol Sci 370(1671). https://doi.org/10.1098/rstb.2014.0148

Karaolis DK, Newstead MW, Zeng X, Hyodo M, Hayakawa Y, Bhan U, Liang H, Standiford TJ (2007) Cyclic di-GMP stimulates protective innate immunity in bacterial pneumonia. Infect Immun 75 (10):4942–4950. https://doi.org/10.1128/IAI.01762-06

Kaufmann E, Sanz J, Dunn JL, Khan N, Mendonca LE, Pacis A, Tzelepis F, Pernet E, Dumaine A, Grenier JC, Mailhot-Leonard F, Ahmed E, Belle J, Besla R, Mazer B, King IL, Nijnik A, Robbins CS, Barreiro LB, Divangahi M (2018) BCG educates hematopoietic stem cells to generate protective innate immunity against tuberculosis. Cell 172(1–2):176–190 e119. https://doi.org/10.1016/j.cell.2017.12.031

Kavanagh H, Noone C, Cahill E, English K, Locht C, Mahon BP (2010) Attenuated Bordetella pertussis vaccine strain BPZE1 modulates allergen-induced immunity and prevents allergic pulmonary pathology in a murine model. Clin Exp Allergy 40(6):933–941. https://doi.org/10.1111/j.1365-2222.2010.03459.x

Kawai T, Akira S (2009) The roles of TLRs, RLRs and NLRs in pathogen recognition. Int Immunol 21 (4):317–337. https://doi.org/10.1093/intimm/dxp017

Kawai T, Akira S (2010) The role of pattern-recognition receptors in innate immunity: update on toll-like receptors. Nat Immunol 11(5):373–384. https://doi.org/10.1038/ni.1863

King KY, Goodell MA (2011) Inflammatory modulation of HSCs: viewing the HSC as a foundation for the immune response. Nat Rev Immunol 11 (10):685–692. https://doi.org/10.1038/nri3062

Kleinnijenhuis J, Quintin J, Preijers F, Joosten LA, Ifrim DC, Saeed S, Jacobs C, van Loenhout J, de Jong D, Stunnenberg HG, Xavier RJ, van der Meer JW, van Crevel R, Netea MG (2012) Bacille Calmette-Guerin induces NOD2-dependent nonspecific protection from reinfection via epigenetic reprogramming of monocytes. Proc Natl Acad Sci U S A 109 (43):17537–17542. https://doi.org/10.1073/pnas.1202870109

Kooijman S, Brummelman J, van Els C, Marino F, Heck AJR, van Riet E, Metz B, Kersten GFA, Pennings JLA, Meiring HD (2018) Vaccine antigens modulate the innate response of monocytes to Al(OH)3. PLoS One 13(5):e0197885. https://doi.org/10.1371/journal.pone.0197885

Krahenbuhl JL, Sharma SD, Ferraresi RW, Remington JS (1981) Effects of muramyl dipeptide treatment on resistance to infection with toxoplasma gondii in mice. Infect Immun 31(2):716–722

Krieg AM (2002) CpG motifs in bacterial DNA and their immune effects. Annu Rev Immunol 20:709–760. https://doi.org/10.1146/annurev.immunol.20.100301.064842

Krieg AM, Yi AK, Matson S, Waldschmidt TJ, Bishop GA, Teasdale R, Koretzky GA, Klinman DM (1995) CpG motifs in bacterial DNA trigger direct B-cell activation. Nature 374(6522):546–549. https://doi.org/10.1038/374546a0

Kurtz J, Franz K (2003) Innate defence: evidence for memory in invertebrate immunity. Nature 425 (6953):37–38. https://doi.org/10.1038/425037a

Ladant D, Ullmann A (1999) Bordatella pertussis adenylate cyclase: a toxin with multiple talents. Trends Microbiol 7(4):172–176

Leef M, Elkins KL, Barbic J, Shahin RD (2000) Protective immunity to Bordetella pertussis requires both B cells and CD4(+) T cells for key functions other than specific antibody production. J Exp Med 191 (11):1841–1852

Leininger E, Roberts M, Kenimer JG, Charles IG, Fairweather N, Novotny P, Brennan MJ (1991) Pertactin, an Arg-Gly-Asp-containing Bordetella pertussis surface protein that promotes adherence of mammalian cells. Proc Natl Acad Sci U S A 88(2):345–349

Levine MI, Sackett MF (1946) Results of BCG immunization in New York City. Am Rev Tuberc 53:517–532

Li R, Lim A, Phoon MC, Narasaraju T, Ng JK, Poh WP, Sim MK, Chow VT, Locht C, Alonso S (2010) Attenuated Bordetella pertussis protects against highly pathogenic influenza A viruses by dampening the cytokine storm. J Virol 84(14):7105–7113. https://doi.org/10.1128/JVI.02542-09

Li S, Rouphael N, Duraisingham S, Romero-Steiner S, Presnell S, Davis C, Schmidt DS, Johnson SE, Milton A, Rajam G, Kasturi S, Carlone GM, Quinn C, Chaussabel D, Palucka AK, Mulligan MJ, Ahmed R, Stephens DS, Nakaya HI, Pulendran B (2014) Molecular signatures of antibody responses derived from a systems biology study of five human vaccines. Nat Immunol 15(2):195–204. https://doi.org/10.1038/ni.2789

Locht C, Bertin P, Menozzi FD, Renauld G (1993) The filamentous haemagglutinin, a multifaceted adhesion produced by virulent Bordetella spp. Mol Microbiol 9 (4):653–660

Locht C, Papin JF, Lecher S, Debrie AS, Thalen M, Solovay K, Rubin K, Mielcarek N (2017) Live attenuated pertussis vaccine BPZE1 protects baboons against Bordetella pertussis disease and infection. J Infect Dis 216(1):117–124. https://doi.org/10.1093/infdis/jix254

Macdonald-Fyall J, Xing D, Corbel M, Baillie S, Parton R, Coote J (2004) Adjuvanticity of native and detoxified adenylate cyclase toxin of Bordetella pertussis towards co-administered antigens. Vaccine 22 (31–32):4270–4281. https://doi.org/10.1016/j.vaccine.2004.04.033

Magalhaes JG, Philpott DJ, Nahori MA, Jehanno M, Fritz J, Le Bourhis L, Viala J, Hugot JP, Giovannini M, Bertin J, Lepoivre M, Mengin-Lecreulx D, Sansonetti PJ, Girardin SE (2005) Murine Nod1 but not its human orthologue mediates innate immune detection of tracheal cytotoxin. EMBO Rep 6(12):1201–1207. https://doi.org/10.1038/sj.embor.7400552

Mann PB, Kennett MJ, Harvill ET (2004) Toll-like receptor 4 is critical to innate host defense in a murine model of bordetellosis. J Infect Dis 189(5):833–836. https://doi.org/10.1086/381898

Manz MG, Boettcher S (2014) Emergency granulopoiesis. Nat Rev Immunol 14(5):302–314. https://doi.org/10.1038/nri3660

Marr N, Novikov A, Hajjar AM, Caroff M, Fernandez RC (2010) Variability in the lipooligosaccharide structure and endotoxicity among Bordetella pertussis strains. J Infect Dis 202(12):1897–1906. https://doi.org/10.1086/657409

Mattoo S, Miller JF, Cotter PA (2000) Role of Bordetella bronchiseptica fimbriae in tracheal colonization and development of a humoral immune response. Infect Immun 68(4):2024–2033

Mbow ML, De Gregorio E, Valiante NM, Rappuoli R (2010) New adjuvants for human vaccines. Curr Opin Immunol 22(3):411–416. https://doi.org/10.1016/j.coi.2010.04.004

McGuirk P, Mills KH (2000) Direct anti-inflammatory effect of a bacterial virulence factor: IL-10-dependent suppression of IL-12 production by filamentous hemagglutinin from Bordetella pertussis. Eur J Immunol 30 (2):415–422. https://doi.org/10.1002/1521-4141(200002)30:2<415::AID-IMMU415>3.0.CO;2-X

McGuirk P, Mahon BP, Griffin F, Mills KH (1998) Compartmentalization of T cell responses following respiratory infection with Bordetella pertussis: hyporesponsiveness of lung T cells is associated with modulated expression of the co-stimulatory molecule CD28. Eur J Immunol 28(1):153–163. https://doi.org/10.1002/(SICI)1521-4141(199801)28:01<153::AID-IMMU153>3.0.CO;2-#

McGuirk P, McCann C, Mills KH (2002) Pathogen-specific T regulatory 1 cells induced in the respiratory tract by a bacterial molecule that stimulates interleukin 10 production by dendritic cells: a novel strategy for evasion of protective T helper type 1 responses by Bordetella pertussis. J Exp Med 195(2):221–231

McKee AS, Munks MW, MacLeod MK, Fleenor CJ, Van Rooijen N, Kappler JW, Marrack P (2009) Alum induces innate immune responses through macrophage and mast cell sensors, but these sensors are not required for alum to act as an adjuvant for specific immunity. J Immunol 183(7):4403–4414. https://doi.org/10.4049/jimmunol.0900164

Mielcarek N, Debrie AS, Raze D, Bertout J, Rouanet C, Younes AB, Creusy C, Engle J, Goldman WE, Locht C (2006) Live attenuated B. pertussis as a single-dose nasal vaccine against whooping cough. PLoS Pathog 2 (7):e65. https://doi.org/10.1371/journal.ppat.0020065

Mills KH, Barnard A, Watkins J, Redhead K (1993) Cell-mediated immunity to Bordetella pertussis: role of Th1 cells in bacterial clearance in a murine respiratory infection model. Infect Immun 61(2):399–410

Misiak A, Leuzzi R, Allen AC, Galletti B, Baudner BC, D'Oro U, O'Hagan DT, Pizza M, Seubert A, Mills KHG (2017a) Addition of a TLR7 agonist to an acellular pertussis vaccine enhances Th1 and Th17 responses and protective immunity in a mouse model. Vaccine 35(39):5256–5263. https://doi.org/10.1016/j.vaccine.2017.08.009

Misiak A, Wilk MM, Raverdeau M, Mills KH (2017b) IL-17-producing innate and pathogen-specific tissue resident memory gammadelta T cells expand in the lungs of Bordetella pertussis-infected mice. J Immunol 198(1):363–374. https://doi.org/10.4049/jimmunol.1601024

Moreno G, Errea A, Van Maele L, Roberts R, Leger H, Sirard JC, Benecke A, Rumbo M, Hozbor D (2013) Toll-like receptor 4 orchestrates neutrophil recruitment into airways during the first hours of Bordetella pertussis infection. Microbes Infect 15(10–11):708–718. https://doi.org/10.1016/j.micinf.2013.06.010

Munoz N, Van Maele L, Marques JM, Rial A, Sirard JC, Chabalgoity JA (2010) Mucosal administration of flagellin protects mice from Streptococcus pneumoniae lung infection. Infect Immun 78(10):4226–4233. https://doi.org/10.1128/IAI.00224-10

Nasso M, Fedele G, Spensieri F, Palazzo R, Costantino P, Rappuoli R, Ausiello CM (2009) Genetically detoxified pertussis toxin induces Th1/Th17 immune response through MAPKs and IL-10-dependent mechanisms. J Immunol 183(3):1892–1899. https://doi.org/10.4049/jimmunol.0901071

Netea MG, van Crevel R (2014) BCG-induced protection: effects on innate immune memory. Semin Immunol 26 (6):512–517. https://doi.org/10.1016/j.smim.2014.09.006

Netea MG, Joosten LA, Latz E, Mills KH, Natoli G, Stunnenberg HG, O'Neill LA, Xavier RJ (2016) Trained immunity: a program of innate immune memory in health and disease. Science 352(6284):aaf1098. https://doi.org/10.1126/science.aaf1098

Obermoser G, Presnell S, Domico K, Xu H, Wang Y, Anguiano E, Thompson-Snipes L, Ranganathan R, Zeitner B, Bjork A, Anderson D, Speake C, Ruchaud E, Skinner J, Alsina L, Sharma M, Dutartre H, Cepika A, Israelsson E, Nguyen P, Nguyen QA, Harrod AC, Zurawski SM, Pascual V, Ueno H, Nepom GT, Quinn C, Blankenship D, Palucka K, Banchereau J, Chaussabel D (2013) Systems scale interactive exploration reveals quantitative and qualitative differences in response to influenza and pneumococcal vaccines. Immunity 38(4):831–844. https://doi.org/10.1016/j.immuni.2012.12.008

O'Garra A (1998) Cytokines induce the development of functionally heterogeneous T helper cell subsets. Immunity 8(3):275–283

Osicka R, Osickova A, Hasan S, Bumba L, Cerny J, Sebo P (2015) Bordetella adenylate cyclase toxin is a unique ligand of the integrin complement receptor 3. elife 4: e10766. https://doi.org/10.7554/eLife.10766

Paik D, Monahan A, Caffrey DR, Elling R, Goldman WE, Silverman N (2017) SLC46 family transporters facilitate cytosolic innate immune recognition of monomeric peptidoglycans. J Immunol 199(1):263–270. https://doi.org/10.4049/jimmunol.1600409

Phongsisay V, Iizasa E, Hara H, Yoshida H (2017) Pertussis toxin targets the innate immunity through DAP12, FcRgamma, and MyD88 adaptor proteins. Immunobiology 222(4):664–671. https://doi.org/10.1016/j.imbio.2016.12.004

Pichichero ME, Casey JR, Almudevar A, Basha S, Surendran N, Kaur R, Morris M, Livingstone AM, Mosmann TR (2016) Functional immune cell differences associated with low vaccine responses in infants. J Infect Dis 213(12):2014–2019. https://doi.org/10.1093/infdis/jiw053

Pillsbury A, Quinn HE, McIntyre PB (2014) Australian vaccine preventable disease epidemiological review series: pertussis, 2006-2012. Commun Dis Intell Q Rep 38(3):E179–E194

Place DE, Muse SJ, Kirimanjeswara GS, Harvill ET (2014) Caspase-1-independent interleukin-1beta is required for clearance of Bordetella pertussis infections and whole-cell vaccine-mediated immunity. PLoS One 9(9):e107188. https://doi.org/10.1371/journal.pone.0107188

Plotkin SA (2014) The pertussis problem. Clin Infect Dis 58(6):830–833. https://doi.org/10.1093/cid/cit934

Powlesland AS, Ward EM, Sadhu SK, Guo Y, Taylor ME, Drickamer K (2006) Widely divergent biochemical properties of the complete set of mouse DC-SIGN-related proteins. J Biol Chem 281(29):20440–20449. https://doi.org/10.1074/jbc.M601925200

Pulendran B (2009) Learning immunology from the yellow fever vaccine: innate immunity to systems vaccinology. Nat Rev Immunol 9(10):741–747. https://doi.org/10.1038/nri2629

Pulendran B (2014) Systems vaccinology: probing humanity's diverse immune systems with vaccines. Proc Natl Acad Sci U S A 111(34):12300–12306. https://doi.org/10.1073/pnas.1400476111

Pulendran B, Tang H, Manicassamy S (2010) Programming dendritic cells to induce T(H)2 and tolerogenic responses. Nat Immunol 11(8):647–655. https://doi.org/10.1038/ni.1894

Queenan AM, Dowling DJ, Cheng WK, Fae K, Fernandez J, Flynn PJ, Joshi S, Brightman SE, Ramirez J, Serroyen J, Wiertsema S, Fortanier A, van den Dobbelsteen G, Levy O, Poolman J (2019) Increasing FIM2/3 antigen-content improves efficacy of Bordetella pertussis vaccines in mice in vivo without altering vaccine-induced human reactogenicity biomarkers in vitro. Vaccine 37(1):80–89. https://doi.org/10.1016/j.vaccine.2018.11.028

Quintin J, Saeed S, Martens JHA, Giamarellos-Bourboulis EJ, Ifrim DC, Logie C, Jacobs L, Jansen T, Kullberg BJ, Wijmenga C, Joosten LAB, Xavier RJ, van der Meer JWM, Stunnenberg HG, Netea MG (2012) Candida albicans infection affords protection against reinfection via functional reprogramming of monocytes. Cell Host Microbe 12(2):223–232. https://doi.org/10.1016/j.chom.2012.06.006

Raeven RH, Brummelman J, Pennings JL, Nijst OE, Kuipers B, Blok LE, Helm K, van Riet E, Jiskoot W, van Els CA, Han WG, Kersten GF, Metz B (2014) Molecular signatures of the evolving immune response in mice following a Bordetella pertussis infection. PLoS One 9(8):e104548. https://doi.org/10.1371/journal.pone.0104548

Raeven RH, van der Maas L, Tilstra W, Uittenbogaard JP, Bindels TH, Kuipers B, van der Ark A, Pennings JL, van Riet E, Jiskoot W, Kersten GF, Metz B (2015) Immunoproteomic profiling of Bordetella pertussis outer membrane vesicle vaccine reveals broad and balanced humoral immunogenicity. J Proteome Res 14(7):2929–2942. https://doi.org/10.1021/acs.jproteome.5b00258

Raeven RH, Brummelman J, Pennings JL, van der Maas L, Tilstra W, Helm K, van Riet E, Jiskoot W, van Els CA, Han WG, Kersten GF, Metz B (2016) Bordetella pertussis outer membrane vesicle vaccine confers equal efficacy in mice with milder inflammatory responses compared to a whole-cell vaccine. Sci Rep 6:38240. https://doi.org/10.1038/srep38240

Raeven RH, Brummelman J, Pennings JLA, van der Maas L, Helm K, Tilstra W, van der Ark A, Sloots A, van der Ley P, van Eden W, Jiskoot W, van Riet E, van Els CA, Kersten GF, Han WG, Metz B (2018) Molecular and cellular signatures underlying superior immunity against Bordetella pertussis upon pulmonary vaccination. Mucosal Immunol 11(3):979–993. https://doi.org/10.1038/mi.2017.81

Rappuoli R (2000) Reverse vaccinology. Curr Opin Microbiol 3(5):445–450

Rappuoli R, Bottomley MJ, D'Oro U, Finco O, De Gregorio E (2016) Reverse vaccinology 2.0: human immunology instructs vaccine antigen design. J Exp Med 213(4):469–481. https://doi.org/10.1084/jem.20151960

Relyveld EH, Bizzini B, Gupta RK (1998) Rational approaches to reduce adverse reactions in man to vaccines containing tetanus and diphtheria toxoids. Vaccine 16(9–10):1016–1023

Renauld-Mongenie G, Cornette J, Mielcarek N, Menozzi FD, Locht C (1996) Distinct roles of the N-terminal and C-terminal precursor domains in the biogenesis of the Bordetella pertussis filamentous hemagglutinin. J Bacteriol 178(4):1053–1060

Ribes S, Meister T, Ott M, Redlich S, Janova H, Hanisch UK, Nessler S, Nau R (2014) Intraperitoneal prophylaxis with CpG oligodeoxynucleotides protects neutropenic mice against intracerebral Escherichia coli K1 infection. J Neuroinflammation 11:14. https://doi.org/10.1186/1742-2094-11-14

Roberts M, Fairweather NF, Leininger E, Pickard D, Hewlett EL, Robinson A, Hayward C, Dougan G, Charles IG (1991) Construction and characterization of Bordetella pertussis mutants lacking the vir-regulated P.69 outer membrane protein. Mol Microbiol 5(6):1393–1404

Rodrigues J, Brayner FA, Alves LC, Dixit R, Barillas-Mury C (2010) Hemocyte differentiation mediates innate immune memory in Anopheles gambiae mosquitoes. Science 329(5997):1353–1355. https://doi.org/10.1126/science.1190689

Ronchi F, Basso C, Preite S, Reboldi A, Baumjohann D, Perlini L, Lanzavecchia A, Sallusto F (2016) Experimental priming of encephalitogenic Th1/Th17 cells requires pertussis toxin-driven IL-1beta production by myeloid cells. Nat Commun 7:11541. https://doi.org/10.1038/ncomms11541

Ross PJ, Lavelle EC, Mills KH, Boyd AP (2004) Adenylate cyclase toxin from Bordetella pertussis synergizes with lipopolysaccharide to promote innate interleukin-10 production and enhances the induction of Th2 and regulatory T cells. Infect Immun 72(3):1568–1579

Ross PJ, Sutton CE, Higgins S, Allen AC, Walsh K, Misiak A, Lavelle EC, McLoughlin RM, Mills KH (2013) Relative contribution of Th1 and Th17 cells in adaptive immunity to Bordetella pertussis: towards the rational design of an improved acellular pertussis vaccine. PLoS Pathog 9(4):e1003264. https://doi.org/10.1371/journal.ppat.1003264

Ryan M, Murphy G, Gothefors L, Nilsson L, Storsaeter J, Mills KH (1997) Bordetella pertussis respiratory infection in children is associated with preferential activation of type 1 T helper cells. J Infect Dis 175(5):1246–1250

Ryan M, McCarthy L, Rappuoli R, Mahon BP, Mills KH (1998a) Pertussis toxin potentiates Th1 and Th2 responses to co-injected antigen: adjuvant action is associated with enhanced regulatory cytokine production and expression of the co-stimulatory molecules B7-1, B7-2 and CD28. Int Immunol 10(5):651–662

Ryan M, Murphy G, Ryan E, Nilsson L, Shackley F, Gothefors L, Oymar K, Miller E, Storsaeter J, Mills KH (1998b) Distinct T-cell subtypes induced with whole cell and acellular pertussis vaccines in children. Immunology 93(1):1–10

Sallusto F (2016) Heterogeneity of human CD4(+) T cells against microbes. Annu Rev Immunol 34:317–334. https://doi.org/10.1146/annurev-immunol-032414-112056

Samore MH, Siber GR (1996) Pertussis toxin enhanced IgG1 and IgE responses to primary tetanus immunization are mediated by interleukin-4 and persist during secondary responses to tetanus alone. Vaccine 14(4):290–297

Scanlon KM, Gau Y, Zhu J, Skerry C, Wall SM, Soleimani M, Carbonetti NH (2014) Epithelial anion transporter pendrin contributes to inflammatory lung pathology in mouse models of Bordetella pertussis infection. Infect Immun 82(10):4212–4221. https://doi.org/10.1128/IAI.02222-14

Schaeffer LM, McCormack FX, Wu H, Weiss AA (2004) Bordetella pertussis lipopolysaccharide resists the bactericidal effects of pulmonary surfactant protein a. J Immunol 173(3):1959–1965

Scheller EV, Melvin JA, Sheets AJ, Cotter PA (2015) Cooperative roles for fimbria and filamentous hemagglutinin in Bordetella adherence and immune modulation. MBio 6(3):e00500–e00515. https://doi.org/10.1128/mBio.00500-15

Schenten D, Medzhitov R (2011) The control of adaptive immune responses by the innate immune system. Adv Immunol 109:87–124. https://doi.org/10.1016/B978-0-12-387664-5.00003-0

Schiavoni I, Fedele G, Quattrini A, Bianco M, Schnoeller C, Openshaw PJ, Locht C, Ausiello CM (2014) Live attenuated B. pertussis BPZE1 rescues the immune functions of respiratory syncytial virus infected human dendritic cells by promoting Th1/Th17 responses. PLoS One 9(6):e100166. https://doi.org/10.1371/journal.pone.0100166

Schnoeller C, Roux X, Sawant D, Raze D, Olszewska W, Locht C, Openshaw PJ (2014) Attenuated Bordetella pertussis vaccine protects against respiratory syncytial virus disease via an IL-17-dependent mechanism. Am J Respir Crit Care Med 189(2):194–202. https://doi.org/10.1164/rccm.201307-1227OC

Schreibelt G, Benitez-Ribas D, Schuurhuis D, Lambeck AJ, van Hout-Kuijer M, Schaft N, Punt CJ, Figdor CG, Adema GJ, de Vries IJ (2010) Commonly used prophylactic vaccines as an alternative for synthetically produced TLR ligands to mature monocyte-derived dendritic cells. Blood 116(4):564–574. https://doi.org/10.1182/blood-2009-11-251884

Schroder NW, Schumann RR (2005) Single nucleotide polymorphisms of toll-like receptors and susceptibility to infectious disease. Lancet Infect Dis 5(3):156–164. https://doi.org/10.1016/S1473-3099(05)01308-3

Solans L, Debrie AS, Borkner L, Aguilo N, Thiriard A, Coutte L, Uranga S, Trottein F, Martin C, Mills KHG, Locht C (2018) IL-17-dependent SIgA-mediated protection against nasal Bordetella pertussis infection by live attenuated BPZE1 vaccine. Mucosal Immunol 11 (6):1753–1762. https://doi.org/10.1038/s41385-018-0073-9

Souwer Y, Groot Kormelink T, Taanman-Kueter EW, Muller FJ, van Capel TMM, Varga DV, Bar-Ephraim YE, Teunissen MBM, van Ham SM, Kuijpers TW, Wouters D, Meyaard L, de Jong EC (2018) Human TH17 cell development requires processing of dendritic cell-derived CXCL8 by neutrophil elastase. J Allergy Clin Immunol 141(6):2286–2289. e2285. https://doi.org/10.1016/j.jaci.2018.01.003

Spensieri F, Fedele G, Fazio C, Nasso M, Stefanelli P, Mastrantonio P, Ausiello CM (2006) Bordetella pertussis inhibition of interleukin-12 (IL-12) p70 in human monocyte-derived dendritic cells blocks IL-12 p35 through adenylate cyclase toxin-dependent cyclic AMP induction. Infect Immun 74(5):2831–2838. https://doi.org/10.1128/IAI.74.5.2831-2838.2006

Spreafico R, Ricciardi-Castagnoli P, Mortellaro A (2010) The controversial relationship between NLRP3, alum, danger signals and the next-generation adjuvants. Eur J Immunol 40(3):638–642. https://doi.org/10.1002/eji.200940039

Stefanelli P, Fazio C, Fedele G, Spensieri F, Ausiello CM, Mastrantonio P (2009) A natural pertactin deficient strain of Bordetella pertussis shows improved entry in human monocyte-derived dendritic cells. New Microbiol 32(2):159–166

Stenger RM, Meiring HD, Kuipers B, Poelen M, van Gaans-van den Brink JA, Boog CJ, de Jong AP, van Els CA (2014) Bordetella pertussis proteins dominating the major histocompatibility complex class II-presented epitope repertoire in human monocyte-derived dendritic cells. Clin Vaccine Immunol 21(5):641–650. https://doi.org/10.1128/CVI.00665-13

Surendran N, Nicolosi T, Pichichero M (2016) Infants with low vaccine antibody responses have altered innate cytokine response. Vaccine 34(47):5700–5703. https://doi.org/10.1016/j.vaccine.2016.09.050

Surendran N, Nicolosi T, Kaur R, Morris M, Pichichero M (2017) Prospective study of the innate cellular immune response in low vaccine responder children. Innate Immun 23(1):89–96. https://doi.org/10.1177/1753425916678471

Takeda A, Hamano S, Yamanaka A, Hanada T, Ishibashi T, Mak TW, Yoshimura A, Yoshida H (2003) Cutting edge: role of IL-27/WSX-1 signaling for induction of T-bet through activation of STAT1 during initial Th1 commitment. J Immunol 170 (10):4886–4890

Takizawa H, Regoes RR, Boddupalli CS, Bonhoeffer S, Manz MG (2011) Dynamic variation in cycling of hematopoietic stem cells in steady state and inflammation. J Exp Med 208(2):273–284. https://doi.org/10.1084/jem.20101643

Tan T, Dalby T, Forsyth K, Halperin SA, Heininger U, Hozbor D, Plotkin S, Ulloa-Gutierrez R, Wirsing von Konig CH (2015) Pertussis across the globe: recent epidemiologic trends from 2000 to 2013. Pediatr Infect Dis J 34(9):e222–e232. https://doi.org/10.1097/INF.0000000000000795

Thorstensson R, Trollfors B, Al-Tawil N, Jahnmatz M, Bergstrom J, Ljungman M, Torner A, Wehlin L, Van Broekhoven A, Bosman F, Debrie AS, Mielcarek N, Locht C (2014) A phase I clinical study of a live attenuated Bordetella pertussis vaccine--BPZE1; a single centre, double-blind, placebo-controlled, dose-escalating study of BPZE1 given intranasally to healthy adult male volunteers. PLoS One 9(1):e83449. https://doi.org/10.1371/journal.pone.0083449

Tonon S, Goriely S, Aksoy E, Pradier O, Del Giudice G, Trannoy E, Willems F, Goldman M, De Wit D (2002) Bordetella pertussis toxin induces the release of inflammatory cytokines and dendritic cell activation in whole blood: impaired responses in human newborns. Eur J Immunol 32(11):3118–3125. https://doi.org/10.1002/1521-4141(200211)32:11<3118::AID-IMMU3118>3.0.CO;2-B

Trinchieri G, Pflanz S, Kastelein RA (2003) The IL-12 family of heterodimeric cytokines: new players in the regulation of T cell responses. Immunity 19 (5):641–644

Trollfors B, Taranger J, Lagergard T, Lind L, Sundh V, Zackrisson G, Lowe CU, Blackwelder W, Robbins JB (1995) A placebo-controlled trial of a pertussis-toxoid vaccine. N Engl J Med 333(16):1045–1050. https://doi.org/10.1056/NEJM199510193331604

Tsang RS, Lau AK, Sill ML, Halperin SA, Van Caeseele P, Jamieson F, Martin IE (2004) Polymorphisms of the fimbria fim3 gene of Bordetella pertussis strains isolated in Canada. J Clin Microbiol 42(11):5364–5367. https://doi.org/10.1128/JCM.42.11.5364-5367.2004

Underhill DM, Goodridge HS (2012) Information processing during phagocytosis. Nat Rev Immunol 12 (7):492–502. https://doi.org/10.1038/nri3244

van Beelen AJ, Zelinkova Z, Taanman-Kueter EW, Muller FJ, Hommes DW, Zaat SA, Kapsenberg ML, de Jong EC (2007) Stimulation of the intracellular bacterial sensor NOD2 programs dendritic cells to promote interleukin-17 production in human memory T cells. Immunity 27(4):660–669. https://doi.org/10.1016/j.immuni.2007.08.013

van Loo IH, Heuvelman KJ, King AJ, Mooi FR (2002) Multilocus sequence typing of Bordetella pertussis based on surface protein genes. J Clin Microbiol 40 (6):1994–2001

van Twillert I, van Gaans-van den Brink JA, Poelen MC, Helm K, Kuipers B, Schipper M, Boog CJ, Verheij TJ, Versteegh FG, van Els CA (2014) Age related differences in dynamics of specific memory B cell populations after clinical pertussis infection. PLoS One 9(1):e85227. https://doi.org/10.1371/journal.pone.0085227

Vandebriel RJ, Hellwig SM, Vermeulen JP, Hoekman JH, Dormans JA, Roholl PJ, Mooi FR (2003) Association of Bordetella pertussis with host immune cells in the mouse lung. Microb Pathog 35(1):19–29

Varney ME, Boehm DT, DeRoos K, Nowak ES, Wong TY, Sen-Kilic E, Bradford SD, Elkins C, Epperly MS, Witt WT, Barbier M, Damron FH (2018) Bordetella pertussis whole cell immunization, unlike acellular immunization, mimics naive infection by driving hematopoietic stem and progenitor cell expansion in mice. Front Immunol 9:2376. https://doi.org/10.3389/fimmu.2018.02376

Villarino Romero R, Hasan S, Fae K, Holubova J, Geurtsen J, Schwarzer M, Wiertsema S, Osicka R, Poolman J, Sebo P (2016) Bordetella pertussis filamentous hemagglutinin itself does not trigger anti-inflammatory interleukin-10 production by human dendritic cells. Int J Med Microbiol 306(1):38–47. https://doi.org/10.1016/j.ijmm.2015.11.003

Vojtova J, Kamanova J, Sebo P (2006) Bordetella adenylate cyclase toxin: a swift saboteur of host defense. Curr Opin Microbiol 9(1):69–75. https://doi.org/10.1016/j.mib.2005.12.011

Vukman KV, Ravida A, Aldridge AM, O'Neill SM (2013) Mannose receptor and macrophage galactose-type lectin are involved in Bordetella pertussis mast cell interaction. J Leukoc Biol 94(3):439–448. https://doi.org/10.1189/jlb.0313130

Wang ZY, Yang D, Chen Q, Leifer CA, Segal DM, Su SB, Caspi RR, Howard ZO, Oppenheim JJ (2006) Induction of dendritic cell maturation by pertussis toxin and its B subunit differentially initiate Toll-like receptor 4-dependent signal transduction pathways. Exp Hematol 34(8):1115–1124. https://doi.org/10.1016/j.exphem.2006.04.025

Wang X, Gray MC, Hewlett EL, Maynard JA (2015) The Bordetella adenylate cyclase repeat-in-toxin (RTX) domain is immunodominant and elicits neutralizing antibodies. J Biol Chem 290(6):3576–3591. https://doi.org/10.1074/jbc.M114.585281

Warfel JM, Merkel TJ (2013) Bordetella pertussis infection induces a mucosal IL-17 response and long-lived Th17 and Th1 immune memory cells in nonhuman primates. Mucosal Immunol 6(4):787–796. https://doi.org/10.1038/mi.2012.117

Warfel JM, Zimmerman LI, Merkel TJ (2014) Acellular pertussis vaccines protect against disease but fail to prevent infection and transmission in a nonhuman primate model. Proc Natl Acad Sci U S A 111 (2):787–792. https://doi.org/10.1073/pnas.1314688110

White MJ, Nielsen CM, McGregor RH, Riley EH, Goodier MR (2014) Differential activation of CD57-defined natural killer cell subsets during recall responses to vaccine antigens. Immunology 142(1):140–150. https://doi.org/10.1111/imm.12239

Wilk MM, Misiak A, McManus RM, Allen AC, Lynch MA, Mills KHG (2017) Lung CD4 tissue-resident memory T cells mediate adaptive immunity induced by previous infection of mice with Bordetella pertussis. J Immunol 199(1):233–243. https://doi.org/10.4049/jimmunol.1602051

Wu V, Smith AA, You H, Nguyen TA, Ferguson R, Taylor M, Park JE, Llontop P, Youngman KR, Abramson T (2016) Plasmacytoid dendritic cell-derived IFNalpha modulates Th17 differentiation during early Bordetella pertussis infection in mice. Mucosal Immunol 9(3):777–786. https://doi.org/10.1038/mi.2015.101

Yan H, KuoLee R, Tram K, Qiu H, Zhang J, Patel GB, Chen W (2009) 3′,5′-Cyclic diguanylic acid elicits mucosal immunity against bacterial infection. Biochem Biophys Res Commun 387(3):581–584. https://doi.org/10.1016/j.bbrc.2009.07.061

Zeddeman A, van Gent M, Heuvelman CJ, van der Heide HG, Bart MJ, Advani A, Hallander HO, Wirsing von Konig CH, Riffelman M, Storsaeter J, Vestrheim DF, Dalby T, Krogfelt KA, Fry NK, Barkoff AM, Mertsola J, He Q, Mooi F (2014) Investigations into the emergence of pertactin-deficient Bordetella pertussis isolates in six European countries, 1996 to 2012. Euro Surveill 19(33):pii: 20881

Zhou H, Wang Y, Lian Q, Yang B, Ma Y, Wu X, Sun S, Liu Y, Sun B (2014) Differential IL-10 production by DCs determines the distinct adjuvant effects of LPS and PTX in EAE induction. Eur J Immunol 44 (5):1352–1362. https://doi.org/10.1002/eji.201343744

Zorzeto TQ, Higashi HG, da Silva MT, Carniel Ede F, Dias WO, Ramalho VD, Mazzola TN, Lima SC, Morcillo AM, Stephano MA, Antonio MA, Zanolli Mde L, Raw I, Vilela MM (2009) Immunogenicity of a whole-cell pertussis vaccine with low lipopolysaccharide content in infants. Clin Vaccine Immunol 16(4):544–550. https://doi.org/10.1128/CVI.00339-08

Adv Exp Med Biol - Advances in Microbiology, Infectious Diseases and Public Health (2019) 1183: 81–98
https://doi.org/10.1007/5584_2019_405
© Springer Nature Switzerland AG 2019
Published online: 19 July 2019

Superior *B. pertussis* Specific CD4+ T-Cell Immunity Imprinted by Natural Infection

Eleonora E. Lambert, Anne-Marie Buisman, and Cécile A. C. M. van Els

Abstract

Pertussis remains endemic in vaccinated populations due to waning of vaccine-induced immunity and insufficient interruption of transmission. Correlates of long-term protection against whooping cough remain elusive but increasing evidence from experimental models indicates that the priming of particular lineages of *B. pertussis* (Bp) specific CD4+ T cells is essential to control bacterial load. Critical hallmarks of these protective CD4+ T cell lineages in animals are suggested to be their differentiation profile as Th1 and Th17 cells and their tissue residency. These features seem optimally primed by previous infection but insufficiently or only partially by current vaccines. In this review, evidence is sought indicating whether infection also drives such superior Bp specific CD4+ T cell lineages in humans. We highlight key features of effector immunity downstream of Th1 and Th17 cell cytokines that explain clearing of primary Bp infections in naïve hosts, and effective prevention of infection in convalescent hosts during secondary challenge. Outstanding questions are put forward that need answers before correlates of human Bp infection-primed CD4+ T cell immunity can be used as benchmark for the development of improved pertussis vaccines.

Keywords

Bordetella pertussis (Bp) · Mechanism of protection · Natural infection · Th1 and Th17 polarized subsets · Tissue residency

Abbreviations

ACV	acellular pertussis vaccine
Bp	*Bordetella pertussis*
FHA	filamentous hemagglutinin
HSPC	hematopoietic stem and progenitor cell
PRN	pertactin
PTX	pertussis toxin
Th	T helper subset
WCV	whole cell pertussis vaccine

E. E. Lambert, A.-M. Buisman, and C. A. C. M. van Els (✉)
Centre for Infectious Disease Control, National Institute for Public Health and the Environment, Bilthoven, The Netherlands
e-mail: nora.lambert@rivm.nl;
annemarie.buisman@rivm.nl; cecile.van.els@rivm.nl

1 Introduction

Bordetella pertussis (Bp) is a gram negative encapsulated coccobacillus that colonizes the mucosa of the upper respiratory tract (URT). It is transmitted via airborne droplets and the average incubation time is 7–10 days. Bp is the causative agent of the disease pertussis or whooping cough, and can lead to severe illness, especially in young infants and newborns due to bronchopneumonia (Masseria et al. 2017). Pertussis is one of the leading causes of infant death due to respiratory infection (Tan et al. 2015; Muloiwa et al. 2018). The disease proceeds in several stages. It starts with a catarrhal phase, in which patients experience malaise and have symptoms of a cold. This catarrhal phase is followed by the paroxysmal phase, which presents with frequent attacks of coughing and characteristic squeaking inhalation due to breathing problems (2–6 weeks) and subsequently by the convalescent phase, which is characterized by a continuous coughing. Full recovery from disease may take weeks up to several months. However, a host immune response is initiated as soon as Bp attaches to the ciliated epithelial cells in the URT and starts colonizing. Here the bacterium is sensed by resident innate immune cells that shape a complex series of interactions with other recruited innate and adaptive cell types, influenced by both pathogen-derived factors and the local inflammatory microenvironment. In primary infections Bp can modulate virtually all aspects of the immune response and avoid early clearance by producing numerous virulence factors and adapting various bacterial life styles, as reviewed extensively (Locht et al. 2011; de Gouw et al. 2011; Higgs et al. 2012; Melvin et al. 2014; Brummelman et al. 2015; Jongerius et al. 2015; Cattelan et al. 2016; Fedele et al. 2017; Dorji et al. 2018; Gestal et al. 2018). Eventually effector immunity develops and the pathogen is cleared from its nasopharyngeal niche within weeks. Hence, patients are contagious and transmit the bacterium only during the catarrhal phase and the first weeks of the paroxysmal phase (Kilgore et al. 2016). In addition to effector immunity, the convalescent host develops long-lived memory immunity, able to react faster and to prevent secondary Bp infections and transmission. Although symptomatic reinfections have been described (Versteegh et al. 2002), most reinfections likely occur unnoticed and result in boosting levels of naturally acquired immunity that may have waned (Van Twillert et al. 2016). In the era before the introduction of pertussis vaccination, immunity against whooping cough was only naturally acquired and, based on its epidemiology, pertussis was considered a childhood disease (Hewlett and Edwards 2005). Then infant pertussis

immunization was introduced in many developed countries in the 40s and 50s, that strongly reduced the mortality and disease load but also shifed its epidemiology to older age groups (Hewlett and Edwards 2005). The first type of pertussis vaccines were whole cell vaccines (WCV), consisting of killed Bp biomass, next to a broad spectrum of bacterial protein antigens these included the reactogenic Bp endotoxin LOS. While WCVs were highly effective, they were associated with side effects and were replaced by safer acellular pertussis vaccines (ACV) in most high income countries since the 90's. ACVs contain one up to five major immunogenic purified proteins of the bacterium that are adjuvanted by Aluminum salts. Despite vaccination, however, the disease reoccurred in many countries and is seen with two- or three-yearly epidemic peaks in the last few decades (Fulton et al. 2016). Various explanations for this resurgence have been postulated, such as waning vaccine-induced immunity, strain adaptation, improved diagnostics and medical awareness (Cherry 2012a, b). Since 2012 epidemiological evidence has accumulated that immune protection induced by ACV is less long-lasting than its WCV-induced counterpart (Klein et al. 2013; Sheridan et al. 2014). This prompted novel interest and research into the differences between correlates of protection against pertussis as induced by the two vaccine types or by natural infection (Diavatopoulos et al. 2018). Infection-primed immunity is regarded to represent the broadest and most durable protective immune response, based on epidemiological data (Wendelboe et al. 2005; Wearing and Rohani 2009). Apart from the multispecificity of the response, interactions with the whole live bacterium within an inflammatory milieu unique for primary infections might underlie the higher effectiveness of naturally-acquired Bp specific immunity.

2   Elusive Correlates of Protection against Pertussis

Basically Bp is regarded a noninvasive pathogen, however the pathogen can also enter and survive inside a number of human cell types (reviewed in (de Gouw et al. 2011)), mostly evidenced for pulmonary alveolar macrophages (Paddock et al. 2008; Bromberg et al. 1991), monocyte-derived macrophages (Friedman et al. 1992; Lamberti et al. 2010), monocytes (Fedele et al. 2005), and respiratory epithelium (Lamberti et al. 2013). This indicates that the host's defense should be able to act against extracellular as well as intracellular Bp. In agreement herewith, both humoral and cellular immune mechanisms contribute to adaptive immune protection against Bp. The presence of serum antibodies against filamentous hemagglutinin (FHA), pertactin (PRN) and/or fimbriae (Fim2 and 3) have been suggested to protect against colonization based on their adhesin-specificity and potential to opsonize or mediate killing of whole Bp bacteria (Storsaeter et al. 1998; Cherry et al. 1998; Thorstensson et al. 2014). Low levels of antibodies to pertussis toxin (PTX) showed some correlation with susceptibility to disease (Taranger et al. 2000; Storsaeter et al. 2003). Yet, unlike other infectious diseases for which clear protective cut-off levels of antigen-specific antibodies have been identified (Plotkin 2010), a level of 20 IU/ml IgG antibodies to PTX is currently used as just an arbitrary level of protection to pertussis.

As a second protective arm of defense against pertussis, cell-mediated immunity by CD4+ T cells is implied. Moreover, CD4+ T-cell immunity imprinted by natural infection is thought to contain crucial signatures that correlate with the long-term protective capacity of the convalescent immune response. Recently, developments in animal models of primary Bp infection and secondary challenge have shed new light on CD4+ T-cell lineages induced by natural infection and the effector mechanisms they may propagate. Here we will review the insights from animal studies regarding superior cell-mediated immunity in the convalescent host and ask what is the evidence hereof in humans. It is discussed which gaps in knowledge should be closed before we understand hallmarks of human superior CD4+ T cell immunity that may be used as benchmark to improve current pertussis vaccines or vaccination strategies.

3 Clearance of *B. pertussis* Requires CD4+ T-Cells with a Distinct Differentiation Profile

In the past few decades, mouse models have been used extensively to address the protective role of cellular immunity to control an infectious respiratory challenge of Bp, generally inoculated via intranasal or aerosol administration (Van Der Ark et al. 2012). Naïve mice typically clear an infectious dose within 4–6 weeks (Mills et al. 1993; Barbic et al. 1997; Raeven et al. 2014) but recovered mice clear a secondary challenge within several days through the presence of primary infection-acquired memory immunity (Mills et al. 1993; Raeven et al. 2016). The role of T-cells and in particular of CD4+ T-cells producing IFNγ in controlling a bacterial load became clear in studies using athymic *nu/nu* mice (Mills et al. 1993) as well as SCID and IFN$\gamma^{-/-}$ mice (Barbic et al. 1997), IFNγR$^{-/-}$ mice (Mahon et al. 1997), or mice depleted of the CD4+ T-cell subset (Leef et al. 2000). Transferred CD4+ (and not CD8+) T-cells, derived from convalescent mice 6 weeks after Bp infection, were able to clear an infectious challenge in sublethally irradiated *nu/nu* recipient mice (Mills et al. 1993). Later Bp specific IL-17-producing T-cells were recognized as an additional T-cell lineage implied in protection, since neutralizing anti-IL-17 Ab significantly reduced the protective efficacy of a WCV-induced immune response against Bp (Higgins et al. 2006). Banus et al. found that Th17 responses, additive to Th1 responses, were also driven by infection (Banus et al. 2008) and Bp-infected IL-17A defective mice were found to have significantly higher bacterial loads in the lungs (Ross et al. 2013). Proof for the protective capacity of the infection-induced Th17 cells came from data showing that adoptive cell transfer of cultured splenic Th1 and Th17 lineage cells from convalescent mice reduced, separately and synergistically, the bacterial loads over the course of a Bp infection (Ross et al. 2013). Bacterial toxins PTX and adenylate cyclase toxin (ACT or CyaA) seem to be involved in triggering protective Th1 and Th17 lineages by virulent Bp, since reduced Th1 as well as Th17 responses are seen in the absence of PTX (Andreasen et al. 2009) and reduced Th17 responses in the absence of ACT (Dunne et al. 2010). However, protective T-cell responses are not absolutely dependent on functional PTX since higher doses of $PTX^{-/-}$ strain or infection with the live attenuated Bp strain BPZE1 expressing a genetically detoxified PTX (and lacking functional tracheal cytotoxin and dermonecrotic toxin) (Mielcarek et al. 2006) can still drive Th17 (Andreasen et al. 2009) or Th1 and Th17 (Feunou et al. 2010a; Kammoun et al. 2012; Solans et al. 2018) type responses, respectively. Infection-primed cell-mediated immune protection seemed long-lasting since adoptive transfer of spleen cells 12 months after intranasal inoculation with BPZE1 protected SCID recipient mice (Feunou et al. 2010b).

Recently, Mills and colleagues applied methodology including *in vivo* labeling of lymphocytes to study local mouse CD4+ T-cells primed by Bp infection (Wilk et al. 2017). Tissue-resident memory T-cells (Trm) are traditionally characterized by the expression of CD69, which inhibits sphingosine-1-phosphate receptor 1 (S1PR1)-mediated egress from tissues (Arnon et al. 2011), and of CD103 (alpha subunit of aEb7 integrin), which docks cells to epithelial E-cadherin (Cepek et al. 1994; Casey et al. 2012). Wilk et al. showed that blocking the capacity of immune CD4+ T-cells to migrate to respiratory tissue through the S1PR1 antagonist, FTY720, abrogated the capacity to rapidly clear bacteria (Wilk et al. 2017) and that clearance of convalescent mice was associated with accumulation of CD4+ T-cells with the CD69 + CD103+ Trm phenotype (and negative for the lymphoid homing receptor CD62L) in lungs and nose. Adoptive transfer of CD4+ T-cells derived from the lungs of 60 days-convalescent mice, but not from vaccinated mice, strongly reduced the bacterial burden in Bp-challenged naïve recipient mice (Wilk et al. 2019). Furthermore, this study indicated that the CD4+ T-cells isolated from lungs, nose and spleen from convalescent mice,

like those from WCV-vaccinated but unlike those from ACV-vaccinated mice, produced IFNγ and IL-17 upon stimulation. Notably, Solans et al. found that intranasal infection with the attenuated BPZE1 strain also primed IFNγ-producing and IL-17-producing CD4+ Trm cells in the lungs (Solans et al. 2018), indicating that full virulence of Bp was not required to steer this type of immunity. This study also revealed a probable link between the IL-17-producing CD4+ Trm cells and the induction of protective local secretory IgA responses. Post-infection co-development of Th17 type and IgA responses was also reported earlier in a systems immunology study using a recent clinical Bp strain (Raeven et al. 2014). Together, the studies in the mouse indicate that CD4+ T-cell lineages primed by natural infection, and to some extent by WCV, rapidly protect against an i.n. challenge dose, in profound contrast to CD4+ T-cells primed by ACV. As reviewed earlier (Brummelman et al. 2015) and elsewhere in this volume (chapter 6, Ausiello et al), ACV induce an alternative functional CD4+ T-cell profile, i.e. in mice dominated by the production of IL-4, IL-5 and IL-13 (Th2) or mixed Th2/Th17 cytokines, and prevent disease but do not protect against nasopharyngeal infection and transmission.

4   Bp Infection in Baboons as a Model to Study Hallmarks of CD4+ T-Cell Lineages Protecting against Colonization, Transmission and Disease

Since 'mice don't cough' the question is always raised as to whether the above findings can be translated to humans. Recently the baboon (*Papio anubis*) was identified as a non-human primate species to have a body temperature close to humans and reproducing typical clinical signs of pertussis (Warfel et al. 2012a). When experimentally challenged with strains representing currently circulating Bp clades, baboons develop leukocytosis, paroxysmal coughing, mucus production and heavy colonization of the airway

(Warfel et al. 2012b; Warfel and Merkel 2014; Naninck et al. 2018; Zimmerman et al. 2018). Importantly, transmission of the bacteria between hosts is observed (Warfel et al. 2012b). Based on these similarities with human (histo)pathology, transmission and immunology, this novel non-human primate model is regarded of high translational value to evaluate pertussis vaccine efficacy and correlates of protection. Merkel and coworkers found that baboons recovered from primary infection did not get colonized nor showed symptoms upon secondary challenge and that this convalescent state was associated with detectable peripheral Bp specific Th1 and Th17 responses, persisting for several years (Warfel and Merkel 2013; Warfel et al. 2014). Furthermore, infected animals had significant induction of IL-17 in the nasopharyngeal mucosa, and enhancement of IL-6, IL-23 and IL-1ß, cytokines responsible for the initiation and proliferation of Th17 immune responses in humans, as well as of several Th17 effector molecules, GCSF, IL-8, MCP-1 and MIP1a. Although the model so far was not designed to look at immune cell recruitment to the URT, the data are suggestive of the presence of a local Th17 type response. Further (indirect) evidence for a local immune response triggered by infection came from the detection of serum IgA to PTX, FHA, and PRN after intranasal/intratracheal inoculation of the attenuated BPZE1 strain in baboons. Since serum IgA responses are usually minimally or not at all induced after intramuscular pertussis vaccination in infants but develop at the respiratory mucosal sites after infection, this finding was interpreted to relate to local induction of IgA by BPZE1 (Locht et al. 2017). Local tissue-resident humoral and cell-mediated immunity remain to be studied in the baboon model.

Hence, from the available animal models, the concept emerges that protective infection-induced immunity against a subsequent Bp re-infection relies on the presence of specific CD4+ T-cells in the URT that produce Th1 and predominantly Th17 cytokines, and that have originally migrated to and remain resident in the respiratory tissues. Yet an important question is whether this paradigm holds true for humans as well.

5 Evidence for Bp Infection-Induced Th1 and Th17 Lineages and Tissue Residency in Humans

5.1 Th1 Lineage

The first studies addressing the cytokine profile of human CD4+ T-cells induced by clinical or asymptomatic Bp infection date back three decades. This was just after the discovery of the first Th1 and Th2 CD4+ T-cell subsets (Mosmann et al. 1986; Mosmann and Sad 1996; Del Prete et al. 1991) but well before the recognition of IL-17 producing CD4+ T cells as a separate lineage (Kolls and Linden 2004; Harrington et al. 2005; Wynn 2005; Park et al. 2005; Bettelli et al. 2007). In a series of 'classical' papers several groups interrogated cytokine responses from cases, ex-cases or naturally exposed individuals without history of pertussis vaccination, after stimulation of PBMC with killed whole bacteria or with vaccine antigen, typically (inactivated) PTX, FHA and/or PRN. Altogether these studies demonstrated that also in humans the Th1 cell subset is activated during Bp infection. In 1991 a first observation of Th1-polarization of Bp specific T-cells concerned a number of PTX-specific CD4+ T-cell clones isolated from a single adult donor with a history of whooping cough, which all were found to secrete IFNγ, some IL-2 but no IL-4 in their supernatants in response to antigen (Peppoloni et al. 1991). To extend these observations using uncloned T-cells, Ryan et al. took advantage of a cohort of acutely Bp-infected or convalescent young children (between 2 months and 8 years of age), all unvaccinated. Fresh PBMC were tested for proliferation and cytokine production after in vitro stimulation with Bp antigen. Most cases and ex-cases showed proliferative responses and production of IFNγ but not of IL-5 (Ryan et al. 1997). Hereafter Ausiello et al. showed, using enriched T-cells from a small sample of healthy adults with no history of pertussis vaccination, antigen-specific induction of gene transcripts for IFNγ and IL-2 but not for IL-4 nor IL-5. At the time, these data were interpreted to reflect natural acquisition of an antigen-specific protective Th1 cytokine response by repeated exposure to Bp in an endemic setting (Ausiello et al. 1998). Also, these investigators described Th1 type Bp specific cell-mediated immunity several years after pertussis diagnosis in unvaccinated children, as opposed to more mixed Th1/Th2 type responses in ACV vaccinated non-infected age-matched controls (Ausiello et al. 2000). Meanwhile in a case report Hafler and Pohl-Koppe (1998) had described a strong proliferative response and IFNγ cytokine secretion to whole killed Bp cells and PT in two clinically infected teenagers, without detectable secretion of the Th2 cytokine IL-4. An important contribution to this series was by Mascart et al., who studied one to 4 months old infants suffering from an acute Bp infection, in comparison to non-immune age-matched controls, prior to their first dose of WCV. High levels of IFNγ in culture supernatants and high numbers of IFNγ spot forming cells were shown in PTX- and FHA-stimulated cultures from the infected infants, which were absent in the control non-immune group. Also, PTX or FHA stimulation did not trigger any Th2 responses, in either group of infants, as evidenced by the absence of IL-13 cytokine release or IL-4 spot forming cells, while all infants could mount Th2 responses to mitogen stimulation (Mascart et al. 2003). Taken together, these pioneering studies indicated that human priming by Bp infection alone, in the absence of any vaccination background, strongly steers Th1 responses. That this seems to be a unique hallmark of natural infection became clear from clinical vaccine studies in the same time frame and thereafter, showing that priming by WCV or ACV vaccination can induce mixed Th1/Th2 type CD4+ T-cell responses, with WCV-induced responses generally being more Th1-dominated (reminiscent of infection-induced responses), and ACV-induced responses being more Th2-dominated (van Twillert et al. 2015; Brummelman et al. 2015; Fedele et al. 2015; Ryan et al. 1998; Mascart et al. 2007; Schure et al. 2012, 2013; van der Lee et al. 2018a) (chapter 6 in this volume, Ausiello C. et al).

5.2 Th17 Lineage

After the discovery of the Th17 T-cell lineage (Kolls and Linden 2004; Wynn 2005; Bettelli et al. 2007) and its importance, in synergy with Th1 cells, to mediate protection in convalescent mice (Dunne et al. 2010; Ross et al. 2013), measurement of Th17 cytokines in human Bp specific T-cell studies became key. Yet, these studies involved vaccine immunogenicity or booster comparisons, or studies in convalescent patients with a history of vaccination (van Twillert et al. 2015; Schure et al. 2013), but none addressed primary Bp infection alone. In cases within 2–3 months after clinical infection, all with a WCV or ACV priming background, we did not observe IL-17 cytokine secretion but instead mixed Th1 and Th2 cytokine responses, when stimulating PBMCs with synthetic peptides representing epitopes from PTX or PRN (Han et al. 2013, 2015), or with naturally presented Bp epitopes from other proteins (Stenger et al. 2014). Also in our group, Schure et al. and Van der Lee et al. described just low levels of IL-17 in Bp antigen-stimulated PBMC cultures from WCV or ACV-primed pediatric cohorts (Schure et al. 2013; Van Der Lee et al. 2018a). Yet in agreement with recent work by da Silva Antunes et al. (2018), we found that compared to ACV-primed adults, Bp antigen-stimulated PBMC cultures from WCV-primed adults, having received an ACV booster vaccination, not only produced increased levels of Th1 and Th2 cytokines, but also of Th17 cytokines (van der Lee et al. 2018b). In line with these age trends, in samples from a prospective pertussis household study (de Greeff et al. 2012) we found significant levels of IL-17 in Bp antigen-stimulated PBMC cultures from infected adults with a history of vaccination, while these were low or non-detectable in cultures from infected pediatric counterparts (Buisman et al., unpublished data). Whether this implies that Th17 responses are less well developed at younger age or that assays may have to be optimized for detection of children's IL-17 responses needs to be verified. It has been shown that naïve T-cells in cord blood fail to differentiate to Th17 cells in co-culture with autologous antigen-presenting cells, despite the presence of polarizing cytokines (de Roock et al. 2013) and that newborns lack circulating Th17 cells (Stoppelenburg et al. 2014). In vitro studies using human monocyte-derived dendritic cells indicated that enzymatically active and inactive PTX and enzymatically active ACT can drive Th17-promoting IL-23 responses (Nasso et al. 2009; Fedele et al. 2010), indirectly supporting the in vivo requirements for these factors in human Th17 lineage differentiation, as evidenced in the mouse (Andreasen 2009; Feunou 2010a; Kammoun 2012; Solans 2018).

5.3 Tissue Residency of CD4+ T-Cells

The discovery of Trm-cells in several tissues has added to the scope of local immunity. Their central role is to respond rapidly to mediate clearance of a pathogen on the site of infection (Mueller and Mackay 2016). The phenotype of human CD4+ Trm-cells resembles that of Trm-cells in mice through constitutive expression of CD69 and CD103 (and absence of lymphoid homing receptors such as CCR7) (Kumar et al. 2017). Although human Trm cells are best defined for the CD8+ T cell subset, CD4+ Trm cells, including in the lung, have been described to play a role in protective immunity and immunopathology (Turner and Farber 2014). Human lung-residing CD4+ Trm cells have been studied for pathogens such as influenza virus and *Mycobacterium tuberculosis* (Walrath and Silver 2011; de Bree et al. 2007; Purwar et al. 2011), but not yet in the context of Bp infection.

5.4 Imprinting of T-Cell Lineage Differentiation

Taken together, there is no dispute about the fact that Bp infection in humans strongly triggers the Th1 lineage, and not the Th2 lineage. The additional involvement of the Th17 lineage, and tissue

residency of both Th1 and Th17 lineages in humans, however, remain unknown due to the absence of IL-17 data in older studies in unvaccinated subjects primed by natural Bp infection. Based on shared (histo)pathology, transmission and immunology features between the baboon model of Bp infection and humans, it can be hypothesized that in addition to Th1 cells, human Th17 T-cells and tissue residency of the T-cell lineages might be involved as well. Lineage differentiation of CD4+ T-cells during immune responses reflects their priming by dendritic cells, which in turn get polarized by innate signals associated with the immunizing event (Sallusto et al. 2018). CD4+ T-cell programming is regulated by early 'signal transducers and activators of transcription' (STAT) pathways and imprinted epigenetically by networks of regulator proteins and permissive or repressive chromatin signatures (O'Shea and Paul 2010). The presence of IFNγ and IL-12 and of IL-6, IL-23 and IL-1ß, are involved in priming of Th1 and Th17 lineages, respectively (Sallusto et al. 2018). Understanding how CD4+ T-cell heterogeneity is formed and whether a degree of lineage plasticity exists in the context of immunity to Bp is not the topic of this review but an important field to investigate.

6 Protective Effector Mechanisms Downstream of Th1 and Th17 Cell Lineages

Bp has many virulence factors mediating evasion of early host clearance mechanisms and prolongation of survival and growth on the respiratory mucosae. Immune modulation includes interference with the production of antimicrobial peptides, the recruitment of phagocytes, complement- and phagocyte-mediated killing, the secretion of chemokines and cytokines, and development of adaptive B- and T-cell responses (Brummelman et al. 2015). It goes beyond the scope of this chapter to cover in detail which initial host-pathogen interactions and host

Primary infection

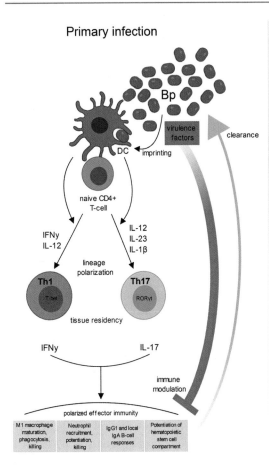

Secondary infection

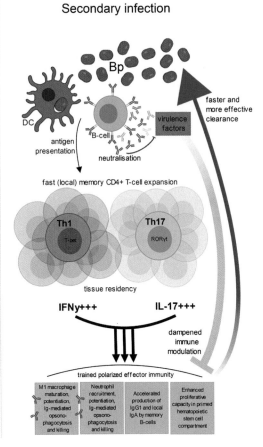

Fig. 1 Model for the role of IFNγ and IL-17 in clearance of Bp infection

Primary infection polarizes towards tissue-resident Th1 and Th17 lineage immunity, resulting in local production of IFNγ and IL-17. These cytokines potentiate macrophage, neutrophil and antibody-mediated mechanisms of clearance. Left panel: clearance in primary infections is significantly delayed by immune modulating virulence factors of Bp. Right panel: Fast, local expansion of infection primed memory responses results in accelerated and more effective Bp clearance during secondary infections.

Accelerated production of IFNγ, IL-17 and antibodies enhance effector immunity, as indicated.
NB not illustrated is the contribution of NK cells and γδ T-cells to local IFNγ and IL-17 production during Bp infection, respectively. Red lines represent inhibiting effects and green arrows represent potentiating effects. Abbreviations: Bp, Bordetella pertussis; DC, dendritic cells; T-bet, T-box expressed in T-cells (transcription factor in Th1 lineage cells); RORγt, retinoic acid receptor-related orphan receptor gamma (transcription factor in Th17 lineage cells). Created with BioRender.com.

responses are implied, first delaying and then promoting the clearance of the pathogen. We here focus on four mechanisms by which effector cytokines produced by the Th lineages associated with infection-primed immunity, i.e. IFNγ and IL-17, can arm innate and adaptive cells to successfully clear primary and prevent secondary infections (illustrated in Fig. 1).

6.1   IFNγ- and IL17-Induced Potentiation of M1 Macrophages

It has been well established that macrophage polarization plays a key role in infectious diseases, and that IFNγ triggers 'classical' activation and differentiation of monocyte precursors into M1 macrophages (Benoit et al. 2008; Arora et al. 2018). Activated M1 macrophages promote enhanced secretion of M1 chemokines, inducible nitric oxide synthase (iNOS) dependent reactive nitrogen intermediates (NO), reactive oxygen species (ROS), high levels of IL-12, IL-23 IL-1β and TNFα, and low levels of IL-10. This creates a pro-inflammatory milieu and NO- and ROS-mediated capacity of the M1 macrophages to kill bacteria engulfed in phagosomes. Bp may end up in phagosomes of phagocytic cells by various pathways including phagocytosis via complement receptors, innate receptors, and, in if antibodies are present, via immunoglobulin receptors (FcRs).

Other cell types than Th1 cells can also produce IFNγ, i.e. CD8+ cytotoxic T-cells, natural killer (NK) cells, antigen-presenting cells (APC) and B cells, and NK cells have indeed been found to be an important early source of IFNγ in the lungs of Bp infected mice, required for the initial polarization of Th1 cells (Byrne et al. 2004). Yet Th1 cells, capable of secreting large amounts of IFNγ, are thought to be the major activators of M1 macrophages. In turn, M1 macrophages secrete large quantities of IL-12 which aids in amplifying Th1 polarization of CD4+ lymphocytes (Martinez and Gordon 2014; Arango Duque and Descoteaux 2014). Macrophage killing of Bp is regarded an important protective mechanism to control Bp infection (Valdez et al. 2016; Bernard et al. 2015). Higgs et al. found that not only IFNγ but also TNFα and IL-17 had potentiating effects on the in vitro bactericidal capacity of murine peritoneal macrophages and a murine alveolar macrophage cell line (Higgins et al. 2006), This suggests that Th1 and Th17 lineage cytokines act in concert to stimulate M1 macrophage function. In fact, this dual M1 potentiating effect may be truly significant to counteract the M2 polarization effect of a high expression ratio of genes encoding the suppressors of cytokine signaling 1 and 3 (SOCS1/SOCS3) (Wilson 2014), observed in macrophages after intracellular infection with Bp (Valdez et al. 2016). Notably, Th2 associated cytokines IL-4 and IL-13 trigger the 'alternative' pathway of macrophage activation. Resultant M2 macrophages secrete high amounts of IL-10 and little IL-12 and IL-23, and lack microbicidal activity (Arora et al. 2018).

6.2 IL-17- Induced Recruitment and Activation of Neutrophils

Neutrophils are the dominant population among granulocytes and play a primary role in the host defense by removing pathogens through phagocytosis. High numbers of neutrophils are maintained in the circulation to facilitate rapid recruitment to infected tissue, mediated by chemotactic host signals and secreted bacterial molecules. Initially, recruitment of neutrophils to lungs after Bp infection is blocked through the action of PTX (Kirimanjeswara et al. 2005; Carbonetti 2015; Locht et al. 2011). Then subsequent host-pathogen interactions initiate neutrophil recruitment, peaking at 10–14 days after inoculation in mice. Th17 cells are known to orchestrate the recruitment of neutrophils to the site of infection, via IL-17A. In IL-17A defective mice neutrophil recruitment and Bp clearance was impaired, compared to wildtype mice (Ross et al. 2013). IL-17 not only mediates neutrophil recruitment but increases their ability to kill phagocytosed Bp through oxygen species generation. Other antimicrobial functions of neutrophils are NET-formation and degranulation (Eby et al. 2015). As discussed by Eby et al., it is likely that Th17 cells are the major source of IL-17 promoting the late Bp clearance by neutrophils. Yet, in parallel with the early polarizing innate IFNγ production by NK cells, early innate IL-17A was found to be produced already 2 h after Bp infection by lung γδ T-cells, promoting a Th17 polarizing milieu (Misiak et al. 2017).

Finally, neutrophils are no longer regarded only as terminally differentiated short-lived cells. Human neutrophils were reported to possess antigen presentation capacity in a MHC class II restricted manner, including the expression of co-stimulatory molecules, and ability to polarize naive T-cells into Th1 and Th17 lineages (Abi Abdallah et al. 2011; Radsak et al. 2000; Lin and Lore 2017). This interaction between neutrophils and Th17 lineage cells has not been described for Bp infection yet.

6.3 Local Cross Talk Between Th17 Cells and IgA-Producing B Cells

In the naïve host, Bp has ample opportunity to modulate host defense mechanisms and clearance is delayed until the adaptive immune response helps to arm the major phagocytes to eliminate Bp at the site of infection. In the convalescent host, memory immune cells can rapidly mount elevated effector T-cell responses and antibody levels, preferably locally in the respiratory tract. IgA has been widely recognized as the antibody class of mucosal immunity. Cytokine patterns in antigen-specific CD4+ T-cells have been shown to instruct B cells to class-switch to particular isotypes. By programming particular T-cell lineages and effector cytokines, pathogens can modulate class-switching in B-cells (Tarlinton and Good-Jacobson 2013; McHeyzer-Williams et al. 2012). Hence pathogen specific T- and B-cell responses are tightly interlinked, with associations between isotype switching in murine B cells to IgG2a (or IgG2c) guided by Th1 cytokine IFNγ, to IgG1 and IgE guided by Th2 cytokine IL-4, and to IgA guided by TGFß and IL-17 in the Th17 cell milieu (Reinhardt et al. 2009; Christensen et al. 2017; Tarlinton and Good-Jacobson 2013). As revealed by Mills et al. in the mouse in addition to specific Th1 cells, Bp infection steers Th17 lineage cells that become tissue-resident in the lungs and nasal cavity shortly after priming (Wilk et al. 2019). Recently, Locht and coworkers showed interlinkage between these tissue-resident Th17 lineage cells and the capacity to produce IL-17 and protective local IgA in mice (Solans et al. 2018; Solans and Locht 2019), and between local Th17 lineage promoting cytokines and serum IgA in baboons (Locht et al. 2017). Hence, T-cell derived IL-17 production in the URT can now be proposed to promote local IgA-switched B cell responses

including rapid boosting of specific IgA levels upon secondary challenge. Local specific antibodies would not only be capable of neutralizing virulence factors, thereby avoiding immune modulation by Bp, but also of enhancing opsonophagocytosis of Bp by local macrophages and neutrophils, thereby accelerating clearance in the convalescent host. Whether a similar link exists between local IFNγ-producing Th1 cells and murine IgG2a (or IgG2c)-switched B-cells (and IgG1-switched B cells in humans) remains to be investigated.

6.4 IFNγ-Induced Training of Myeloid Cells and Hematopoietic Stem and Progenitor Cells

Recently, Varney et al. suggested an important mechanism downstream of IFNγ, explaining how the more durable immune protection could work in hosts recovered from primary Bp infection or immunized with WCV, as opposed to the shorter duration of protection seen in ACV-immunized hosts. It was demonstrated in mice that infection- and WCV-induced immunity impacted responsiveness at the level of Hematopoietic Stem and Progenitor Cells (HSPC) in the bone marrow, especially pointing at myeloid preparedness and rapid expansion of HSPCs and tissue homing upon reinfection (Varney et al. 2018). Gene set enrichment analyses demonstrated that, like WCV-immunized but unlike ACV-immunized mice, Bp-infected mice exhibited unique gene signatures that suggested roles for IFNγ-induced gene expression. Mice exhibiting an IFNγ-priming milieu had relatively large myeloid proportions in the spleen as well as enhanced gene expression in HSPCs, regarding processes such as survival, cell renewal, autophagy and antigen processing and presentation. In line with some of these findings, Raeven et al. already found clusters of genes typically expressed in lungs of mice recovered from primary Bp infection, indicating enhanced activity of 'trained' innate immune cells and involving antigen processing and presentation or MHC signalling (Raeven et al. 2016; Raeven et al. 2017). Programming at the HSPC level was first described in a model of priming with the live attenuated Bacillus Calmette Guerin strain (BCG) of Mycobacterium tuberculosis bovis (MTB) (Kaufmann et al. 2018), indicating that IFNγ-induced training of innate cells and HSPCs is a biological reproducible and relevant phenomenon.

Clearly, IFNγ and IL-17 are master regulators in various effector mechanisms downstream of infection-primed Bp specific Th1 and Th17 cell responses. As these cells likely also produce other cytokines or soluble mediators and engage in many cell-cell interactions, additional mechanisms contributing to the durable type of Bp specific protective immunity should not be excluded.

7 Conclusions & Future Directions

Pertussis remains a public health problem despite vaccination, affecting all age groups and especially vulnerable infants. To forward the development of future pertussis vaccines with a longer duration of protection and interrupting transmission, correlates of protection against pertussis should be better understood. Over the past three decades, the mouse and baboon models of Bp infection have been exploited to elucidate why convalescent hosts primed by infection are better protected than hosts immunized with pertussis vaccines. The basis seems to lie in the imprinting of CD4+ Th1 and Th17 cell lineages, their tissue residency in the lung and nasal tissue and the steering role of their effector cytokines on mechanisms such as phagocytic clearing, differentiation of IgA B-cell responses and training of myeloid cells and HSPC. Natural infection-acquired immunity in unvaccinated humans has also been found to be associated with the most durable protection against pertussis, based on epidemiological evidence. In view of shared features between human and baboon Bp infection biology, it can be hypothesized that similar CD4+ T-cell lineages and mechanisms play a role as

well. However only few aspects of human infection-primed CD4+ T-cell immunity have been explored to date.

Several challenges lie ahead to increase our understanding of the role of CD4+ T-cells in human protective immunity against pertussis, using natural infection as a benchmark. Questions that need answers include:

– Can we detect Bp specific CD4+ T-cell responses in the peripheral blood of unvaccinated humans after natural infection or (likely) exposure, and can functional differentiation of especially Th17 lineages be confirmed? Are these associated with IgA B-cell responses?
– Can human Bp specific CD4+ T-cells be found with a tissue-resident phenotype and a Th1/Th17 lineage profile in cell suspensions obtained from bronchoalveolar lavages or lung surgery biopsies from unvaccinated naturally exposed subjects?
– Can infection-primed human CD4+ T-cells be enriched and in depth explored for candidate biomarkers of durable protective Bp immunity? What are the defining immunological signatures of these cells at the proteome, transcriptome and/or epigenome level?
– Can we develop assays to immunomonitor candidate biomarkers that predict durable protective CD4+ T-cell immunity to Bp in future clinical registration studies?

To address these questions blood and preferably mucosal samples from pertussis cases, representing all ages, without vaccination history are pivotal. Except for older age groups recruiting non-vaccinated younger cases, born after the implementation of immunization programs, may pose a problem. Also, assays to interrogate the Th lineage differentiation of human CD4+ T-cells in depth should be designed with care. First of all, Bp antigen-specificity should be well-defined and perhaps antigen panels need to be optimized for infection-primed immunity. Second, in view of their low frequency, enrichment of Bp specific CD4+ T-cells prior to analysis is likely required, either based on in vitro activation markers (da Silva 2018) or ex vivo labeling with MHC class II tetramers (Han et al. 2015). Third, depending on the question, methodology to in-depth analyse the enriched Bp specific CD4+ T-cells for candidate biomarkers of protective immunity may be selected at different systems' levels, including multiparameter flow cytometry, CyTOF, RNAseq, genome-wide STAT binding or ATAQ-seq, where possible at the single cell level. Associations between human Th17 lineage immunity and levels of Bp specific IgA determined in serum and mucosal samples such as nasal lining fluid and lung lavages should be addressed. Also a systems approach including other immune cells like myeloid cell types and HSPC could reveal wider mechanisms of imprinting of Bp infection-induced immune protection in humans.

Altogether this will help to understand deeper hallmarks of Bp infection-primed CD4+ T-cell immunity that could be used as benchmark to guide development and implementation of improved pertussis vaccines with a longer duration of protection against disease, and preventing transmission.

- It is currently unknown how the Bp specific CD4+ T-cell lineages acquired by natural infection are programmed exactly and to what extent lineage imprinting is reversible
- A great body of evidence discussed in this review comes from animal models, but it remains to be investigated whether findings can be translated to human immunity
- Although studying infection-induced immunity in unvaccinated cohorts is challenging due to the rarity of these cohorts and confounding factors such as age and unknown (subclinical) repeated exposures, it should be given more priority
- Studying human naturally-acquired cell-mediated mechanisms will help to understand immunity that is essential to combat Bp infections and guide or accelerate pertussis vaccine development

Key Issues

- Immunity induced by natural Bp infection is superior as it has higher durability compared to vaccine-induced immunity based on epidemiological data
- Based on animal models, infection-induced superior CD4+ T-cell immunity is characterized by Th1 and Th17 lineages that enhance macrophage killing capacity and neutrophil recruitment and function
- Bp infection in animal models induces tissue-resident memory CD4+ T-cells in lung and nose. These play an important role within the local mucosal immune response, including stimulation of IgA class switching
- Bp infection seems associated with IFN-γ-induced training of innate cells and HSPCs

Compliance with Ethical Standards EL is supported by the PERISCOPE project, funded by the Innovative Medicines Initiative 2 Joint Undertaking under grant agreement No 115910, with support from the European Union's Horizon 2020 research and innovation program and EFPIA, and BMGF.

References

Abi Abdallah DS, Egan CE, Butcher BA, Denkers EY (2011) Mouse neutrophils are professional antigen-presenting cells programmed to instruct Th1 and Th17 T-cell differentiation. Int Immunol 23 (5):317–326. https://doi.org/10.1093/intimm/dxr007

Andreasen C, Powell DA, Carbonetti NH (2009) Pertussis toxin stimulates IL-17 production in response to Bordetella pertussis infection in mice. PLoS One 4 (9):e7079

Arango Duque G, Descoteaux A (2014) Macrophage cytokines: involvement in immunity and infectious diseases. Front Immunol 5:491. https://doi.org/10.3389/fimmu.2014.00491

Arnon TI, Xu Y, Lo C, Pham T, An J, Coughlin S, Dorn GW, Cyster JG (2011) GRK2-dependent S1PR1 desensitization is required for lymphocytes to overcome their attraction to blood. Science (New York, NY) 333(6051):1898–1903. https://doi.org/10.1126/science.1208248

Arora S, Dev K, Agarwal B, Das P, Syed MA (2018) Macrophages: their role, activation and polarization in pulmonary diseases. Immunobiology 223 (4–5):383–396. https://doi.org/10.1016/j.imbio.2017.11.001

Ausiello CM, Lande R, la Sala A, Urbani F, Cassone A (1998) Cell-mediated immune response of healthy adults to Bordetella pertussis vaccine antigens. J Infect Dis 178(2):466–470

Ausiello CM, Lande R, Urbani F, Di Carlo B, Stefanelli P, Salmaso S, Mastrantonio P, Cassone A (2000) Cell-mediated immunity and antibody responses to Bordetella pertussis antigens in children with a history of pertussis infection and in recipients of an acellular pertussis vaccine. J Infect Dis 181(6):1989–1995. https://doi.org/10.1086/315509

Banus S, Stenger RM, Gremmer ER, Dormans JA, Mooi FR, Kimman TG, Vandebriel RJ (2008) The role of toll-like receptor-4 in pertussis vaccine-induced immunity. BMC Immunol 9:21. https://doi.org/10.1186/1471-2172-9-21

Barbic J, Leef MF, Burns DL, Shahin RD (1997) Role of gamma interferon in natural clearance of Bordetella pertussis infection. Infect Immu 65(12):4904–4908

Benoit M, Desnues B, Mege JL (2008) Macrophage polarization in bacterial infections. J Immunol (Baltimore, Md: 1950) 181(6):3733–3739

Bernard NJ, Finlay CM, Tannahill GM, Cassidy JP, O'Neill LA, Mills KH (2015) A critical role for the TLR signaling adapter Mal in alveolar macrophage-mediated protection against Bordetella pertussis. Mucosal Immunol 8(5):982–992. https://doi.org/10.1038/mi.2014.125

Bettelli E, Oukka M, Kuchroo VK (2007) T H-17 cells in the circle of immunity and autoimmunity. Nat Immunol 8(4):345–350

Bromberg K, Tannis G, Steiner P (1991) Detection of Bordetella pertussis associated with the alveolar macrophages of children with human immunodeficiency virus infection. Infect Immunol 59 (12):4715–4719

Brummelman J, Wilk MM, Han WGH, van Els CACM, Mills KHG (2015) Roads to the development of improved pertussis vaccines paved by immunology. Pathog Dis 73(8):ftv067. https://doi.org/10.1093/femspd/ftv067

Byrne P, McGuirk P, Todryk S, Mills KH (2004) Depletion of NK cells results in disseminating lethal infection with Bordetella pertussis associated with a reduction of antigen-specific Th1 and enhancement of Th2, but not Tr1 cells. Eur J Immunol 34 (9):2579–2588. https://doi.org/10.1002/eji.200425092

Carbonetti NH (2015) Contribution of pertussis toxin to the pathogenesis of pertussis disease. Pathog Dis 73(8): ftv073. https://doi.org/10.1093/femspd/ftv073

Casey KA, Fraser KA, Schenkel JM, Moran A, Abt MC, Beura LK, Lucas PJ, Artis D, Wherry EJ, Hogquist K, Vezys V, Masopust D (2012) Antigen-independent differentiation and maintenance of effector-like resident memory T cells in tissues. J Immunol (Baltimore, Md: 1950) 188(10):4866–4875. https://doi.org/10.4049/jimmunol.1200402

Cattelan N, Dubey P, Arnal L, Yantorno OM, Deora R (2016) Bordetella biofilms: a lifestyle leading to persistent infections. Pathog Dis 74(1):ftv108. https://doi.org/10.1093/femspd/ftv108

Cepek KL, Shaw SK, Parker CM, Russell GJ, Morrow JS, Rimm DL, Brenner MB (1994) Adhesion between epithelial cells and T lymphocytes mediated by E-cadherin and the αEβ7 integrin. Nature 372 (6502):190–193

Cherry JD (2012a) Epidemic pertussis in 2012--the resurgence of a vaccine-preventable disease. N Engl J Med 367(9):785–787. https://doi.org/10.1056/NEJMp1209051

Cherry JD (2012b) Why do pertussis vaccines fail? Pediatrics 129(5):968–970. https://doi.org/10.1542/peds.2011-2594

Cherry JD, Gornbein J, Heininger U, Stehr K (1998) A search for serologic correlates of immunity to Bordetella pertussis cough illnesses. Vaccine 16 (20):1901–1906

Christensen D, Mortensen R, Rosenkrands I, Dietrich J, Andersen P (2017) Vaccine-induced Th17 cells are established as resident memory cells in the lung and promote local IgA responses. Mucosal Immunol 10 (1):260–270. https://doi.org/10.1038/mi.2016.28

da Silva AR, Babor M, Carpenter C, Khalil N, Cortese M, Mentzer AJ, Seumois G, Petro CD, Purcell LA, Vijayanand P, Crotty S, Pulendran B, Peters B, Sette A (2018) Th1/Th17 polarization persists following whole-cell pertussis vaccination despite repeated acellular boosters. J Clin Invest 128(9):3853–3865. https://doi.org/10.1172/JCI121309

de Bree GJ, Daniels H, Schilfgaarde M, Jansen HM, Out TA, van Lier RA, Jonkers RE (2007) Characterization of CD4+ memory T cell responses directed against common respiratory pathogens in peripheral blood and lung. J Infect Dis 195(11):1718–1725. https://doi.org/10.1086/517612

de Gouw D, Diavatopoulos DA, Bootsma HJ, Hermans PW, Mooi FR (2011) Pertussis: a matter of immune modulation. FEMS Microbiol Rev 35(3):441–474. https://doi.org/10.1111/j.1574-6976.2010.00257.x

de Greeff SC, de Melker HE, Westerhof A, Schellekens JF, Mooi FR, van Boven M (2012) Estimation of household transmission rates of pertussis and the effect of cocooning vaccination strategies on infant pertussis. Epidemiology 23:852–860 de Roock S, Stoppelenburg AJ, Scholman R, Hoeks SB, Meerding J, Prakken BJ, Boes M (2013) Defective TH17 development in human neonatal T cells involves reduced RORC2 mRNA content. J Allergy Clin Immunol 132(3):754–756. e753

Del Prete GF, De Carli M, Mastromauro C, Biagiotti R, Macchia D, Falagiani P, Ricci M, Romagnani S (1991) Purified protein derivative of Mycobacterium tuberculosis and excretory-secretory antigen (s) of Toxocara canis expand in vitro human T cells with stable and opposite (type 1 T helper or type 2 T helper) profile of cytokine production. J Clin Invest 88(1):346–350

Diavatopoulos DA, Mills KH, Kester KE, Kampmann B, Silerova M, Heininger U, van Dongen JJ, van der Most RG, Huijnen MA, Siena E (2018) PERISCOPE: road towards effective control of pertussis. Lancet Infect Dis 19(5):e179–e186

Dorji D, Mooi F, Yantorno O, Deora R, Graham RM, Mukkur TK (2018) Bordetella Pertussis virulence factors in the continuing evolution of whooping cough vaccines for improved performance. Med Microbiol Immunol 207(1):3–26. https://doi.org/10.1007/s00430-017-0524-z

Dunne A, Ross PJ, Pospisilova E, Masin J, Meaney A, Sutton CE, Iwakura Y, Tschopp J, Sebo P, Mills KH (2010) Inflammasome activation by adenylate cyclase toxin directs Th17 responses and protection against Bordetella pertussis. J Immunol (Baltimore, Md: 1950) 185(3):1711–1719. https://doi.org/10.4049/jimmunol.1000105

Eby JC, Hoffman CL, Gonyar LA, Hewlett EL (2015) Review of the neutrophil response to Bordetella pertussis infection. Pathog Dis 73(9):ftv081

Fedele G, Stefanelli P, Spensieri F, Fazio C, Mastrantonio P, Ausiello CM (2005) Bordetella pertussis-infected human monocyte-derived dendritic cells undergo maturation and induce Th1 polarization and interleukin-23 expression. Infect Immun 73(3):1590–1597

Fedele G, Spensieri F, Palazzo R, Nasso M, Cheung GYC, Coote JG, Ausiello CMJ (2010) Bordetella pertussis commits human dendritic cells to promote a Th1/Th17 response through the activity of adenylate cyclase toxin and MAPK-pathways. PLoS one 5(1):e8734

Fedele G, Cassone A, Ausiello CM (2015) T-cell immune responses to Bordetella pertussis infection and vaccination. Pathog Dis 73(7):ftv051. https://doi.org/10.1093/femspd/ftv051

Fedele G, Schiavoni I, Adkins I, Klimova N, Sebo P (2017) Invasion of dendritic cells, macrophages and neutrophils by the Bordetella adenylate cyclase toxin: a subversive move to fool host immunity. Toxins 9 (10):293. https://doi.org/10.3390/toxins9100293

Feunou P, Bertout J, Locht C (2010a) T-and B-cell-mediated protection induced by novel, live attenuated pertussis vaccine in mice. Cross protection against parapertussis. PloS one 5(4):e10178

Feunou PF, Kammoun H, Debrie AS, Mielcarek N, Locht C (2010b) Long-term immunity against pertussis induced by a single nasal administration of live attenuated B. pertussis BPZE1. Vaccine 28 (43):7047–7053. https://doi.org/10.1016/j.vaccine.2010.08.017

Friedman RL, Nordensson K, Wilson L, Akporiaye E, Yocum D (1992) Uptake and intracellular survival of Bordetella pertussis in human macrophages. Infect Immun 60(11):4578–4585

Fulton TR, Phadke VK, Orenstein WA, Hinman AR, Johnson WD, Omer SB (2016) Protective effect of contemporary pertussis vaccines: a systematic review and meta-analysis. Clin Infect Dis 62(9):1100–1110. https://doi.org/10.1093/cid/ciw051

Gestal M, Whitesides L, Harvill E (2018) Integrated signaling pathways mediate Bordetella immunomodulation, persistence, and transmission. Trends Microbiol 27(2):118–130

Hafler JP, Pohl-Koppe A (1998) The cellular immune response to Bordetella pertussis in two children with whooping cough. Eur J Med Res 3(11):523–526

Han WG, van Twillert I, Poelen MC, Helm K, van de Kasssteele J, Verheij TJ, Versteegh FG, Boog CJ, van Els CA (2013) Loss of multi-epitope specificity in memory CD4(+) T cell responses to B. pertussis with age. PLoS One 8(12):e83583. https://doi.org/10.1371/journal.pone.0083583

Han WG, Helm K, Poelen MM, Otten HG, van Els CA (2015) Ex vivo peptide-MHC II tetramer analysis reveals distinct end-differentiation patterns of human pertussis-specific CD4(+) T cells following clinical infection. Clin Immunol (Orlando, Fla) 157 (2):205–215. https://doi.org/10.1016/j.clim.2015.02.009

Harrington LE, Hatton RD, Mangan PR, Turner H, Murphy TL, Murphy KM, Weaver CT (2005) Interleukin 17-producing CD4+ effector T cells develop via a lineage distinct from the T helper type 1 and 2 lineages. Nat Immunol 6(11):1123–1132. https://doi.org/10.1038/ni1254

Hewlett EL, Edwards KM (2005) Pertussis—not just for kids. N Engl J Med 352(12):1215–1222

Higgins SC, Jarnicki AG, Lavelle EC, Mills KH (2006) TLR4 mediates vaccine-induced protective cellular immunity to Bordetella pertussis: role of IL-17-producing T cells. J Immunol (Baltimore, Md: 1950) 177 (11):7980–7989

Higgs R, Higgins S, Ross P, Mills K (2012) Immunity to the respiratory pathogen Bordetella pertussis. Mucosal Immunol 5(5):485–500

Jongerius I, Schuijt TJ, Mooi FR, Pinelli E (2015) Complement evasion by Bordetella pertussis: implications for improving current vaccines. J Mol Med 93 (4):395–402

Kammoun H, Feunou PF, Foligne B, Debrie A-S, Raze D, Mielcarek N, Locht C (2012) Dual mechanism of protection by live attenuated Bordetella pertussis BPZE1 against Bordetella bronchiseptica in mice. Vaccine 30 (40):5864–5870

Kaufmann E, Sanz J, Dunn JL, Khan N, Mendonça LE, Pacis A, Tzelepis F, Pernet E, Dumaine A, Grenier J-C (2018) BCG educates hematopoietic stem cells to generate protective innate immunity against tuberculosis. Cell 172(1–2):176–190. e119

Kilgore PE, Salim AM, Zervos MJ, Schmitt H-J (2016) Pertussis: microbiology, disease, treatment, and prevention. Clin Microbiol Rev 29(3):449–486

Kirimanjeswara GS, Agosto LM, Kennett MJ, Bjornstad ON, Harvill ET (2005) Pertussis toxin inhibits neutrophil recruitment to delay antibody-mediated clearance of Bordetella pertussis. J Clin Invest 115 (12):3594–3601. https://doi.org/10.1172/jci24609

Klein NP, Bartlett J, Fireman B, Rowhani-Rahbar A, Baxter R (2013) Comparative effectiveness of acellular versus whole-cell pertussis vaccines in teenagers. Pediatrics 131(6):e1716–e1722. https://doi.org/10.1542/peds.2012-3836

Kolls JK, Linden A (2004) Interleukin-17 family members and inflammation. Immunity 21(4):467–476. https://doi.org/10.1016/j.immuni.2004.08.018

Kumar BV, Ma W, Miron M, Granot T, Guyer RS, Carpenter DJ, Senda T, Sun X, Ho S-H, Lerner H (2017) Human tissue-resident memory T cells are defined by core transcriptional and functional signatures in lymphoid and mucosal sites. Cell Rep 20(12):2921–2934

Lamberti YA, Hayes JA, Vidakovics MLP, Harvill ET, Rodriguez ME (2010) Intracellular trafficking of Bordetella pertussis in human macrophages. Infect Immun 78(3):907–913

Lamberti Y, Gorgojo J, Massillo C, Rodriguez ME (2013) Bordetella pertussis entry into respiratory epithelial cells and intracellular survival. Pathog Dis 69 (3):194–204

Leef M, Elkins KL, Barbic J, Shahin RD (2000) Protective immunity to Bordetella pertussis requires both B cells and CD4+ T cells for key functions other than specific antibody production. J Exp Med 191(11):1841–1852

Lin A, Lore K (2017) Granulocytes: new members of the antigen-presenting cell family. Front Immunol 8:1781. https://doi.org/10.3389/fimmu.2017.01781

Locht C, Coutte L, Mielcarek N (2011) The ins and outs of pertussis toxin. FEBS J 278(23):4668–4682. https://doi.org/10.1111/j.1742-4658.2011.08237.x

Locht C, Papin JF, Lecher S, Debrie AS, Thalen M, Solovay K, Rubin K, Mielcarek N (2017) Live attenuated pertussis vaccine BPZE1 protects baboons against Bordetella pertussis disease and infection. J Infect Dis 216(1):117–124. https://doi.org/10.1093/infdis/jix254

Mahon BP, Sheahan BJ, Griffin F, Murphy G, Mills KH (1997) Atypical disease after Bordetella pertussis respiratory infection of mice with targeted disruptions of interferon-gamma receptor or immunoglobulin mu chain genes. J Exp Med 186(11):1843–1851

Martinez FO, Gordon S (2014) The M1 and M2 paradigm of macrophage activation: time for reassessment. F1000prime Rep:6–13. https://doi.org/10.12703/p6-13

Mascart F, Verscheure V, Malfroot A, Hainaut M, Piérard D, Temerman S, Peltier A, Debrie A-S, Levy J, Del Giudice G (2003) Bordetella pertussis infection in 2-month-old infants promotes type 1 T cell responses. J Immunol (Baltimore, Md: 1950) 170 (3):1504–1509

Mascart F, Hainaut M, Peltier A, Verscheure V, Levy J, Locht C (2007) Modulation of the infant immune responses by the first pertussis vaccine administrations. Vaccine 25(2):391–398. https://doi.org/10.1016/j.vaccine.2006.06.046

Masseria C, Martin CK, Krishnarajah G, Becker LK, Buikema A, Tan TQ (2017) Incidence and burden of pertussis among infants less than 1 year of age. Pediatr Infect Dis J 36(3):e54

McHeyzer-Williams M, Okitsu S, Wang N, McHeyzer-Williams L (2012) Molecular programming of B cell memory. Nat Rev Immunol 12(1):24–34

Melvin JA, Scheller EV, Miller JF, Cotter PA (2014) Bordetella pertussis pathogenesis: current and future challenges. Nat Rev Microbiol 12(4):274–288

Mielcarek N, Debrie AS, Raze D, Bertout J, Rouanet C, Younes AB, Creusy C, Engle J, Goldman WE, Locht C (2006) Live attenuated B. pertussis as a single-dose nasal vaccine against whooping cough. PLoS Pathog 2(7):e65. https://doi.org/10.1371/journal.ppat.0020065

Mills K, Barnard A, Watkins J, Redhead K (1993) Cell-mediated immunity to Bordetella pertussis: role of Th1 cells in bacterial clearance in a murine respiratory infection model. Infect Immun 61(2):399–410

Misiak A, Wilk MM, Raverdeau M, Mills KH (2017) IL-17-producing innate and pathogen-specific tissue resident memory gammadelta T cells expand in the lungs of Bordetella pertussis-infected mice. J Immunol (Baltimore, Md: 1950) 198(1):363–374. https://doi.org/10.4049/jimmunol.1601024

Mosmann TR, Sad S (1996) The expanding universe of T-cell subsets: Th1, Th2 and more. Immunol Today 17 (3):138–146

Mosmann TR, Cherwinski H, Bond MW, Giedlin MA, Coffman RL (1986) Two types of murine helper T cell clone. I. Definition according to profiles of lymphokine activities and secreted proteins. J Immunol (Baltimore, Md: 1950) 136(7):2348–2357

Mueller SN, Mackay LK (2016) Tissue-resident memory T cells: local specialists in immune defence. Nat Rev Immunol 16(2):79–89

Muloiwa R, Wolter N, Mupere E, Tan T, Chitkara A, Forsyth KD, von König C-HW, Hussey GJV (2018) Pertussis in Africa: findings and recommendations of the Global Pertussis Initiative (GPI). Vaccine 36 (18):2385–2393

Naninck T, Coutte L, Mayet C, Contreras V, Locht C, Le Grand R, Chapon C (2018) In vivo imaging of bacterial colonization of the lower respiratory tract in a baboon model of Bordetella pertussis infection and transmission. Sci Rep 8(1):12297

Nasso M, Fedele G, Spensieri F, Palazzo R, Costantino P, Rappuoli R, Ausiello CM (2009) Genetically detoxified pertussis toxin induces Th1/Th17 immune response through MAPKs and IL-10-dependent mechanisms. J Immunol (Baltimore, Md: 1950) 183 (3):1892–1899

O'Shea JJ, Paul WEJS (2010) Mechanisms underlying lineage commitment and plasticity of helper. CD4+ T cells 327(5969):1098–1102

Paddock CD, Sanden GN, Cherry JD, Gal AA, Langston C, Tatti KM, Wu K-H, Goldsmith CS, Greer PW, Montague JL (2008) Pathology and pathogenesis of fatal Bordetella pertussis infection in infants. Clin Infect Dis 47(3):328–338

Park H, Li Z, Yang XO, Chang SH, Nurieva R, Wang Y-H, Wang Y, Hood L, Zhu Z, Tian Q (2005) A distinct lineage of CD4 T cells regulates tissue inflammation by producing interleukin 17. Nat Immunol 6 (11):1133–1141

Peppoloni S, Nencioni L, Di Tommaso A, Tagliabue A, Parronchi P, Romagnani S, Rappuoli R, De Magistris M (1991) Lymphokine secretion and cytotoxic activity of human CD4+ T-cell clones against Bordetella pertussis. Infect Immun 59(10):3768–3773

Plotkin SA (2010) Correlates of protection induced by vaccination. Clin Vaccine Immunol 17(7):1055–1065

Purwar R, Campbell J, Murphy G, Richards WG, Clark RA, Kupper TS (2011) Resident memory T cells (T (RM)) are abundant in human lung: diversity, function, and antigen specificity. PLoS One 6(1):e16245. https://doi.org/10.1371/journal.pone.0016245

Radsak M, Iking-Konert C, Stegmaier S, Andrassy K, Hansch GM (2000) Polymorphonuclear neutrophils as accessory cells for T-cell activation: major histocompatibility complex class II restricted antigen-dependent induction of T-cell proliferation. Immunology 101(4):521–530

Raeven RH, Brummelman J, Pennings JL, Nijst OE, Kuipers B, Blok LE, Helm K, van Riet E, Jiskoot W, van Els CA (2014) Molecular signatures of the evolving immune response in mice following a Bordetella pertussis infection. PLoS One 9(8):e104548

Raeven RH, Brummelman J, van der Maas L, Tilstra W, Pennings JL, Han WG, van Els CA, van Riet E, Kersten GF, Metz B (2016) Immunological signatures after Bordetella pertussis infection demonstrate importance of pulmonary innate immune cells. PLoS One 11 (10):e0164027

Raeven RH, Pennings JL, van Riet E, Kersten GF, Metz B (2017) Meta-analysis of pulmonary transcriptomes from differently primed mice identifies molecular signatures to differentiate immune responses following Bordetella pertussis challenge. J Immunol Res 2017:8512847. https://doi.org/10.1155/2017/8512847

Reinhardt RL, Liang HE, Locksley RM (2009) Cytokine-secreting follicular T cells shape the antibody repertoire. Nat Immunol 10(4):385–393. https://doi.org/10.1038/ni.1715

Ross PJ, Sutton CE, Higgins S, Allen AC, Walsh K, Misiak A, Lavelle EC, McLoughlin RM, Mills KH (2013) Relative contribution of Th1 and Th17 cells in adaptive immunity to Bordetella pertussis: towards the rational design of an improved acellular pertussis vaccine. PLoS Pathog 9(4):e1003264. https://doi.org/10.1371/journal.ppat.1003264

Ryan M, Murphy G, Gothefors L, Nilsson L, Storsaeter J, Mills KH (1997) Bordetella pertussis respiratory infection in children is associated with preferential activation of type 1 T helper cells. J Infect Dis 175 (5):1246–1250

Ryan M, Murphy G, Ryan E, Nilsson L, Shackley F, Gothefors L, Oymar K, Miller E, Storsaeter J, Mills KH (1998) Distinct T-cell subtypes induced with whole cell and acellular pertussis vaccines in children. Immunology 93(1):1–10

Sallusto F, Cassotta A, Hoces D, Foglierini M, Lanzavecchia A (2018) Do memory CD4 T cells keep their cell-type programming: plasticity versus fate commitment? T-cell heterogeneity, plasticity, and selection in humans. Cold Spring Harb Perspect Biol 10(3). https://doi.org/10.1101/cshperspect.a029421

Schure RM, Hendrikx LH, de Rond LG, Ozturk K, Sanders EA, Berbers GA, Buisman AM (2012) T-cell responses before and after the fifth consecutive acellular pertussis vaccination in 4-year-old Dutch children. Clin Vaccine Immunol 19(11):1879–1886. https://doi.org/10.1128/CVI.00277-12

Schure RM, Hendrikx LH, de Rond LG, Ozturk K, Sanders EA, Berbers GA, Buisman AM (2013) Differential T- and B-cell responses to pertussis in acellular vaccine-primed versus whole-cell vaccine-primed children 2 years after preschool acellular booster vaccination. Clin Vaccine Immunol 20(9):1388–1395. https://doi.org/10.1128/CVI.00270-13

Sheridan SL, Frith K, Snelling TL, Grimwood K, McIntyre PB, Lambert SB (2014) Waning vaccine immunity in teenagers primed with whole cell and acellular pertussis vaccine: recent epidemiology. Expert Rev Vaccines 13(9):1081–1106. https://doi.org/10.1586/14760584.2014.944167

Solans L, Locht C (2019) The role of mucosal immunity in pertussis. Front Immunol 9(3068). https://doi.org/10.3389/fimmu.2018.03068

Solans L, Debrie A-S, Borkner L, Aguiló N, Thiriard A, Coutte L, Uranga S, Trottein F, Martín C, Mills KH (2018) IL-17-dependent SIgA-mediated protection against nasal Bordetella pertussis infection by live attenuated BPZE1 vaccine. Mucosal Immunol 11 (6):1753–1762

Stenger RM, Meiring HD, Kuipers B, Poelen M, van Gaans-van den Brink JA, Boog CJ, de Jong AP, van Els CA (2014) Bordetella pertussis proteins dominating the major histocompatibility complex class II-presented epitope repertoire in human monocyte-derived dendritic cells. Clin Vaccine Immunol 21(5):641–650. https://doi.org/10.1128/cvi.00665-13

Stoppelenburg AJ, de Roock S, Hennus MP, Bont L, Boes M (2014) Elevated Th17 response in infants undergoing respiratory viral infection. Am J Pathol 184(5):1274–1279

Storsaeter J, Hallander HO, Gustafsson L, Olin P (1998) Levels of anti-pertussis antibodies related to protection after household exposure to Bordetella pertussis. Vaccine 16(20):1907–1916

Storsaeter J, Hallander HO, Gustafsson L, Olin P (2003) Low levels of antipertussis antibodies plus lack of history of pertussis correlate with susceptibility after household exposure to Bordetella pertussis. Vaccine 21(25–26):3542–3549

Tan T, Dalby T, Forsyth K, Halperin SA, Heininger U, Hozbor D, Plotkin S, Ulloa-Gutierrez R, Von König CHW (2015) Pertussis across the globe: recent epidemiologic trends from 2000 to 2013. Pediatr Infect Dis J 34(9):e222–e232

Taranger J, Trollfors B, Lagergård T, Sundh V, Bryla DA, Schneerson R, Robbins JB (2000) Correlation between pertussis toxin IgG antibodies in postvaccination sera and subsequent protection against pertussis. J Infect Dis 181(3):1010–1013

Tarlinton D, Good-Jacobson K (2013) Diversity among memory B cells: origin, consequences, and utility. Science (New York, NY) 341(6151):1205–1211

Thorstensson R, Trollfors B, Al-Tawil N, Jahnmatz M, Bergstrom J, Ljungman M, Torner A, Wehlin L, Van Broekhoven A, Bosman F, Debrie AS, Mielcarek N, Locht C (2014) A phase I clinical study of a live attenuated Bordetella pertussis vaccine--BPZE1; a single centre, double-blind, placebo-controlled, dose-escalating study of BPZE1 given intranasally in healthy adult male volunteers. PLoS One 9(1): e83449. https://doi.org/10.1371/journal.pone.0083449

Turner DL, Farber DL (2014) Mucosal resident memory CD4 T cells in protection and immunopathology. Front Immunol 5:331

Valdez HA, Oviedo JM, Gorgojo JP, Lamberti Y, Rodriguez ME (2016) Bordetella pertussis modulates human macrophage defense gene expression. Pathog Dis 74(6):ftw073. https://doi.org/10.1093/femspd/ftw073

Van Der Ark AA, Hozbor DF, Boog CJ, Metz B, Van Den Dobbelsteen GP, Van Els CA (2012) Resurgence of pertussis calls for re-evaluation of pertussis animal models. Expert Rev Vaccines 11(9):1121–1137

van der Lee S, Hendrikx LH, Sanders EAM, Berbers GAM, Buisman A-M (2018a) Whole-cell or acellular pertussis primary immunizations in infancy determines adolescent cellular immune profiles. Front Immunol 9 (51). https://doi.org/10.3389/fimmu.2018.00051

van der Lee S, van Rooijen DM, de Zeeuw-Brouwer M-L, Bogaard MJM, van Gageldonk PGM, Marinovic AB, Sanders EAM, Berbers GAM, Buisman A-M (2018b) Robust humoral and cellular immune responses to pertussis in adults after a first acellular booster vaccination. Front Immunol 9(681). https://doi.org/10.3389/fimmu.2018.00681

van Twillert I, Han WG, van Els CA (2015) Waning and aging of cellular immunity to Bordetella pertussis. Pathog Dis 73(8):ftv071

Van Twillert I, Marinović AAB, Kuipers B, Berbers GA, van der Maas NA, Verheij TJ, Versteegh FG, Teunis PF, van Els CA (2016) The use of innovative two-component cluster analysis and Serodiagnostic cut-off methods to estimate prevalence of pertussis reinfections. PLoS One 11(2):e0148507

Varney ME, Boehm DT, DeRoos K, Nowak ES, Wong TY, Sen-Kilic E, Bradford SD, Elkins C, Epperly MS, Witt WT, Barbier M, Damron FH (2018) Bordetella pertussis whole cell immunization, unlike acellular immunization, mimics Naïve infection by driving hematopoietic stem and progenitor cell expansion in mice. Front Immunol 9(2376). https://doi.org/10.3389/fimmu.2018.02376

Versteegh F, Schellekens J, Nagelkerke A, Roord J (2002) Laboratory-confirmed reinfections with Bordetella pertussis. Acta Paediatr 91(1):95–97

Walrath JR, Silver RF (2011) The alpha4beta1 integrin in localization of Mycobacterium tuberculosis-specific T helper type 1 cells to the human lung. Am J Respir Cell Mol Biol 45(1):24–30. https://doi.org/10.1165/rcmb.2010-0241OC

Warfel J, Merkel T (2013) Bordetella pertussis infection induces a mucosal IL-17 response and long-lived Th17 and Th1 immune memory cells in nonhuman primates. Mucosal Immunol 6(4):787–796

Warfel JM, Merkel TJ (2014) The baboon model of pertussis: effective use and lessons for pertussis vaccines. Expert Rev Vaccines 13(10):1241–1252

Warfel JM, Beren J, Kelly VK, Lee G, Merkel TJ (2012a) Nonhuman primate model of pertussis. Infect Immun 80 (4):1530–1536. https://doi.org/10.1128/IAI.06310-11

Warfel JM, Beren J, Merkel TJ (2012b) Airborne transmission of Bordetella pertussis. J Infect Dis 206 (6):902–906. https://doi.org/10.1093/infdis/jis443

Warfel JM, Zimmerman LI, Merkel TJ (2014) Acellular pertussis vaccines protect against disease but fail to prevent infection and transmission in a nonhuman primate model. PNAS 111(2):787–792

Wearing HJ, Rohani P (2009) Estimating the duration of pertussis immunity using epidemiological signatures. PLoS Pathog 5(10):e1000647. https://doi.org/10.1371/journal.ppat.1000647

Wendelboe AM, Van Rie A, Salmaso S, Englund JA (2005) Duration of immunity against pertussis after natural infection or vaccination. Pediatr Infect Dis J 24(5 Suppl):S58–S61

Wilk MM, Misiak A, McManus RM, Allen AC, Lynch MA, Mills KHG (2017) Lung CD4 tissue-resident memory T cells mediate adaptive immunity induced by previous infection of mice with Bordetella pertussis. J Immunol (Baltimore, Md: 1950) 199(1):233–243. https://doi.org/10.4049/jimmunol.1602051

Wilk MM, Borkner L, Misiak A, Curham L, Allen AC, Mills KH (2019) Immunization with whole cell but not acellular pertussis vaccines primes CD4 TRM cells that sustain protective immunity against nasal colonization with Bordetella pertussis. Emerg Microbes Infect 8 (1):169–185

Wilson HM (2014) SOCS proteins in macrophage polarization and function. Front Immunol 5:357

Wynn TA (2005) TH-17: a giant step from TH1 and TH2. Nat Immunol 6:1069–1070. https://doi.org/10.1038/ni1105-1069

Zimmerman LI, Papin JF, Warfel J, Wolf RF, Kosanke SD, Merkel TJ (2018) Histopathology of Bordetella pertussis in the Baboon Model. Infect Immun 86(11): e00511–e00518

Adv Exp Med Biol - Advances in Microbiology, Infectious Diseases and Public Health (2019) 1183: 99–113
https://doi.org/10.1007/5584_2019_406
© Springer Nature Switzerland AG 2019
Published online: 25 July 2019

Human Immune Responses to Pertussis Vaccines

Clara M. Ausiello, Françoise Mascart, Véronique Corbière, and Giorgio Fedele

Abstract

Pertussis still represents a major cause of morbidity and mortality worldwide. Although vaccination is the most powerful tool in preventing pertussis and despite nearly 70 years of universal childhood vaccination, incidence of the disease has been rising in the last two decades in countries with high vaccination coverage. Two types of vaccines are commercially available against pertussis: whole-cell pertussis vaccines (wPVs) introduced in the 1940s and still in use especially in low and middle-income countries; less reactogenic acellular pertussis vaccines (aPVs), licensed since the mid-1990s.

In the last years, studies on pertussis vaccination have highlighted significant gaps and major differences between the two types of vaccines in the induction of protective anti-pertussis immunity in humans. This chapter will discuss the responses of the immune system to wPVs and aPVs, with the aim to enlighten critical points needing further efforts to reach a good level of protection in vaccinated individuals.

Keywords

Anti-pertussis immunity · Bordetella pertussis (Bp) · Immunization strategies · Mechanisms of protection · Pertussis vaccines

1 Pertussis Vaccination

Historical reports mention a pertussis reminiscent disease as far back as the twelfth century (Weston 2012) but pathogen isolation only occurred in 1906 by Bordet and Gengou (Bordet and Gengou 1906). The first attempts to use whole-cell killed bacteria to develop a pertussis vaccine were made a few years after Bordet and Gengou studies (Lapidot and Gill 2016).

Routine immunization with whole-cell pertussis vaccines (wPVs) started in the late 1940s in the United States, using a wPV combined with diphtheria and tetanus (DTwPV, trivalent). Immunization campaigns were successful, with pertussis cases falling from 115,000–270,000 annually prior to the vaccine era to 1200–4000 annually during the 1980s (Cherry et al. 1988). Despite the high efficacy, DTwPVs showed high reactogenicity and their use was associated with serious systemic reactions, including convulsions and encephalopathies, due to the pertussis

C. M. Ausiello and G. Fedele (✉)
Department of Infectious Diseases, Istituto Superiore di Sanità, Rome, Italy
e-mail: giorgio.fedele@iss.it

F. Mascart
Laboratory of Vaccinology and Mucosal Immunity, Universitè Libre de Bruxelles, Brussels, Belgium

Immunobiology Clinic, Hôpital Erasme, Université Libre de Bruxelles (U.L.B.), Brussels, Belgium

V. Corbière
Laboratory of Vaccinology and Mucosal Immunity, Universitè Libre de Bruxelles, Brussels, Belgium

component (Cherry 1996; Jefferson et al. 2003). In 1970s and 1980s safety concerns regarding wPVs raised (Miller et al. 1981; Cody et al. 1981; Gangarosa et al. 1998). For this reason, pertussis vaccination programs were suspended in Japan and Sweden (Sato et al. 1984; Romanus et al. 1987), while in several other countries pertussis vaccine acceptance was greatly reduced (Cherry et al. 1988; Gangarosa et al. 1998; Gonfiantini et al. 2014).

Concerns about the safety of wPVs prompted the development of acellular pertussis vaccines (aPVs). These are subunit vaccines composed of 1–5 purified *B. pertussis* antigens. All aPVs contain the pertussis toxin (PT), believed to be the major virulence factor and target of protective immune responses. Other antigens included in aPVs formulations are the filamentous hemagglutinin (FHA), the pertactin (PRN) and the Fimbriae (Fim2 and Fim3) (Pichichero 1996).

In 1986, the first placebo-controlled trial of an acellular vaccine was carried out in Sweden, selected since at that time it was one of the few countries in Europe that did not administer wPVs routinely to infants (Ad Hoc Group for the Study of Pertussis Vaccines 1988). After this first study, others were performed using aPVs of different formulation and different protocols; a summary is shown in Table 1. The trials that ultimately led to the licensure and adoption of aPVs were those conducted in Sweden and Italy. In both trials, DTaPVs were compared to DTwPV and placebo arms, using a blinded, randomized scheme, with culture or serology confirmed clinical pertussis as the primary endpoint (Gustafsson et al. 1996; Greco et al. 1996). Table 2 summarize a few details of the Italian aPV efficacy trial. The vaccine efficacy study was conduct in about 15,000 infants, humoral response was assayed in about 1,500 infants and T-cell response was tested in about 150 infants. Vaccine efficacy, humoral and T-cell responses were followed in a subgroup of aPV vaccinated children till 33 months of age (Salmaso et al. 1998).

Either the Swedish and the Italian clinical trials showed that, compared with wPVs, aPVs have improved tolerability and safety and induce higher concentrations of antibodies against PT,

and proved that the efficacy of aPVs is higher than wPV (Gustaffson et al. 1996; Greco et al. 1996). Unfortunately, the lot of wPV used in the Swedish and the Italian trials, produced by Connaught Laboratories, was less efficacious than expected (Table 2). This probably led to an over-evaluation of aPVs efficacy. In other trials where aPVs were compared to other wPVs preparations, as in the Senegal trial, the wPV showed a better efficacy than the aPV (Simondon et al. 1997).

The vaccine efficacy trials performed in the 1990s marginally investigated crucial parameters of vaccination, such as the duration of protection, the type of immunity evoked or the ability to prevent transmission of infection. These aspects were investigated in depth in follow-up studies. In particular, they were intensified by the observation that the disease was resurging even in countries with high vaccination coverage (Black 1997; Bancroft et al. 2016; van der Lee et al. 2018a, b, c). Most of these studies highlighted straight different responses between wPVs and aPVs, mainly related to the induction of a different type of anti-pertussis immunity.

2 B-cell Immune Responses to Pertussis Vaccination

2.1 Humoral Immune Response after Primary Immunization

Studies on the humoral response to pertussis antigens are crucial in the search of correlates of protection induced by vaccination. In principle, antibodies that can either neutralize the toxic effect of PT and/or prevent the attachment of *B. pertussis* to cells of the upper and lower respiratory tract may provide protection. Primary immunization of children with a pertussis vaccine usually involves a three-dose schedule given in the first 2–11 months of life. In some European countries, a fourth dose is given at 15–18 months of age to complete the primary vaccination schedule (https://ecdc.europa.eu/en/immunisation-vaccines/EU-vaccination-schedules). Figure 1 shows primary vaccination schedules in different

Table 1 Efficacy trials of acellular pertussis vaccine

Study year	Study location	Design and methods	Number of participants	Comments
1985 (Ad Hoc Group for the Study of Pertussis Vaccines 1988)	Sweden	Double blind placebo controlled (compared two Japanese aPV)	3801	No wPV control group 2-dose schedule
1990 (Simondon et al. 1997)	Senegal	Double blind household contact (DTaPV/DTwPV)	4181	No placebo control 3 dose schedule
1991 (Trollfors et al. 1995)	Sweden	Double blind placebo controlled (compared DT/DTaPV)	3450	No wPV control 3-dose schedule
1992 (Gustafsson et al. 1996)	Sweden	Double blind placebo controlled (two-compenent DTaPV/five component DTaPV/DTwPV/DT)	24,336	wPV control (Connaught) 3-dose schedule
1992 (Greco et al. 1996)	Italy	Double blind placebo controlled (DTaPV/DTwPV/DT)	14,751	wPV control (Connaught) 3-dose schedule

Modified from: Lapidot R and Gill CJ (2016)

Table 2 Vaccine efficacy and immunogenicity in the Italian efficacy trials of acellular pertussis vaccine

Vaccine	Nr. of children	Vaccine efficacy (95% CI)	anti-PT IgG 1 month IU/ml (95% CI) N = 1275	anti-PT IgG 15 months IU/ml (95% CI) N = 1275	T-cell proliferation 1 month % of positive response N = 142
aPV (SmithKline and Beecham)	4481	84 (76–89)	51.3 (47.9–57.9)	2.7 (2.4–3.0)	55%
aPV (Chiron Biocine)	4452	84 (76–90)	94.4 (88.8–100.3)	4.5 (4.0–5.0)	83%
wPV (Connaught)	4348	36.1 (14–52)	1.3 (1.1–1.2)	1.1 (1.1–1.2)	46%

From: Greco et al. (1996), Cassone et al. (1997), and Giuliano et al. (1998)

European, Asia-Pacific, African and American countries.

It is known that vaccination with wPV induces specific anti-PT, anti-FHA and anti-PRN immunoglobulin G (IgG) since the first dose (Steinhoff et al. 1995; Pereira et al. 2010), unless in prematurely born infants (Mascart et al. 2018). The induction of higher IgG levels by aPVs compared to wPVs was stated by a study comparing the immunogenicity of 13 different aPVs with a licensed wPV (Edwards et al. 1995). Worth of note, the same study allowed concluding that, particularly for PT, vaccine immunogenicity seems to depend on factors other than antigen concentration, possibly including antigen derivation and formulation. In this regard, it was found that aPVs containing a genetically inactivated PT were responsible of a higher anti-PT IgG response (Edwards et al. 1995;

	Months																	
	1	2	3	4	5	6	7	8	9	10	11	12	13	14	15	16	17	18
Belgium[a]		aPV	aPV	aPV											aPV			
France[a]		aPV		aPV							aPV							
Italy[a]			aPV		aPV						aPV							
Netherlands[a]	aPV		aPV	aPV							aPV							
Poland[a, #]		wPV		wPV		wPV										wPV		
UK[a]		aPV	aPV	aPV														
China[b]			aPV	aPV	aPV													
Japan[b]			aPV	aPV	aPV	aPV												aPV
Nigeria[b]		wPV	wPV	wPV														
Senegal[b]		wPV	wPV	wPV														
Argentina[b]		wPV		wPV		wPV												wPV
Canada[b]		aPV		aPV		aPV												aPV
USA[c]		aPV		aPV		aPV									aPV	aPV	aPV	aPV
Australia[d]		aPV		aPV		aPV												aPV

Fig. 1 Examples of primary vaccination schedules reccomemded in different European, Asia-Pacific, African and American countries
Source: [a]ECDC European Vaccine Scheduler; [b]WHO vaccine-preventable diseases: monitoring system. 2018 global summary; [c]CDC's Advisory Committee on Immunization Practices (ACIP); [d]The Melbourne Vaccine Education Centre (MVEC)
[#]aPV are available on the private market; it has been estimated that in 2013 aPV represented 60% of all vaccines used for primary pertussis vaccination in Poland (U. Heininger, et al. PLoS One, 11 (2016), p. e0155949)

Cassone et al. 1997; Giuliano et al. 1998). It should be considered that antigen concentrations are lower in wPVs compared to aPVs, in particular for PT, and that a limited number of purified antigens are present in aPVs. Therefore, the first explanation for lower anti-PT, anti-FHA and anti-PRN antibody titers induced by wPVs compared to aPVs relay to a lower antigen concentration, whereas significant antibody titers were detected in response to a whole-cell *B. pertussis* lysate (Mascart et al. 2018).

The protective implications of humoral responses induced by vaccination are not well understood since clear serological correlates of vaccine-mediated protection are missing. In fact, although some evidences have suggested that antibody response against PT, PRN, and Fimbriae may be associated with protection (Storsaeter et al. 1998; Taranger et al. 2000), the immunogenicity studies performed within the clinical trials did not demonstrate a satisfactory correlation between the levels of antibodies to the vaccine antigens and vaccine efficacy (Ad Hoc Group for the Study of Pertussis Vaccines 1988; Giuliano et al. 1998).

A key point to be considered is that humoral immune responses to pertussis vaccination are of short duration. Follow-up studies on the persistence of the serological response to primary immunization with DTaPV showed a marked decline of IgG level against vaccine antigens approximately after 15–20 months from the last dose (Table 2) (Giuliano et al. 1998; Huang et al. 1996; Hallander et al. 2009).

2.2 Humoral Immune Response after Booster Vaccination

After the introduction of aPVs, infections and disease caused by *B. pertussis* among older children and adults in immunized populations were increasingly recognized (Cromer et al. 1993; He et al. 1994; Black 1997), indicating that the vaccine-induced immunity was waning below the protective level in these age groups. In addition, several household studies and investigations of outbreaks had shown that older family members constitute an important reservoir for spread of infection to susceptible infants (Nelson 1978; Mertsola et al. 1983; Long et al. 1990). These observations suggested the need for booster immunizations of older children and adults, also with the goal of preventing transmission of *B. pertussis* from these age groups to infants. Less reactogenic aPVs seemed to be suitable not only for primary immunization but also for boosting of preschool children. A study by Hallander and colleagues predicted that 65 months after the third dose of a primary vaccination at 2, 4 and 6 months, anti-PT IgG would have been below the detection level in 50% of the vaccinated children (Hallander et al. 2005). Starting from this and similar observations, public health authorities and strategic advisory group of experts started to recommend a pre-school booster immunization at 5–6 years of age in order to maintain an adequate level of immune protection.

Studies on the induction of humoral immunity after vaccine boosters pointed out their importance in restoring antibody levels (Schure et al. 2013; Aase et al. 2014; Carollo et al. 2014). However, it is becoming apparent that, similarly to primary immunization, boost vaccination tends to decline over time. Recently, it was shown that after the booster dose at around 4 years of age, antibodies to PT became undetectable in 49% of children at the 5-year follow-up visit (Voysey et al. 2016). Increasing incidence of the disease in older age-groups, the need to reduce the risk of spreading the infection to unprotected younger infants, and the rapid decline of antibody levels, prompted for the introduction of vaccine booster doses also for adolescents and adults, using vaccines with a reduced antigen content (Tdap) (Halperin 2001; Campins-Martí et al. 2001; Zepp et al. 2011). Studies evaluating the persistence of humoral responses after the booster vaccine dose in adolescents and adults have shown a decline over time of pertussis-specific antibodies that, nevertheless, are usually maintained at greater than pre-immunization levels for several years after the receipt of the last booster dose (Edelman et al. 2004; Edelman et al. 2007; Le et al. 2004).

2.3 B-cell Memory Response to Pertussis Vaccination

In the search of effective correlates of protection, several studies assessed the induction of B-cell memory immune responses to pertussis antigens following vaccination, since these cells can propagate a booster response rapidly enough to outpace pathogenesis of *B. pertussis* (Pichichero 2009). The results obtained indicate that, despite the rapid antibody decay, long-term memory B-cell responses are induced by vaccination and that memory B-cells, in addition to antibodies, may contribute to protection against pertussis. (Hendrikx et al. 2011; Schure et al. 2013; Carollo et al. 2014; Jahnmatz et al. 2014). In particular, in wPV-primed Dutch children the levels of specific memory B-cells increased at 3, 4, 6 and 9 years of age, and could be detected in vaccinated children whose antibody levels had already waned (Hendrikx et al. 2011). In an Italian study, still >80% of aP vaccinated children presented a positive B-cell memory response 5 years after aPV priming (Carollo et al. 2014). The crucial role of memory B-cells response in protection has been demonstrated by a recent study showing that the low levels of pre-formed serum antibodies are insufficient for protection and that memory B cells play a major role in the adult defense (Marcellini et al. 2017).

3 T-Cell Immune Responses to Pertussis Vaccination

3.1 T-Cell Immune Response after Primary Immunization

During the safety and efficacy trials conducted in the 1990s, immunogenicity studies focused on the induction of pertussis-specific antibodies while the interest in studying the T-cell immune response to vaccination was limited. However, during the Italian trial, studies were performed in a small percentage of infants to assess the induction of T-cell responses by pertussis vaccines, measured as pertussis-specific T-cell proliferation and T helper (Th) type cytokines expression (Cassone et al. 1997; Ausiello et al. 1997). The results showed that aPVs were better inducers of T-cell immune responses than the wPVs, (Cassone et al. 1997) (Table 2). However, as underlined previously, the wPV lot used in the trial was less efficacious than expected. Follow-up studies showed that vaccine-induced T-cell proliferation persisted, in contrast to the rapid decline in antibody levels. In fact, 14 months after the last immunization, anti-PT IgG titers fell to low or undetectable values, while T-cell responses substantially persisted (Table 2) (Cassone et al. 1997). The authors proposed that persistence of T-cell immunity against pertussis could be boosted by exposure to natural infection (Cassone et al. 1997; Ausiello et al. 1997, 1999; Cassone et al. 2000).

The profile of Th cells cytokines produced after antigenic stimulation in wPV or aPV vaccinated individuals was evaluated in the same subgroup of infants. A key difference was evidenced, indeed aPV vaccination induced both a Th type 1 and type 2 cytokine profile, marked by the production of Interferon-gamma and Interleukin 5, activating a cell-mediated immune response against intracellular pathogens or a humoral immune response against extracellular pathogens, respectively. On the contrary, the wPV induced a Th type 1 pattern only (Ausiello et al. 1997). Following this first study, many others highlighted the crucial mismatch between

aPVs and wPVs induced T-cell immune response. In fact, wP vaccination induces Th1 polarized responses, whereas aP vaccination is followed by a predominant Th2 response, that could change from a mixed Th2/Th1 to a robust Th1 profile following a natural booster or a vaccine booster at 15 months of age (Zepp et al. 1996; Ryan et al. 1998; He et al. 1998; Ausiello et al. 1997, 1999; Mascart et al. 2007; Edwards and Berbers 2014; Mascart et al. 2018).

More recently, in studies performed mainly in animal models, it was shown that both the aPVs and wPVs induce the expansion of another Th subset, Th17 cells, activated to fight extracellular bacteria (Ross et al. 2013; Warfel and Merkel 2013). Overall, it is now clear that natural infection and immunization with wPVs induces a similar pattern of Th1/Th17 response while aPVs induce a Th2/Th17 response (Ross et al. 2013; Warfel et al. 2014). The role of CD4+ T-helper cells in mediating immunity against natural infection is reviewed in depth by Lambert and colleagues in this issue (see chapter 5 of this volume).

3.2 T-Cell Immune Response after Booster Immunization

Several studies have been performed on the persistence of vaccine induced T-cell response and the effect of vaccine booster doses. The results on the importance of booster immunizations in enhancing T-cell responses to pertussis antigens are somewhat contrasting. In some studies, an enhancing effect was recorded. Tran Minh et al. (1999) and Edelman et al. (2004) evaluated pertussis-specific T-cell responses in adolescents. At one month and three years after the aPV boost, T-cell responses were higher than those observed before the boost.

Other studies, on the contrary, did not highlight an enhancing effect. A fourth dose given at 13–16 months of age, to complete the primary vaccination schedule had no major effect on antigen-induced cytokine production neither in full-term born infants nor in preterm infants, but it allowed maintaining significant immune responses in the same infants tested before and after the fourth dose (Dirix et al. 2009; Vermeulen et al. 2013). According to Schure and colleagues, an increase in cytokine production was missed after a boost vaccination in children primed with aPV, whereas it was not the case for wP-vaccinated children (Schure et al. 2012a). The same research group reported that in 9 years-old children, T-cell responses did not increase after a second aPV booster (Schure et al. 2012b). Poor effect of vaccinal boost was confirmed by another study evaluating T-cell immunity in children 5 years after primary vaccination with two aPVs. A positive T-cell response, evaluated in terms of proliferation and IFN-γ positive CD4+ T cells, was present only in 36.8% of vaccinees (Palazzo et al. 2016a). PT-specific proliferation was higher in children tested before than after the preschool vaccine booster dose (Palazzo et al. 2016a). Similarly, only a marginal effect of a pre-school booster dose on the proportions of FHA- and PT-induced IFN-gamma-containing CD4+ T lymphocytes was observed in Belgium (Mascart et al. 2018). However, the effect in children of a booster dose on T-cell immune responses may also be restricted to Th2-type cytokine production as reported after an aPV booster administrated in aPV primed children (Ryan et al. 2000).

Despite lack of immediate boosting effect on antigen-specific Th1-type responses, pertussis-specific T-cell immunity increases during the 5 year following the booster at 4 years of age (Schure et al. 2012a). The authors conclude that this phenomenon is probably due to natural boosting caused by the high circulation of *B. pertussis*. This might explain, at least in part, the persistence of protection against pertussis in aPV recipients despite a substantial waning of both antibodies and T-cell responses induced by the primary immunization.

All these studies indicated a probable overestimation of the duration of immunity induced by aPVs introduced in the mid-nineties of the last century, due, in part, to an asymptomatic natural booster in countries with high *B. pertussis* circulation. Very few studies investigated the effect of a booster dose administrated in adults in view of the rapid waning of the aPV-induced immune responses. However, preliminary data suggest that booster dose administrated in adults is not associated with an enhancement of specific T-cell immune responses (Mascart et al. 2018). Quite remarkably, review of data from an observational, cross-sectional study performed in the Netherlands, comprising pertussis patients of various ages, suggested that T-cell responsiveness tends to diminish with age (van Twillert et al. 2015).

3.3 T-cell Memory Response to Pertussis Vaccination

In the search of new parameters to assess the level and duration of protection after vaccination or infection, pertussis-specific memory T-cell populations were assessed in humans. Several data showed that pertussis-specific T-cell responses in infants after aPV primary vaccination were mainly restricted to central memory and effector memory T-cell subsets (Sharma and Pichichero 2012; Smits et al. 2013; Palazzo et al. 2016a). However, a vaccine boost had no specific effect on the frequency of memory subsets expansion (Schure et al. 2012b; Smits et al. 2013; Palazzo et al. 2016a). Hence, a correlation between the percentage of the different T memory subsets and duration of protection from pertussis appears to be still elusive.

The induction of CD8+ T-cell response during *B. pertussis* infection was analyzed in details by Mascart's group (Mascart et al. 2003; Dirix et al. 2012). In CD8+ cells, an expansion of effector memory T-cells was observed leading to assume that pertussis-specific CD8+ T memory cells contribute to protection against pertussis (Rieber et al. 2011; Dirix et al. 2012; de Rond et al. 2015).

3.4 T follicular Helper Cells

An important cellular population involved in the development and maintenance of B cell responses, which have not been investigated yet in pertussis field, is the T follicular helper cells (Tfh). Germinal centers Tfh cells instruct neighboring B lymphocytes to undergo differentiation into memory B cells and plasma cells secreting affinity matured class-switched immunoglobulins (Crotty 2014). Upon recall of the antigen, memory Tfh cells will help part of the B memory cells to differentiate quickly into antibody secreting plasma cells, providing an initial rapid boost of the antibody response (MacLennan et al. 2003). Tfh cells have been initially described in the germinal centers of secondary lymphoid tissues, but circulating Tfh (cTfh) can be detected in the blood and are considered as a memory compartment of germinal center Tfh cells (Morita et al. 2011) with the capacity of rapid and efficient secondary immune responses. cTfh are categorized in distinct subsets which share properties with Th1, Th2 or Th17 cells depending on the combination of surface markers expression (Ueno 2016) and will contribute to the production of different Ig class and subclass (Morita et al. 2011; Locci et al. 2013).

cTfh cells have been associated with protective role in human infectious disease (Locci et al. 2013; Obeng-Adjei et al. 2015; Kumar et al. 2014; Slight et al. 2013; Farooq et al. 2016) and vaccines (Bentebibel et al. 2013; Pallikkuth et al. 2012; Spensieri et al. 2013). Therefore, it is conceivable that Tfh cells play a role also in immunity to pertussis. Long-term specific memory B cells are induced by pertussis vaccines. However, to generate efficient secondary immune responses, Tfh cells are key drivers. The quality of the Tfh cells response induced by pertussis vaccine might influence the type of memory B cell response and the quality of the recall response, and need therefore to be investigated.

4   Different Immune Responses to Different Pertussis Vaccines

There is a rather large consensus for a more rapid waning of protective immunity in aPV than in wPV recipients (Plotkins 2013; Edwards and Berbers 2014; Acosta et al. 2015). Moreover, it is known that teenagers who received wPVs in childhood are more protected than those who received aPVs (Klein et al. 2013; Witt et al. 2013). Rieber and colleagues published a first study focusing on differences in long-term immunity and booster immune response to pertussis antigens between adolescents who previously had received DTaPV or DTwPV. The authors found that subjects who received primary wP vaccination responded with higher IgG-PT titers to the adolescent Tdap booster than those immunized with primary aP vaccination (Rieber et al. 2008). A more recent study, comparing pertussis-specific humoral responses after aP booster vaccination of 4-year-old children who had been vaccinated in the primary series with wPVs or aPVs, showed that the preschool aPV booster at 4 years of age resulted in significantly higher pertussis-specific IgG antibody levels in aPV-primed children than those in wPV-primed children, which remained higher for at least 2 years post-booster (Schure et al. 2013). A follow-up study showed that the pre-adolescent Tdap booster vaccination induced lower vaccine antigen-specific humoral and B memory cell responses in aPV-primed compared with wPV-primed children, suggesting that aPV primed children may experience faster humoral and B memory cells waning (van der Lee et al. 2018a), confirming the result of Rieber et al. (2008). Studies on wPV- or aPV-primed children allowed to demonstrate a different profile of the humoral immune response associated with primary immunization, with high proportions of specific IgG4 in some aPV-primed children, an antibody response associated to a Th2 profile (van der Lee et al. 2018c).

wPV or aPV priming can also determine the outcome of T-cell responses. A study by Smits and colleagues in 9–11 years-old children showed that wPV-primed children have longer lasting Th1-type immune responses than aPV-primed children (Smits et al. 2013). Indeed, even if the time from the last booster vaccine was significantly longer in wPV-compared to aPV-vaccinated children, the T-cell proliferative capacity in response to antigenic stimulation was comparable, and more children had a detectable cytokine response after wPV-compared to aPV-vaccination (Smits et al.

2013). Most interestingly, the influence of pertussis priming vaccines on adult T-cell responses after a Tdap booster vaccination has a key role in skewing the immune profile of vaccine recipients. Indeed, in wPV primed individuals, the T-cell response is Th-1 polarized, while IL-5 is dominant in aPV primed individuals. This differential pattern is maintained after booster vaccination up to several decades after the original aPV/wPV priming (Bancroft et al. 2016). These findings suggest that childhood aPV versus wPV vaccination induces functionally different T-cell responses to pertussis that become fixed and are unchanged even upon boosting. This view was confirmed by a recent study analyzing pertussis-specific memory CD4+ T-cell responses. The authors found a Th2 versus Th1/Th17 differential polarization as a function of childhood vaccination with aPV or wPV, respectively. These differences appeared to be T-cell specific, since equivalent increases of antibody titers and plasmablasts after aPV boost were seen in both groups (da Silva et al. 2018).

Differences in the capacity to induce protective responses by primary or booster vaccination due to differences in aPV components have been reported (Vermeulen et al. 2013; Koepke et al. 2014; Carollo et al. 2014; Palazzo et al. 2016a). Factors causing this differential behavior may include antigenic formulation and concentration, adjuvant content and the PT inactivation process. Specifically, it was conceivable that the milder inactivation of vaccine antigens was responsible for a better T epitope preservation and an induction of a more sustained T-cell proliferative response (Palazzo et al. 2016a). On the contrary, vaccines formulated using antigens adsorbed onto a higher content of aluminum hydroxide better preserved the antibody responses (Carollo et al. 2014).

5 Immune Responses to Pertussis Maternal Immunization

Pertussis-related morbidity and mortality disproportionately affects young infants (Van Hoek et al. 2013), those less than 4 months of age being particularly vulnerable to infection. Vaccination during pregnancy to boost maternal antibody levels and enhance infant passive immunization by IgG placental transfer was therefore considered. This approach was shown to be safe (Campbell et al. 2018; Halperin et al. 2018) and effective to prevent infant pertussis especially during the first 2 months of life (Baxter et al. 2017). Pertussis vaccination during pregnancy was therefore recommended in different countries, the World Health Organization (WHO) considering it as the most cost-effective additional strategy for preventing disease in young infants from birth until protection provided by the first infants immunizations (WHO 2015). The United States were the first to advise in 2011, that pertussis vaccine be administrated to pregnant women in the third trimester, and in 2012, this advice was updated to recommend vaccination in every pregnancy (CDC 2013). A number of countries further introduced maternal Tdap vaccination during pregnancy, starting by Argentina, followed by United Kingdom; Australia, Belgium, Spain (Campbell et al. 2018). There remains however considerable variation between national immunization recommendations. Some countries still recommend the administration of a vaccine booster soon after delivery even if there is now agreement that this cocooning strategy is costly, difficult to implement, and providing uncertain effectiveness (Blain et al. 2016).

The rationale for pertussis vaccination during pregnancy is to provide passive protection for newborn infants by *B. pertussis* antibodies transferred from mother to infant across the placenta, although there is no clear immunological correlate of protection for pertussis. The efficiency of antibody transfer through placenta is dependent on maternal antibody levels, placental function, absence of maternal co-infections that diminish transfer, and IgG subclass induced by vaccine antigens (Kachikis and Englund 2016). As *B. pertussis* antigens present in the acellular vaccines used for booster administration in adults are proteins, they induce IgG1 antibodies which are transported quite efficiently across the placenta, an active transport being mediated by the neonatal receptor for the constant region of immunoglobulin (FcRn) (Roopenian and Akilesh

2007). The concentration of PT-specific antibodies in the cord blood are higher when mothers are immunized during the second trimester or early in the third trimester of gestation as compared to mothers immunized later in the third trimester (Eberhardt et al. 2016; Abu Raya et al. 2015). The highest concentrations of anti-PT antibodies in neonates at birth were observed when the mothers were vaccinated within the window of 27 through 30 weeks of pregnancy whereas the antibody titers declined thereafter (Healy et al. 2018). In these conditions, anti-PT antibody concentrations remained detectable at a substantial level until the initiation of the primary vaccines series in infants, reducing the risk of pertussis-related mortality and morbidity.

The avidity of umbilical cord IgG is reported by some authors to be higher in case of maternal immunization at 27–30 weeks of gestation (Eberhardt et al. 2016), whereas others reported no difference in the pertussis specific antibody avidity when the women are immunized before 27 weeks until at 31–36 weeks of gestation (Maertens et al. 2015).

This recommended strategy has resulted in a 91% reduction in pertussis in infants 3 months or younger in the United Kingdom (Amirthalingam et al. 2014). There is however some concern about the possible impact of the maternal IgG antibodies on the early life immunity. Several studies indicate that high maternally derived pertussis antibody titers can have a suppressive effect on infant responses to primary immunization against pertussis, mostly in case of infant vaccination with the wPV. A smaller and transient inhibitory effect on infant antibody response against pertussis was in contrast observed in case of acellular pertussis vaccination of infants, and globally, the clinical significance of this blunting effect has not yet been assessed (Abu-Raya et al. 2017). The effect of maternally derived antibodies on specific cellular immune responses was only very little investigated in humans but studies performed in animals suggest that T-cell responses would be unaffected.

6   Conclusions

Pertussis outbreaks are recorded even in countries with high vaccination coverage. Resurgence of the disease could be attributed not only to insufficient vaccine uptake, but also to suboptimal protection and waning of vaccine-induced immunity. Data from mathematical modelling (Althouse and Scarpino 2015) and animal experimental models (Warfel et al. 2014) show that even though the aPV is capable of preventing serious symptoms, it does not prevent bacterial colonization (Warfel et al. 2016). Therefore, despite vaccination, people could still transmit the bacteria. This is a possible explanation for the continuing circulation of the pathogen in aPV using countries (Huygen et al. 2014; Palazzo et al. 2016b; Moriuchi et al. 2017).

Maternal immunization has proven safe and effective in limiting severe and deadly pertussis in young infants, thus it should be supported, especially during the outbreak period. Nevertheless, further research efforts are needed to fill knowledge gaps.

It is clear that improvements in aPVs or development of new approaches, like the mucosal administration of an attenuated *B. pertussis* strain, needs an enhanced understanding of the correlates of protection against the disease and of the mechanisms that could induce durable, highly effective immunity.

References

Aase A, Herstad TK, Jørgensen SB, Leegaard TM, Berbers G, Steinbakk M, Aaberge I (2014) Anti-pertussis antibody kinetics following DTaP-IPV booster vaccination in Norwegian children 7-8 years of age. Vaccine 32(45):5931–5936

Abu Raya B, Bamberger E, Almog M, Peri R, Srugo I, Kessel A (2015) Immunization of pregnant women against pertussis: the effect of timing on antibody avidity. Vaccine 33:1948–1952

Abu-Raya B, Edwards KM, Scheifele DW, Halperin SA (2017) Pertussis and influenza immunisation during pregnancy: a land- scape review. Lancet Infect Dis 17 (7):e209–e222

Acosta AM, DeBolt C, Tasslimi A, Lewis M, Stewart LK, Misegades LK, Messonnier NE, Clark TA, Martin SW, Patel M (2015) Tdap vaccine effectiveness in adolescents during the 2012 Washington State pertussis epidemic. Pediatrics 135(6):981–989. https://doi.org/10.1542/peds.2014-3358

Ad Hoc Group for the Study of Pertussis Vaccines (1988) Placebo-controlled trial of two acellular pertussis vaccines in Sweden--protective efficacy and adverse events. Lancet 1(8592):955–960

Althouse BM, Scarpino SV (2015) Asymptomatic transmission and the resurgence of Bordetella pertussis. BMC Med 13:146–158. https://doi.org/10.1186/s12916-015-0382-8

Amirthalingam G, Andrews N, Campbell H, Ribeiro S, Kara E, Donegan K, Fry NK, Miller E, Ramsay M (2014) Effectiveness of maternal pertussis vaccination in England: an observational study. Lancet 384 (9953):1521–1528. https://doi.org/10.1016/S0140-6736(14)60686-60683

Ausiello CM, Urbani F, la Sala A, Lande R, Cassone A (1997) Vaccine- and antigen-dependent type 1 and type 2 cytokine induction after primary vaccination of infants with whole-cell or acellular pertussis vaccines. Infect Immun 65(6):2168–2174

Ausiello CM, Lande R, Urbani F, la Sala A, Stefanelli P, Salmaso S, Mastrantonio P, Cassone A (1999) Cell-mediated immune responses in four-year-old children after primary immunization with acellular pertussis vaccines. Infect Immun 67:4064–4071

Bancroft T, Dillon MB, da Silva AR, Paul S, Peters B, Crotty S, Lindestam Arlehamn CS, Sette A (2016) Th1 versus Th2 T cell polarization by whole-cell and acellular childhood pertussis vaccines persists upon re-immunization in adolescence and adulthood. Cell Immunol 304–305:35–43

Baxter R, Barlett J, Fireman B, Lewis E, Klein NP (2017) Effectiveness of vaccination during pregnancy to prevent infant pertussis. Pediatrics 139(5):pii: e20164091. https://doi.org/10.1542/peds.2016-4091

Bentebibel SE, Lopez S, Obermoser G, Schmitt N, Mueller C, Harrod C, Flano E, Mejias A, Albrecht RA, Blankenship D, Xu H, Pascual V, Banchereau J, Garcia-Sastre A, Palucka AK, Ramilo O, Ueno H (2013) Induction of ICOS+CXCR3+CXCR5+ TH cells correlates with antibody responses to influenza vaccination. Sci Transl Med 5(176):176ra32. https://doi.org/10.1126/scitranslmed.3005191

Black S (1997) Epidemiology of pertussis. Pediatr Infect Dis J 16(Supplement):S85–S89

Blain AE, Lewis M, Banerjee E, Kudish K, Liko J, McGuire S, Selvage D, Watt J, Martin SW, Skoff TH (2016) An assessment of the cocooning strategy for preventing infant pertussis- United Stated, 2011. Clin Infect Dis 63(suppl 4):S221–S226

Bordet J, Gengou O (1906) Le microbe de la coqueluche. Ann Inst Pasteur (Paris) 2:731–741

Campbell H, Gupta S, Dolan G, Kapadia S, Kumar Singh A, Andrews N, Amirthalingam G (2018) Review of vaccination in pregnancy to prevent pertussis in early infancy. J Med Microbiol 67 (10):1426–1456

Campins-Martí M, Cheng HK, Forsyth K, Guiso N, Halperin S, Huang LM, Mertsola J, Oselka G, Ward J, Wirsing von König CH, Zepp F, International Consensus Group on Pertussis Immunisation International Consensus Group on Pertussis Immunisation (2001) Recommendations are needed for adolescent and adult pertussis immunization: rationale and strategies for consideration. Int Consensus Group Pertussis Immunisation Vaccine 20:641–646

Carollo M, Pandolfi E, Tozzi AE, Buisman AM, Mascart F, Ausiello CM (2014) Humoral and B-cell memory responses in children five years after pertussis acellular vaccine priming. Vaccine 32(18):2093–2099. https://doi.org/10.1016/j.vaccine.2014.02.005

Cassone A, Ausiello CM, Urbani F, Lande R, Giuliano M, La Sala A, Piscitelli A, Salmaso S (1997) Cell-mediated and antibody responses to Bordetella pertussis antigens in children vaccinated with acellular or whole-cell pertussis vaccines. The Progetto Pertosse-CMI Working Group. Arch Pediatr Adolesc Med 151:283–289

Cassone A, Mastrantonio P, Ausiello CM (2000) Are only antibody levels involved in the protection against pertussis in acellular pertussis vaccine recipients? J Infect Dis 182(5):1575–1577

Centers for Disease Control and Prevention (CDC) (2013) Updated recommendations for use of tetanus toxoid, reduced diphtheria toxoid, and acellular pertussis vaccine (Tdap) in pregnant women--Advisory Committee on Immunization Practices (ACIP), 2012. MMWR Morb Mortal Wkly Rep 62(7):131–135

Cherry JD (1996) Historical review of pertussis and the classical vaccine. J Infect Dis 174:S259–S263

Cherry JD, Brunnel P, Golden G (1988) Report of the task force on pertussis and pertussis immunization. Pediatrics 81:S933–S984

Cody CL, Baraff LJ, Cherry JD, Marcy SM, Manclark CR (1981) Nature and rates of adverse reactions associated with DTP and DT immunizations in infants and children. Pediatrics 68:650–660

Cromer BA, Goydos J, Hackell J, Mezzatesta J, Dekker C, Mortimer EA (1993) Unrecognized pertussis infection in adolescents. Am J Dis Child 147:575–577

Crotty S (2014) Follicular helper cell differentiation, function, and roles in disease. Immunity 41:529–542

da Silva AR, Babor M, Carpenter C, Khalil N, Cortese M, Mentzer AJ, Seumois G, Petro CD, Purcell LA, Vijayanand P, Crotty S, Pulendran B, Peters B, Sette AJ (2018) Th1/Th17 polarization persists following whole-cell pertussis vaccination despite repeated acellular boosters. Clin Invest 128(9):3853–3865

de Rond L, Schure RM, Öztürk K, Berbers G, Sanders E, van Twillert I, Carollo M, Mascart F, Ausiello CM, van Els CA, Smits K, Buisman AM (2015) Identification of pertussis specific effector memory T-cells in preschool children. Clin Vaccine Immunol 22:561–569

Dirix V, Verscheure V, Goetghebuer T, Hainaut M, Debrie AS, Locht C, Mascart F (2009) Cytokine and antibody profiles in 1-year-old children vaccinated with either acellular or whole-cell pertussis vaccine during infancy. Vaccine 27(43):6042–6047. https://doi.org/10.1016/j.vaccine.2009.07.075

Dirix V, Verscheure V, Vermeulen F, De Schutter I, Goetghebuer T, Locht C, Mascart F (2012) Both CD4 (+) and CD8(+) lymphocytes participate in the IFN-gamma response to filamentous hemagglutinin from Bordetella pertussis in infants, children, and adults. Clin Dev Immunol 2012:795958. https://doi.org/10.1155/2012/795958

Eberhardt CS, Blanchard-Rohner G, Lemaitre B, Boukrid M, Combecure C, Othenin-Girard V, Chilin A, Petre J, de Tejada BM, Siegrist CA (2016) Maternal immunization earlier in pregnancy maximizes antibody transfer and expected infant seropositivity against pertussis. Clin Infect Dis 62:829–836

Edelman KJ, He Q, Makinen JP, Haanpera MS, Tran Minh NN, Schuerman L, Wolter J, Mertsola JA (2004) Pertussis-specific cell-mediated and humoral immunity in adolescents 3 years after booster immunization with acellular pertussis vaccine. Clin Infect Dis 39 (2):179–185

Edelman K, He Q, Mäkinen J, Sahlberg A, Haanperä M, Schuerman L, Wolter J, Mertsola J (2007) Immunity to pertussis 5 years after booster immunization during adolescence. J Clin Infect Dis 44(10):1271–1277

Edwards KM, Berbers GAM (2014) Immune responses to pertussis vaccines and disease. J Infect Dis 209:S10–S15

Edwards KM, Meade BD, Decker MD, Reed GF, Rennels MB, Steinhoff MC, Anderson EL, Englund JA, Pichichero ME, Deloria MA (1995) Comparison of 13 acellular pertussis vaccines: overview and serologic response. Pediatrics 96(3 Pt 2):548–557

Farooq F, Beck K, Paolino KM, Phillips R, Waters NC, Regules JA, Bergmann-Leitner ES (2016) Circulating follicular T helper cells and cytokine profile in humans following vaccination with the rVSV-ZEBOV Ebola vaccine. Sci Rep 6:27944. https://doi.org/10.1038/srep27944

Gangarosa EJ, Galazka AM, Wolfe CR, Phillips LM, Gangarosa RE, Miller E, Chen RT (1998) Impact of anti-vaccine movements on pertussis control: the untold story. Lancet 351:356–361

Giuliano M, Mastrantonio P, Giammanco A, Piscitelli A, Salmaso S, Wassilak SG (1998) Antibody responses and persistence in the two years after immunization with two acellular vaccines and one whole-cell vaccine against pertussis. J Pediatr 132(6):983–988

Gonfiantini MV, Carloni E, Gesualdo F, Pandolfi E, Agricola E, Rizzuto E, Iannazzo S, Ciofi Degli Atti ML, Villani A, Tozzi AE (2014) Epidemiology of pertussis in Italy: disease trends over the last century. Euro Surveill 19(40):20921–20929

Greco D, Salmaso S, Mastrantonio P, Giuliano M, Tozzi AE, Anemona A, Ciofi Degli Atti ML, Giammanco A, Panei P, Blackwelder WC, Klein DL, Wassilak SG (1996) A controlled trial of two acellular vaccines and one whole-cell vaccine against pertussis. Progetto Pertosse Working Group. N Engl J Med 334(6):341–348

Gustafsson L, Hallander HO, Olin P, Reizenstein E, Storsaeter J (1996) A controlled trial of a two-component acellular, a five-component acellular, and a whole-cell pertussis vaccine. N Engl J Med 334 (6):349–355

Hallander HO, Gustafsson L, Ljungman M, Storsaeter J (2005) Pertussis antitoxin decay after vaccination with DTPa: Response to a first booster dose 3 1/2 – 6 1/2 years after the third vaccine dose. Vaccine 23 (46–47):5359–5364

Hallander HO, Ljungman M, Storsaeter J, Gustafsson L (2009) Kinetics and sensitivity of ELISA IgG pertussis antitoxin after infection and vaccination with Bordetella pertussis in young children. APMIS 117:797–807

Halperin SA (2001) Pertussis immunization for adolescents: what are we waiting for? Paediatr Child Health 6(4):184–186

Halperin SA, Langley JM, Ye L, MacKinnon-Cameron D, Elsherif M, Allen VM, Smith B, Halperin BA, McNeil SA, Vanderkooi OG, Dwinnell S, Wilson RD, Tapiero B, Boucher M, Le Saux N, Gruslin A, Vaudry W, Chandra S, Dobson S, Money D (2018) A randomized controlled trial of the safety and immunogenicity of tetanus, diphtheria, and acellular pertussis vaccine immunization during pregnancy and subsequent infant immune response. Clin Infect Dis 67(7):1063–1071. https://doi.org/10.1093/cid/ciy244

He Q, Viljanen MK, Nikkari S, Lyytikäinen R, Mertsola J (1994) Outcomes of Bordetella pertussis infection in different age groups of an immunized population. J Infect Dis 17:873–877

He Q, Tran Minh NN, Edelman K, Viljanen MK, Arvilommi H, Mertsola J (1998) Cytokine mRNA expression and proliferative responses induced by pertussis toxin, filamentous hemagglutinin, and pertactin of Bordetella pertussis in the peripheral blood mononuclear cells of infected and immunized school children and adults. Infect Immun 66:3796–3801

Healy CM, Rench MA, Swaim LS, Smith O, Sangi-Haghpeykar H, Mathis MH, Martin MD, Baker CJ (2018) Association between third-trimester Tdap immunization and neonatal pertussis antibody concentrations. JAMA 320(14):1464–1470

Heininger U, André P, Chlibek R, Kristufkova Z, Kutsar K, Mangarov A, Mészner Z, Nitsch-Osuch A, Petrović V, Prymula R, Usonis V, Zavadska D (2016) Comparative epidemiologic characteristics of pertussis in 10 Central and Eastern European Countries, 2000-2013. PLoS One 11(6):e0155949. https://doi.org/10.1371/journal.pone.0155949

Hendrikx LH, Oztürk K, de Rond LG, Veenhoven RH, Sanders EA, Berbers GA, Buisman AM (2011) Identifying long-term memory B-cells in vaccinated children despite waning antibody levels specific for Bordetella pertussis proteins. Vaccine 29 (7):1431–1437

Huang LM, Lee CY, Lin TY, Chen JM, Lee PI, Hsu CY (1996) Responses to primary and a booster dose of acellular, component, and whole-cell pertussis vaccines initiated at 2 months of age. Vaccine 14 (9):916–922

Huygen K, Rodeghiero C, Govaerts D, Leroux-Roels I, Melin P, Reynders M, Van Der Meeren S, Van Den Wijngaert S, Pierard D (2014) Bordetella pertussis seroprevalence in Belgian adults aged 20-39 years, 2012. Epidemiol Infect 142(4):724–728

Jahnmatz M, Ljungman M, Netterlid E, Jenmalm MC, Nilsson L, Thorstensson R (2014) Pertussis-specific memory B-cell and humoral IgG responses in adolescents after a fifth consecutive dose of acellular pertussis vaccine. Clin Vaccine Immunol 21 (9):1301–1308. https://doi.org/10.1128/CVI.00280-14

Jefferson T, Rudin M, DiPietrantonj C (2003) Systematic review of the effects of pertussis vaccines in children. Vaccine 21:2003–2014

Kachikis A, Englund JA (2016) Maternal immunization: optimizing protection for the mother and infants. J Inf Secur 72(suppl):S83–S90

Klein NP, Bartlett J, Fireman B, Rowhani-Rahbar A, Baxter R (2013) Comparative effectiveness of acellular versus whole-cell pertussis vaccines in teenagers. Pediatrics 131:e1716–e1722. https://doi.org/10.1542/peds.2012-3836

Koepke R, Eickhoff JC, Ayele RA, Petit AB, Schauer SL, Hopfensperger DJ, Conway JH, Davis JP (2014) Estimating the effectiveness of tetanus-diphtheria-acellular pertussis vaccine (Tdap) for preventing pertussis: evidence of rapidly waning immunity and difference in effectiveness by Tdap brand. J Infect Dis 210:942–953

Kumar NP, Sridhar R, Hanna LE, Banurekha VV, Nutman TB, Babu S (2014) Decreased frequencies of circulating CD4+ T follicular helper cells associated with diminished plasma IL-21 in active pulmonary tuberculosis. PLoS One 9:e111098. https://doi.org/10.1371/journal.pone.0111098

Lapidot R, Gill CJ (2016) The 2016 pertussis resurgence: putting together the pieces of the puzzle. Trop Dis Travel Med Vaccines 2:26. https://doi.org/10.1186/s40794-016-0043-8. eCollection

Le T, Cherry JD, Chang SJ, Knoll MD, Lee ML, Barenkamp S, Bernstein D, Edelman R, Edwards KM, Greenberg D, Keitel W, Treanor J, Ward JI, APERT Study (2004) Immune responses and antibody decay after immunization of adolescents and adults with an acellular pertussis vaccine: the APERT Study. J Infect Dis 190:535–544

Locci MH-DC, Landais E, Wu J, Kroenke MA, Arlehamn CL, Su LF, Cubas R, Davis MM, Sette A, Haddad EK, International AIDS Vaccine Initiative Protocol C Principal Investigators, Poignard P, Crotty S (2013) Human circulating PD-1+CXCR3-CXCR5+ memory Tfh cells are highly functional and correlate with broadly neutralizing HIV antibody responses. Immunity 39:758–769

Long SS, Welkon CJ, Clark JL (1990) Widespread silent transmission of pertussis in families: antibody correlates of infection and symptomatology. J Infect Dis 161:480–486

MacLennan IC, Toellner KM, Cunningham AF, Serre K, Sze DM, Zúñiga E, Cook MC, Vinuesa CG (2003) Extrafollicular antibody responses. Immunol Rev 194:8–18

Maertens K, Hoang TH, Caboré RN, Leuridan E (2015) Avidity of maternal pertussis antibodies after vaccination during pregnancy. Vaccine 33:5489

Marcellini V, Piano Mortari E, Fedele G, Gesualdo F, Pandolfi E, Midulla F, Leone P, Stefanelli P, Tozzi AE, Carsetti R, Pertussis Study Group (2017) Protection against pertussis in humans correlates to elevated serum antibodies and memory B cells. Front Immunol 8:1158

Mascart F, Verscheure V, Malfroot A, Hainaut M, Piérard D, Temerman S, Peltier A, Debrie AS, Levy J, Del Giudice G, Locht C (2003) Bordetella pertussis infection in 2-month-old infants promotes type 1 T cell responses. J Immunol 170:1504–1509

Mascart F, Hainaut M, Peltier A, Verscheure V, Levy J, Locht C (2007) Modulation of the infant immune responses by the first pertussis vaccine administrations. Vaccine 25(2):391–398

Mascart F, DIrix V, Locht C (2018) Chapter 7 The human immune responses to pertussis and pertussis vaccines. In: Rohani P, Scarpino S (eds) Pertussis: epidemiology, immunology, and evolution. Oxford University Press, Oxford

Mertsola J, Ruuskanen O, Eerola E, Viljanen MK (1983) Intrafamilial spread of pertussis. J Pediatr 103:359–363

Miller DL, Ross EM, Alderslade R, Bellman MH, Rawson NS (1981) Pertussis immunisation and serious acute neurological illness in children. Br Med J (Clin Res Ed) 282(6276):1595–1599

Morita R, Schmitt N, Bentebibel SE, Ranganathan R, Bourdery L, Zurawski G, Foucat E, Dullaers M, Oh S, Sabzghabaei N, Lavecchio EM, Punaro M, Pascual V, Banchereau J, Ueno H (2011) Human blood CXCR5(+)CD4(+) T cells are counterparts of T follicular cells and contain specific subsets that differentially support antibody secretion. Immunity 34:108–121

Moriuchi T, Otsuka N, Hiramatsu Y, Shibayama K, Kamachi K (2017) A high seroprevalence of antibodies to pertussis toxin among Japanese adults: qualitative and quantitative analyses. PLoS One 12(7):e0181181. https://doi.org/10.1371/journal.pone.0181181

Nelson JD (1978) The changing epidemiology of pertussis in young infants: the role of adults as reservoirs of infection. Am J Dis Child 132:371–373

Obeng-Adjei N, Portugal S, Tran TM, Yazew TB, Skinner J, Li S, Jain A, Felgner PL, Doumbo OK, Kayentao K, Ongoiba A, Traore B, Crompton PD (2015) Circulating Th1-cell-type Tfh cells that exhibit impaired B cell help are preferentially activated during acute malaria in children. Cell Rep 13:425–439

Palazzo R, Carollo M, Bianco M, Fedele G, Schiavoni I, Pandolfi E, Villani A, Tozzi AE, Mascart F, Ausiello CM (2016a) Persistence of T-cell immune response induced by two acellular pertussis vaccines in children five years after primary vaccination. New Microbiol 39 (1):35–47

Palazzo R, Carollo M, Fedele G, Rizzo C, Rota MC, Giammanco A, Iannazzo S, Ausiello CM (2016b) Sero-Epidemiology Working Group. Evidence of increased circulation of Bordetella pertussis in the Italian adult population from seroprevalence data (2012-2013). J Med Microbiol 65(7):649–657

Pallikkuth S, Parmigiani A, Silva SY, George VK, Fischl M, Pahwa R, Pahwa S (2012) Impaired peripheral blood T-follicular helper cell function in HIV-infected nonresponders to the 2009 H1N1/09 vaccine. Blood 120:985–993. https://doi.org/10.1182/blood-2011-12-396648

Pereira A, Pietro Pereira AS, Silva CL, de Melo RG, Lebrun I, Sant'Anna OA, Tambourgi DV (2010) Antibody response from whole-cell pertussis vaccine immunized Brazilian children against different strains of Bordetella pertussis. Am J Trop Med Hyg 82:678–682

Pichichero ME (1996) Acellular pertussis vaccines. Towards an improved safety profile. Drug Saf 15 (5):311–324

Pichichero ME (2009) Booster vaccinations: can immunologic memory outpace disease pathogenesis? Pediatrics 124:1633–1641

Plotkins SA (2013) Complex correlates of protection after vaccination. Clin Infect Dis 56:1458–1465

Rieber N, Graf A, Belohradsky BH, Hartl D, Urschel S, Riffelmann M, Wirsing von König CH, Liese J (2008) Differences of humoral and cellular immune response to an acellular pertussis booster in adolescents with a whole cell or acellular primary vaccination. Vaccine 26:6929–6935

Rieber N, Graf A, Hartl D, Urschel S, Belohradsky BH, Liese J (2011) Acellular pertussis booster in adolescents induces Th1 and memory CD8+ T cell immune response. PLoS One 6:e17271. https://doi.org/10.1371/journal.pone.0017271

Romanus V, Jonsell R, Bergquist SO (1987) Pertussis in Sweden after the cessation of general immunization in 1979. Pediatr Infect Dis J 6:364–371

Roopenian DC, Akilesh S (2007) FcRn: the neonatal fc receptor comes of age. Nat Rev Immunol 7 (9):715–725

Ross PJ, Sutton CE, Higgins S, Allen AC, Walsh K, Misiak A, Lavelle EC, McLoughlin RM, Mills KH (2013) Relative contribution of Th1 and Th17 cells in adaptive immunity to Bordetella pertussis: towards the rational design of an improved acellular pertussis vaccine. PLoS Pathog 9:e1003264

Ryan M, Murphy G, Ryan E, Nilsson L, Shackley F, Gothefors L, Oymar K, Miller E, Storsaeter J, Mills KH (1998) Distinct T-cell subtypes induced with whole cell and acellular pertussis vaccines in children. Immunology 93:1–10

Ryan EJ, Nilsson L, Kjellman N, Gothefors L, Mills KH (2000) Booster immunization of children with an acellular pertussis vaccine enhances Th2 cytokine production and serum IgE responses against pertussis toxin but not against common allergens. Clin Exp Immunol 121(2):193–200

Salmaso S, Mastrantonio P, Wassilak SG, Giuliano M, Anemona A, Giammanco A, Tozzi AE, Ciofi Degli Atti ML, Greco D (1998) Persistence of protection through 33 months of age provided by immunization in infancy with two three-component acellular pertussis vaccines. Stage II Working Group. Vaccine 16 (13):1270–1275

Sato Y, Kimura M, Fukumi H (1984) Development of a pertussis component vaccine in Japan. Lancet 1:122–126

Schure RM, Hendrikx LH, de Rond LG, Oztürk K, Sanders EA, Berbers GA, Buisman AM (2012a) T-cell responses before and after the fifth consecutive acellular pertussis vaccination in 4-year-old Dutch children. Clin Vaccine Immunol 19:1879–1886

Schure RM, de Rond L, Ozturk K, Hendrikx L, Sanders E, Berbers G, Buisman AM (2012b) Pertussis circulation has increased T-cell immunity during childhood more than a second acellular booster vaccination in Dutch children 9 years of age. PLoS One 7:e41928. https://doi.org/10.1371/journal.pone.0041928

Schure RM, Hendrikx LH, de Rond LG, Oztürk K, Sanders EA, Berbers GA, Buisman AM (2013) Differential T- and B-cell responses to pertussis in acellular vaccine-primed versus whole-cell vaccine-primed children 2 years after preschool acellular booster vaccination. Clin Vaccine Immunol 20(9):1388–1395

Sharma SK, Pichichero ME (2012) Functional deficits of pertussis-specific CD4+ T cells in infants compared to adults following DTaP vaccination. Clin Exp Immunol 169:281–291

Simondon F, Preziosi MP, Yam A, Kane CT, Chabirand L, Iteman I, Sanden G, Mboup S, Hoffenbach A, Knudsen K et al (1997) A randomized double-blind trial comparing a two-component acellular to a whole-cell pertussis vaccine in Senegal. Vaccine 15 (15):1606–1612

Slight SR, Rangel-Moreno J, Gopal R, Lin Y, Fallert Junecko BA, Mehra S, Selman M, Becerril-Villanueva E, Baquera-Heredia J, Pavon L, Kaushal D, Reinhart TA, Randall TD, Khader SA (2013) CXCR5+ T helper cells mediate protective immunity against tuberculosis. J Clin Invest 123:712–726

Smits K, Pottier G, Smet J, Dirix V, Vermeulen F, De Schutter I, Carollo M, Locht C, Ausiello CM, Mascart F (2013) Different T cell memory in preadolescents after whole-cell or acellular pertussis vaccination. Vaccine 32:111–118

Spensieri F, Borgogni E, Zedda L, Bardelli M, Buricchi F, Volpini G, Fragapane E, Tavarini S, Finco O, Rappuoli R, Del Giudice G, Galli G, Castellino F (2013) Human circulating influenza-CD4+ ICOS1 +IL-21+ T cells expand after vaccination, exert helper function, and predict antibody responses. Proc Natl Acad Sci U S A 110:14330–14335

Steinhoff MC, Reed GF, Decker MD, Edwards KM, Englund JA, Pichichero ME, Rennels MB, Anderson EL, Deloria MA, Meade BD (1995) A randomized comparison of reactogenicity and immunogenicity of two whole-cell pertussis vaccines. Pediatrics 96(3 Pt 2):567–570

Storsaeter J, Hallander HO, Gustafsson L, Olin P (1998) Levels of anti-pertussis antibodies related to protection after household exposure to Bordetella pertussis. Vaccine 16:1907–1916

Taranger J, Trollfors B, Lagergård T, Sundh V, Bryla DA, Schneerson R, Robbins JB (2000) Correlation between pertussis toxin IgG antibodies in postvaccination sera and subsequent protection against pertussis. J Infect Dis 181(3):1010–1013

Tran Minh NN, He Q, Ramalho A, Kaufhold A, Viljanen MK, Arvilommi H, Mertsola J (1999) Acellular vaccines containing reduced quantities of pertussis antigens as a booster in adolescents. Pediatrics 104:e70

Trollfors B, Taranger J, Lagergård T, Lind L, Sundh V, Zackrisson G, Lowe CU, Blackwelder W, Robbins JB (1995) A placebo-controlled trial of a pertussis-toxoid vaccine. N Engl J Med 333(16):1045–1050

Ueno H (2016) T follicular helper cells in human autoimmunity. Curr Opin Immunol 43:24–31

van der Lee S, Hendrikx LH, Sanders EAM, Berbers GAM, Buisman AM (2018a) Whole-cell or acellular pertussis primary immunizations in infancy determines adolescent cellular immune profiles. Front Immunol 9:51

van der Lee S, Sanders EAM, Berbers GAM, Buisman AM (2018b) Whole-cell or acellular pertussis vaccination in infancy determines IgG subclass profiles to DTaP booster vaccination. Vaccine 36(2):220–226

van der Lee S, van Rooijen DM, de Zeeuw-Brouwer ML, Bogaard MJM, van Gageldonk PGM, Marinovic AB, Sanders EAM, Berbers GAM, Buisman AM (2018c) Robust humoral and cellular immune responses to pertussis in adults after a first acellular booster vaccination. Front Immunol 9:681

Van Hoek AJ, Campbell H, Amirthalingam G, Andrews N, Miller E (2013) The number of deaths among infants under oe year of age in England with pertussis: results of a capture/recapture analysis for the period 2001 to 2011. Euro Surveill 18(9):pii: 20414

van Twillert I, Han WG, van Els CA (2015) Waning and aging of cellular immunity to Bordetella pertussis. Pathog Dis 73(8):ftv071

Vermeulen F, Dirix V, Verscheure V, Damis E, Vermeylen D, Locht C, Mascart F (2013) Persistence at one year of age of antigen-induced cellular immune responses in preterm infants vaccinated against whooping cough: comparison of three different vaccines and effect of a booster dose. Vaccine 31:1981–1986

Voysey M, Kandasamy R, Yu LM, Baudin M, Sadorge C, Thomas S, John T, Pollard AJ (2016) The predicted persistence and kinetics of antibody decline 9 years after pre-school booster vaccination in UK children. Vaccine 34(35):4221–4228. https://doi.org/10.1016/j.vaccine.2016.06.051

Warfel JM, Merkel TJ (2013) Bordetella pertussis infection induces a mucosal IL-17 response and long-lived Th17 and Th1 immune memory cells in nonhuman primates. Mucosal Immunol 6(4):787–796

Warfel JM, Zimmerman LI, Merkel TJ (2014) Acellular pertussis vaccines protect against disease but fail to prevent infection and transmission in a nonhuman primate model. Proc Natl Acad Sci U S A 111:787–792. https://doi.org/10.1073/pnas.1314688110

Warfel JM, Zimmerman LI, Merkel TJ (2016) Comparison of three whole-cell pertussis vaccines in the baboon model of pertussis. Clin Vaccine Immunol 23 (1):47–54. https://doi.org/10.1128/CVI.00449-15

Weston R (2012) Whooping cough: a brief history to the 19th century. Can Bull Med Hist 29:329–349

WHO Position paper on Vaccines against Pertussis – September 2015. www.who.int/immunization/documents/positionpapers

Witt MA, Arias L, Katz PH, Truong ET, Witt DJ (2013) Reduced risk of pertussis among persons ever vaccinated with whole cell pertussis vaccine compared to recipients of acellular pertussis vaccines in a large US cohort. Clin Infect Dis 56(9):1248–1254

Zepp F, Knuf M, Habermehl P, Schmitt JH, Rebsch C, Schmidtke P, Clemens R, Slaoui M (1996) Pertussis-specific cell-mediated immunity in infants after vaccination with a tricomponent acellular pertussis vaccine. Infect Immun 64:4078–4084

Zepp F, Heininger U, Mertsola J, Bernatowska E, Guiso N, Roord J, Tozzi AE, Van Damme P (2011) Rationale for pertussis booster vaccination throughout life in Europe. Lancet Infect Dis 11:557–570

Adv Exp Med Biol - Advances in Microbiology, Infectious Diseases and Public Health (2019) 1183: 115–126
https://doi.org/10.1007/5584_2019_407
© Springer Nature Switzerland AG 2019
Published online: 21 August 2019

New Pertussis Vaccines: A Need and a Challenge

Daniela Hozbor

Abstract

Effective diphtheria, tetanus toxoids, whole-cell pertussis (wP) vaccines were used for massive immunization in the 1950s. The broad use of these vaccines significantly reduced the morbidity and mortality associated with pertussis. Because of reports on the induction of adverse reactions, less-reactogenic acellular vaccines (aP) were later developed and in many countries, especially the industrialized ones, the use of wP was changed to aP. For many years, the situation of pertussis seemed to be controlled with the use of these vaccines, however in the last decades the number of pertussis cases increased in several countries. The loss of the immunity conferred by the vaccines, which is faster in the individuals vaccinated with the acellular vaccines, and the evolution of the pathogen towards geno/phenotypes that escape more easily the immunity conferred by the vaccines were proposed as the main causes of the disease resurgence. According to their composition of few immunogens, the aP vaccines seem to be exerting a greater selection pressure on the circulating bacterial population causing the prevalence of bacterial isolates defective in the expression of vaccine antigens. Under this context, it is clear that new vaccines against pertussis should be developed. Several vaccine candidates are in preclinical development and few others have recently completed phaseI/phaseII trials. Vaccine candidate based on OMVs is a promising candidate since appeared overcoming the major weaknesses of current aP-vaccines. The most advanced development is the live attenuated-vaccine BPZE1 which has successfully completed a first-in-man clinical trial.

Keywords

Acellular vaccine · *Bordetella pertussis* · Epidemiology · Pertussis · Whole-cell vaccines

1 Current Pertussis Vaccines

Pertussis, also known as whooping cough, is a highly contagious respiratory disease mainly caused by *Bordetella pertussis*, a Gram-negative bacterium. This disease that causes uncontrollable violent coughing, affects all ages, being the most vulnerable the infants under 6 months of age (Stefanelli et al. 2017). The best way to prevent pertussis is to get vaccinated. The first experimentations with vaccines began after Jules Bordet and Octave Gengou of the Pasteur Institute of Brussels identified the etiological agent in 1906; these vaccines were made from killed whole-cell

D. Hozbor (✉)
Laboratorio VacSal. Instituto de Biotecnología y Biología Molecular, Departamento de Ciencias Biológicas, Facultad de Ciencias Exactas, Universidad Nacional de La Plata y CCT-La Plata, CONICET, La Plata, Argentina
e-mail: hozbor@biol.unlp.edu.ar

B. pertussis. In ensuing years, such type of vaccine (whole-cell vaccine, wP) was used in children in different countries. Thorvald Madsen was the first to describe the use of a wP vaccine on a large scale (Madsen 1933). Madsen's vaccine successfully controlled two outbreaks in the Faroe Islands, however some deaths within 48 h of immunization were reported (Madsen 1933). Noteworthy at that time physicians used the vaccine as either a therapeutic or a prophylactic formulation and in both cases the vaccine was given in three injections intramuscularly or subcutaneously with intervals of three to 4 days (Madsen 1933). Madsen T. in his work summarized some reports that concluded sic...*if the vaccine is given early in the catarrhal stage the vaccine will have a good effect; the later the vaccine is given in the convulsive stage, the less effect can be expected. This appears from the reports of most of the Danish officers of Health and also is the consensus of the Danish pediatric society* (Madsen 1933). Louis Sauer of Northwestern University Medical School, Chicago, described minor reactions to a whole-cell pertussis vaccine being used in the United States as an adjuvanted combined vaccine (Sauer 1948). Pearl Kendrick of the State of Michigan Health Department further refined wP vaccines. She and Grace Eldering combined this improved killed vaccine with diphtheria and tetanus toxoids to produce the diphtheria-tetanus-pertussis (DTP) and used it in children (Kendrick 1936). The Committee on Infectious Diseases of the American Academy of Pediatrics suggested in 1944 and recommended in 1947 the routine use of pertussis vaccine in the form of the DTP combination. The use of this vaccine was then expanded to other countries. The coverages of pertussis vaccine were improved when the Expanded Program on Immunization (EPI) was established in 1974. The mission of the EPI is to develop and expand immunization programs throughout the world. In particular, in 1977, the goal was set to make immunization against diphtheria, pertussis, tetanus, poliomyelitis, measles and tuberculosis available to every child in the world by 1990. The massive pertussis vaccination dramatically reduced the morbidity and mortality associated with the disease (Table 1). After this important achievement in the control of the disease, unfortunately, doubts about the safety of

wP vaccines began to arise and this led to a decrease in the acceptance of this type of formulation by the population and even in some countries its use was rejected (Klein 2014; Romanus et al. 1987). The first published reports on irreversible brain damage after whole-cell pertussis vaccination was described by Brody and Sorley. These reports led to the first warnings that pertussis vaccine should not be administered to those with a known neurologic disorder (Brody and Sorley 1947). In Great Britain, concerns on the safety of this vaccine were widely publicized in the popular press and because of that the proportion of children vaccinated against pertussis diminished (Kulenkampff et al. 1974). The adverse reactions ranged from local reactions (redness, swelling, and pain at the injection site) to systemic reactions (fever, persistent crying and, in rare cases encephalopathy) were reported in other countries (Klein 2014; Romanus et al. 1987). Concerns about safety finally led to the development of component (acellular) pertussis vaccines that are associated with a lower frequency of adverse reactions (Sato and Sato 1985; Edwards and Karzon 1990). These second-generation of pertussis vaccines, referred to as aP vaccines, are constituted of purified *B. pertussis* antigens combined with diphtheria and tetanus toxoids. The first acellular vaccine that was developed in Japan in 1970 consisted of two proteins: pertussis toxin (PTx) and filamentous haemagglutinin (FHA) (Sato and Sato 1985). Field trials showed that component vaccine was as effective as and produced less side-effects than did conventional whole-cell vaccine (Sato et al. 1984). The vaccine has been used for mass immunization in Japan since 1981 and was highly effective in preventing pertussis disease. In 1994 the efficacy for two, three-component acellular, pertussis vaccines containing inactivated PTx, FHA, and pertactin (PRN), and one five-component acellular pertussis vaccine containing the same components plus fimbriae 2 and 3 was compared with a UK whole-cell vaccine (Olin et al. 1997). This study demonstrated that the wP vaccine and the five-component aP vaccine had similar efficacy against culture-confirmed typical pertussis, defined by at least 21 days of paroxysmal cough. The authors also found that the three-component acellular vaccine was less effective than the five-

Table 1 Number of reported cases of pertussis and type of pertussis vaccine used in different regions of the world (data extracted from WHO public information)

	1980	2000	2017	Percentage of countries that use whole cell or acellular pertussis vaccines in the primary doses
African Region (Algeria, Angola, Benin, Botswana, Burkina Faso, Burundi, Cameroon, Cape Verde, Central African Republic, Chad, Comoros, Congo, Côte d'Ivoire, Democratic Republic of the Congo, Equatorial Guinea, Eritrea, Eswatini, Ethiopia, Gabon, Gambia, Ghana, Guinea, Guinea-Bissau, Kenya, Lesotho, Liberia, Madagascar, Malawi, Mali, Mauritania, Mauritius, Mozambique, Namibia, Niger, Nigeria, Rwanda, Sao Tome and Principe, Senegal, Seychelles, Sierra Leone, South Africa, South Sudan, Togo, Uganda, United Republic of Tanzania, Zambia, Zimbabwe)	367,961	52,008	7082	
Region of the Americas (Antigua and Barbuda, Argentina, Bahamas, Barbados, Belize, Bolivia, Brazil, Canada, Chile, Colombia, Costa Rica, Cuba, Dominica, Dominican Republic, Ecuador, El Salvador, Grenada, Guatemala, Guyana, Haiti, Honduras, Jamaica, Mexico, Nicaragua, Panama, Paraguay, Peru, Saint Kitts and Nevis, Saint Lucia, Saint Vincent and the Grenadines, Suriname, Trinidad and Tobago, United States of America, Uruguay, Venezuela)	123,734	18,888	10,237	
Eastern Mediterranean Region (Afghanistan, Bahrain, Djibouti, Egypt, Iran (Islamic Republic of), Iraq, Jordan, Kuwait, Lebanon, Libyan Arab Jamahiriya, Morocco, Oman, Pakistan, Qatar, Saudi Arabia, Somalia, Sudan, Syrian Arab Republic, Tunisia, United Arab Emirates, Yemen)	171,631	2112	2012	
European Region (Albania, Andorra, Armenia, Austria, Azerbaijan, Belarus, Belgium, Bosnia and Herzegovina, Bulgaria, Croatia, Cyprus, Czech Republic, Denmark, Estonia, Finland, France, Georgia, Germany, Greece, Hungary, Iceland, Ireland, Israel, Italy, Kazakhstan, Kyrgyzstan, Latvia, Lithuania, Luxembourg, Malta, Monaco, Montenegro, Netherlands, Norway, Poland, Portugal, Republic of Moldova, Romania, Russian Federation, San Marino, Serbia, Slovakia, Slovenia, Spain, Sweden, Switzerland, Tajikistan, The former Yugoslav Republic of Macedonia, Turkey, Turkmenistan, Ukraine, United Kingdom of Great Britain and Northern Ireland, Uzbekistan)	90,546	53,675	63,037	

(continued)

Table 1 (continued)

	1980	2000	2017	Percentage of countries that use whole cell or acellular pertussis vaccines in the primary doses
South-East Asia Region (Bangladesh, Bhutan, Democratic People's Republic of Korea, India, Indonesia, Maldives, Myanmar, Nepal, Sri Lanka, Thailand, Timor-Leste)	399,310	38,510	33,976	
Western Pacific Region (Australia, Brunei Darussalam, Cambodia, China, Cook Islands, Fiji, Japan, Kiribati, Lao People's Democratic Republic, Malaysia, Marshall Islands, Micronesia (Federated States of), Mongolia, Nauru, New Zealand, Niue, Palau, Papua New Guinea, Philippines, Republic of Korea, Samoa, Singapore, Solomon Islands, Tonga, Tuvalu, Vanuatu, Viet Nam)	829,173	25,282	27,624	

component-vaccine and the whole-cell vaccines against culture-confirmed pertussis when all cases irrespective of the duration of severity of cough, were included in the analysis (Olin et al. 1997). Thus, though there was no compelling evidence to support that wP vaccines should not be used, the aP vaccines began to be broadly accepted because of their lower reactogenicity, especially in industrialized countries where wP vaccines of the primary series (3 doses in infancy) was replaced by aP vaccine (Table 1). Currently, US and most of the EU countries use only aP vaccines (Table 1). The aP formulations restored people's confidence in *pertussis*-containing vaccines, and the infection was controlled for several years. Notwithstanding, during the last decades the epidemiology of pertussis has changed (Clark 2014; Tan et al. 2015) with several major outbreaks occurring, the incidence of which not only indicated a waning immunity but also demonstrated that the wP vaccines gave children a longer lasting immunity than aP (Klein et al. 2013; Witt et al. 2012; Sheridan et al. 2012). Furthermore, the risk of pertussis was increased in schoolchildren and adolescents vaccinated exclusively with aP compared to those receiving at least one wP dose (Witt et al. 2013; Sheridan et al. 2012). This difference could result from the weaker immune response induced by aP

vaccines (Mills et al. 2014): while aP vaccines mainly induce a Th2-skewed response (Ryan et al. 1998), wP vaccines induce a robust Th1 profile and the proliferation of respiratory tissue-resident memory CD4 T cells (Brummelman et al. 2015; Wilk and Mills 2018). Therefore, the aP vaccine induced immunity shows a more rapid decay and possibly a reduced impact on transmission compared with currently available wP vaccines (Tartof et al. 2013; McGirr et al. 2013). In addition to the waning of immunity induced by vaccination, in particular with aP vaccines (Koepke et al. 2014; McGirr and Fisman 2015), pathogen adaptation to escape vaccine induced immunity (King et al. 2001; Mooi et al. 2001; Mäkelä 2000; David et al. 2004; He et al. 2003; Bottero et al. 2007; Gzyl et al. 2004; Bowden et al. 2016), and the failure of pertussis vaccines, in particular aP vaccines, to prevent infection and spread of *B. pertussis* were also proposed to explain the resurgence of the disease. Regarding pathogen evolution, the first reports were related to polymorphism in genes coding for proteins included in the vaccine (PRN and PTx among others) (Mooi et al. 1998) and later in the pertussis toxin promoter (*ptx*P) (Advani et al. 2011; Kallonen et al. 2012). Recently, there has been an increase in *B. pertussis* isolates that do not produce some of the vaccine antigens (Lam et al. 2014; Barkoff et al. 2019). It has been proposed that the loss of this vaccine antigen probably provides a selective advantage for bacterial survival in populations vaccinated with aP vaccines (Martin et al. 2015). Commercial aP vaccines containing PTx, PRN and FHA are not as effective as expected in controlling the infection caused by the recent circulating bacteria that do not express PRN (Hegerle et al. 2014). Moreover, recently it was demonstrated in a mixed infection mouse model that PRN deficient *B. pertussis* strain colonizes the respiratory tract of aP immunized mice more effectively than the PRN positive strain (Safarchi et al. 2015).

Under this context, in 2015 the Strategic Advisory Group of Experts on immunization expressed concerns regarding the resurgence of pertussis in certain industrialized countries despite high aP-vaccine coverage (Meeting of the Strategic Advisory Group of Experts on immunization 2015). The switch from wP to aP for primary infant immunization was proposed as, at least partially responsible for that resurgence (Table 1, see reported cases of European Region). The World Health Organization (WHO) therefore recommended that the switch be considered only if, in the national immunization schedules, large numbers of doses including several boosters can be assured. Countries currently using aP vaccines may continue using them, but should consider the need for additional booster doses and strategies to prevent early-childhood mortality upon pertussis resurgence. In fact, the WHO published a position paper on this subject and wrote the following:

A switch from wP to aP vaccines for primary infant immunization should only be considered if the inclusion in the national immunization schedules of additional periodic booster or maternal immunization can be assured and sustained (Pertussis vaccines: WHO position paper, August 2015— Recommendations 2016).

National programmes currently using aP vaccine may continue using this vaccine but should consider the need for additional booster doses and strategies to prevent early childhood mortality such as maternal immunization in case of resurgence of pertussis (Pertussis vaccines: WHO position paper, August 2015— Recommendations 2016).

2 New Pertussis Vaccines

Pertussis vaccines are currently on the agenda due to the worrying increase of pertussis cases detected in different countries. There are an estimated 24.1 million cases of the disease and approximately 160,700 deaths occurring worldwide every year in children younger than 5 years of age (Yeung et al. 2017). It is very clear that the non-use of the current pertussis vaccines would lead to an even more challenging epidemiological scenario and for this reason the current vaccine administration and surveillance of the disease should be improved while new vaccines are being developed. The development of a new pertussis vaccine is a difficult task to achieve since no absolute correlate for protection exists, however there are enough data from animal models and human studies showing that although antibodies may mediate protection, Th1 and Th17 cellular responses and tissue resident memory (T_{RM}) response are responsible for long-lasting protection (Mills et al. 2014). To induce or drive a Th1, Th17 and T_{RM} response, different approaches have already been proposed (Allen and Mills 2014; Mielcarek et al. 2006; Dias et al. 2013). In the next section, the main approaches used so far for the development of new vaccines are discussed.

3 Live Attenuated Vaccine

The most advanced novel pertussis vaccine candidate is that developed by Locht et al. in Lille, France (Thorstensson et al. 2014; Mielcarek et al. 2010; Feunou et al. 2010; Skerry et al. 2009). This vaccine candidate, referred as BPZE1, and consisting in a live attenuated bacterial strain, (Locht 2014) was shown to be immunogenic and protective in mice and baboons after intranasal administration (Locht 2016, 2017). In mice a single nasal administration of BPZE1, but not a high dose of current commercial aP vaccine, induced *B. pertussis*-specific secretory IgA in the nasal cavity, and transfer of the nasal IgA was able to protect recipient mice against nasal colonization after *B. pertussis* challenge (Solans and Locht 2018). Though no protection experiments have yet been performed with BPZE1 against circulating bacteria, other interesting findings have already reported. It was detected that BPZE1 vaccine was able to induce $CD4^+CD69^+CD103^+$ T_{RM} cells in the nasal mucosa of mice, and these cells produced high levels of IL-17 and appreciable levels of IFN-γ. Thus, BPZE1 protects mice against nasal infection by virulent *B. pertussis* via an IL-17-dependent and sIgA-mediated mechanism (Solans and Locht 2018; Fedele et al. 2011). Moreover, recently a double-blind, placebo-controlled, dose-escalating study of BPZE1 given intranasally for the first time to human volunteers was performed as the first trial of a live attenuated bacterial vaccine against pertussis. In this study, 12 subjects per dose group received different quantities of colony-forming units as droplets with half of the dose in each nostril and 12 subjects received the diluent (control group) (Thorstensson et al. 2014). Local and systemic safety and immune responses were assessed during 6 months, and nasopharyngeal colonization with BPZE1 was determined with repeated cultures during the first 4 weeks after vaccination. In this trial, the vaccine candidate was found safe in young human adults, able to transiently colonize the human nasopharynx, and to induce antibodies to PTx, FHA, PRN and fimbriae after a single nasal administration (Thorstensson et al. 2014). This vaccine candidate is currently entering a clinical phase II trial.

4 Less Reactogenic Whole Cell Vaccine

The major cause of wP vaccine reactions is associated to the endotoxin which is a lipo-oligosaccharide (LOS) and because of that attempts were made to detoxify wP vaccines. Researchers at the Institute Butantan in São Paulo, Brazil, diminished the endotoxicity of the wP vaccine by performing a chemical extraction of LOS from the outer membrane (Dias et al. 2013). Chemical extraction of LOS resulted in a

significant decrease in endotoxin content without affecting the integrity of the product. This development, however, raises doubts because with the LOS extraction the adjuvant capacity associated with this molecule would also be decreasing. Other alternative strategies to LOS removal are being sought, specifically a consortium of researchers proposed to work on structural changes of the molecule (on the LipidA) in order to retain de beneficial effects induced by the molecule but eliminating its reactogenicity. The results on this strategy have not yet been disclosed.

5 Acellular Pertussis Vaccines Containing Recombinant Inactivated Pertussis Toxin

The safety and superior immunogenicity of 9 K/129G genetically detoxified PTx (rPT) was demonstrated long time ago (Rappuoli 1999; Podda et al. 1993). Under this context, BioNet-Asia developed a new rPT-expressing *B. pertussis* strain (Buasri et al. 2012). This strain generated increased amounts of rPT compared to wild type strain and strains used in vaccine production and the purified rPT did not show any toxicity (Buasri et al. 2012). Thus, Bionet formulated a new acellular vaccine containing the recombinant genetically detoxified Pertussis Toxin (PTgen), FHA and PRN and presented the results of the first clinical study of this recombinant aP vaccine formulated alone or in combination with tetanus and diphtheria toxoids. For the phase I/II trial, 60 subjects (20 per each vaccine group) were enrolled and included in the safety analysis. This first-in-human study showed that BioNet's PTgen-containing vaccine has a similar reactogenicity and safety profile than the Adacel® acellular vaccine. Moreover, the high immunogenicity of PTgen in adults was demonstrated Sirivichayakul et al. (2016). The results were consistent with previous studies that demonstrated high and sustained efficacy of rPT-containing aP vaccines in infants (Seubert et al. 2014). Recent findings on the ability of rPT-containing acellular vaccine to induce memory response make a significant difference with current acellular vaccines that include chemically detoxified components in terms of long-term protection. Specifically, the authors reported that the boosting of aP-primed adolescents with recombinant-aP induced higher anti-PTx and PTx-neutralizing responses than the current aP vaccine and increased PTx-specific memory B cells (Blanchard Rohner et al. 2018). These new acellular vaccines can thus overcome one of the weaknesses of current acellular vaccines: the rapid loss of induced immunity. However, it remains to study the protection capacity of this vaccine against current circulating bacteria and the selection pressure that this type of vaccine would exert on the circulating bacterial population. This last aspect, in principle, would not be solved with the recombinant acellular vaccine, since it is constituted by the same few immunogens as the current acellular vaccines.

6 New Antigens and Adjuvants for aP Formulations

The incorporation of novel antigens derived from *B. pertussis* to improve the current aP vaccines has also been explored. The *B. pertussis* adenylate cyclase toxin (Cheung et al. 2006), the serum-resistance autotransporter protein BrkA (Marr et al. 2008) and the iron-regulated *B. pertussis* proteins (Alvarez Hayes et al. 2013) among others, have been proposed as a protective antigen. Though none of these antigens alone offered significant protection against *B. pertussis* infection in an intranasal challenge model, when combined with acellular pertussis vaccine, they conferred improved protection over the acellular vaccine alone. The combination of all these immunogens together with the current acellular vaccines could be an attractive proposal to reduce the selection pressure of the current acellular vaccines by offering a greater number of epitopes.

Improvements of the acellular vaccines could also be achieved by using novel adjuvants for pertussis. Combination of aP vaccine with adjuvants that are able to drive Th1 and Th17 responses would be expected to enhance

protection. Cyclic di-GMP, MF59 emulsions, the combination of aluminium hydroxide with the TLR-4 agonist monophosphoryl lipid A, have been shown to enhance Th1 type immune responses however the impact in protection of these adjuvants was not deeply investigated (Geurtsen et al. 2007; Allen et al. 2018). The *B. pertussis* lipoprotein BP1569, a TLR-2 agonist that activates murine dendritic cells and macrophages has recently been shown to possess adjuvant properties (Dunne et al. 2015). Recently it was reported that this protein in combination with c-di-GMP synergistically induces the production of IFN-β, IL-12 and IL-23, and maturation of dendritic cells (Allen et al. 2018). Parenteral immunization of mice with an experimental aP vaccine formulated with this combined adjuvant promoted Th1 and Th17 responses and conferred protection against lung infection with *B. pertussis*. Interestingly, intranasal immunization with this vaccine induced potent *B. pertussis*-specific Th17 responses and IL-17-secreting respiratory tissue-resident memory (T_{RM}) CD4 T cells, and conferred a high level of protection against nasal colonization (sterilizing immunity) as well as lung infection. Furthermore, long-term protection against nasal colonization with *B. pertussis* was observed. This formulation would thus prolong the duration of the protective response but it is not clear that it is capable of overcoming the deficiencies of the current acellular vaccines against the circulating bacterial population. More research must be done in this regard.

7 Outer Membrane Vesicles as Vaccine Candidates Against *B. pertussis* Infections

All Gram-negative bacteria that have been investigated so far are able to naturally release spherical structures originated from the outer membrane (referred to as outer membrane vesicles, OMVs). Although OMVs formation seems to be a common feature of Gram-negative bacteria, the knowledge of their biogenesis and biological roles remains limited. OMVs naturally contain multiple native surface-exposed antigens as well as immunostimulatory molecules. Based on their aforementioned immunogenic potency and on positive examples of the OMV-derived vaccines against *Neisseria meningitides* serogroup *B*, we initiated several studies over the last years to analyze the potential of OMVs derived from *Bordetella pertussis* as vaccine candidates (Hozbor et al. 1999; Roberts et al. 2008; Asensio et al. 2011). We characterized the composition of the pertussis nanoparticles at >200 protein components—including the virulence factors PT, PRN, fimbriae, FHA, and adenylate-cyclase (Hozbor 2016). The presence of a high number of immunogens in the vaccine formulation is essential since they may avoid the high selective pressure conferred by a single or a few protective-vaccine antigens. To date, we have obtained almost 50 batches of *B. pertussis*–derived OMVs with robust results. Our OMV-based vaccine is safe and exhibits an adequate protection capacity against different *B. pertussis* genetic backgrounds, including those not expressing the vaccine antigen PRN (Gaillard et al. 2014).

The OMVs derived from *B. pertussis* represent an attractive acellular pertussis vaccine candidate (Hozbor 2016; Ormazabal et al. 2014; Asensio et al. 2011; Roberts et al. 2008) not only because of its safety and ability to induce protective Th1, Th17 cells (Mills et al. 1993; Ryan et al. 1997; Raeven et al. 2014; Warfel and Merkel 2013; Ross et al. 2013) and T_{RM} cells, but because it contains a greater number of immunogens in conformations close to those found in pathogen, when compared with the current aP vaccines (Hozbor 2016; Advani et al. 2011). Consistent with previous reports (Hegerle et al. 2014; Safarchi et al. 2015), we found that immunization with commercial aP vaccine does not protect against PRN deficient isolate as effectively as against *B. pertussis* Tohama strain (PRN+). Since the PRN deficient isolate is not isogenic to *B. pertussis* Tohama strain (PRN+) and contains polymorphisms at other loci that may affect the fitness of these bacteria, we have also examined the protection of the OMV based vaccine against a PRN defective mutant derived from *B. pertussis* Tohama strain. We found that the commercial aP

vaccine but not the OMV based vaccine exhibits lower level of protection against the PRN deficient strain when compared with the parental PRN(+) positive strain. These results clearly showed the impact of the absence of PRN expression in the effectiveness of aP vaccine against *B. pertussis* when comparisons are made on strains that contain the same genetic background (submitted manuscript).

The results obtained here clearly showed that the OMVs vaccine is more effective than a current commercial aP vaccine against PRN deficient strains. Therefore, the OMV formulation appears as an attractive vaccine candidate that could replace the current aP without causing concern on the reactogenicity associated with wP vaccines because of the proven safety of the OMVs vaccines (Bottero et al. 2016). Since major limitations of the current aP are their strong selection pressure exerted on the circulating bacterial population and their failure to induce sustained protective immunity, the OMV-based vaccine, that contains high number of antigens and that induces INF-γ and IL17-secreting T_{RM} cells, has the potential to replace the current aP vaccine.

Conflict of Interest Statement The authors declare that the research was conducted in the absence of any commercial or financial relationships that could be construed as a potential conflict of interest.

Funding This study was funded by a grant from the ANCPyT (PICT 2014-3617, PICT 2012- 2719), CONICET and FCE-UNLP (Argentina) grants to DFH. DFH is member of the Scientific Career of CONICET. The funders had no role in study design, data collection and analysis, decision to publish, or preparation of the manuscript.

References

Advani A, Gustafsson L, Ahren C, Mooi FR, Hallander HO (2011) Appearance of Fim3 and ptxP3-Bordetella pertussis strains, in two regions of Sweden with different vaccination programs. Vaccine 29(18):3438–3442. https://doi.org/10.1016/j.vaccine.2011.02.070

Allen AC, Mills KH (2014) Improved pertussis vaccines based on adjuvants that induce cell-mediated immunity. Expert Rev Vaccines 13(10):1253–1264. https://doi.org/10.1586/14760584.2014.936391

Allen AC, Wilk MM, Misiak A, Borkner L, Murphy D, Mills KHG (2018) Sustained protective immunity against Bordetella pertussis nasal colonization by intranasal immunization with a vaccine-adjuvant combination that induces IL-17-secreting TRM cells. Mucosal Immunol 11:1763–1776. https://doi.org/10.1038/s41385-018-0080-x

Alvarez Hayes J, Erben E, Lamberti Y, Principi G, Maschi F, Ayala M, Rodriguez ME (2013) Bordetella pertussis iron regulated proteins as potential vaccine components. Vaccine 31(35):3543–3548. https://doi.org/10.1016/j.vaccine.2013.05.072

Asensio CJ, Gaillard ME, Moreno G, Bottero D, Zurita E, Rumbo M, van der Ley P, van der Ark A, Hozbor D (2011) Outer membrane vesicles obtained from Bordetella pertussis Tohama expressing the lipid A deacylase PagL as a novel acellular vaccine candidate. Vaccine 29(8):1649–1656. https://doi.org/10.1016/j.vaccine.2010.12.068

Barkoff AM, Mertsola J, Pierard D, Dalby T, Hoegh SV, Guillot S, Stefanelli P, van Gent M, Berbers G, Vestrheim D, Greve-Isdahl M, Wehlin L, Ljungman M, Fry NK, Markey K, He Q (2019) Pertactin-deficient Bordetella pertussis isolates: evidence of increased circulation in Europe, 1998 to 2015. Euro Surveill 24(7). https://doi.org/10.2807/1560-7917.ES.2019.24.7.1700832

Blanchard Rohner G, Chatzis O, Chinwangso P, Rohr M, Grillet S, Salomon C, Lemaitre B, Boonrak P, Lawpoolsri S, Clutterbuck E, Poredi IK, Wijagkanalan W, Spiegel J, Pham HT, Viviani S, Siegrist CA (2018) Boosting teenagers with acellular pertussis vaccines containing recombinant or chemically inactivated pertussis toxin: a randomized clinical trial. Clin Infect Dis 68:1213–1222. https://doi.org/10.1093/cid/ciy594

Bottero D, Gaillard ME, Fingermann M, Weltman G, Fernandez J, Sisti F, Graieb A, Roberts R, Rico O, Rios G, Regueira M, Binsztein N, Hozbor D (2007) Pulsed-field gel electrophoresis, pertactin, pertussis toxin S1 subunit polymorphisms, and surfaceome analysis of vaccine and clinical Bordetella pertussis strains. Clin Vaccine Immunol 14(11):1490–1498. https://doi.org/10.1128/CVI.00177-07

Bottero D, Gaillard ME, Zurita E, Moreno G, Martinez DS, Bartel E, Bravo S, Carriquiriborde F, Errea A, Castuma C, Rumbo M, Hozbor D (2016) Characterization of the immune response induced by pertussis OMVs-based vaccine. Vaccine 34(28):3303–3309. https://doi.org/10.1016/j.vaccine.2016.04.079

Bowden KE, Weigand MR, Peng Y, Cassiday PK, Sammons S, Knipe K, Rowe LA, Loparev V, Sheth M, Weening K, Tondella ML, Williams MM (2016) Genome structural diversity among 31 Bordetella pertussis isolates from two recent U.S. Whooping Cough Statewide Epidemics. mSphere 1(3). https://doi.org/10.1128/mSphere.00036-16

Brody M, Sorley RG (1947) Neurologic complications following the administration of pertussis vaccine. N Y State J Med 47(9):1016

Brummelman J, Wilk MM, Han WG, van Els CA, Mills KH (2015) Roads to the development of improved pertussis vaccines paved by immunology. Pathog Dis 73(8):ftv067. https://doi.org/10.1093/femspd/ftv067

Buasri W, Impoolsup A, Boonchird C, Luengchaichawange A, Prompiboon P, Petre J, Panbangred W (2012) Construction of Bordetella pertussis strains with enhanced production of genetically-inactivated Pertussis Toxin and Pertactin by unmarked allelic exchange. BMC Microbiol 12:61. https://doi.org/10.1186/1471-2180-12-61

Cheung GY, Xing D, Prior S, Corbel MJ, Parton R, Coote JG (2006) Effect of different forms of adenylate cyclase toxin of Bordetella pertussis on protection afforded by an acellular pertussis vaccine in a murine model. Infect Immun 74(12):6797–6805. https://doi.org/10.1128/IAI.01104-06

Clark TA (2014) Changing pertussis epidemiology: everything old is new again. J Infect Dis 209(7):978–981. https://doi.org/10.1093/infdis/jiu001

David S, van Furth R, Mooi FR (2004) Efficacies of whole cell and acellular pertussis vaccines against Bordetella parapertussis in a mouse model. Vaccine 22(15–16):1892–1898. https://doi.org/10.1016/j.vaccine.2003.11.005

Dias WO, van der Ark AA, Sakauchi MA, Kubrusly FS, Prestes AF, Borges MM, Furuyama N, Horton DS, Quintilio W, Antoniazi M, Kuipers B, van der Zeijst BA, Raw I (2013) An improved whole cell pertussis vaccine with reduced content of endotoxin. Hum Vaccin Immunother 9(2):339–348

Dunne A, Mielke LA, Allen AC, Sutton CE, Higgs R, Cunningham CC, Higgins SC, Mills KH (2015) A novel TLR2 agonist from Bordetella pertussis is a potent adjuvant that promotes protective immunity with an acellular pertussis vaccine. Mucosal Immunol 8(3):607–617. https://doi.org/10.1038/mi.2014.93

Edwards KM, Karzon DT (1990) Pertussis vaccines. Pediatr Clin N Am 37(3):549–566

Fedele G, Bianco M, Debrie AS, Locht C, Ausiello CM (2011) Attenuated Bordetella pertussis vaccine candidate BPZE1 promotes human dendritic cell CCL21-induced migration and drives a Th1/Th17 response. J Immunol 186(9):5388–5396. https://doi.org/10.4049/jimmunol.1003765

Feunou PF, Kammoun H, Debrie AS, Mielcarek N, Locht C (2010) Long-term immunity against pertussis induced by a single nasal administration of live attenuated B. pertussis BPZE1. Vaccine 28(43):7047–7053. https://doi.org/10.1016/j.vaccine.2010.08.017

Gaillard ME, Bottero D, Errea A, Ormazabal M, Zurita ME, Moreno G, Rumbo M, Castuma C, Bartel E, Flores D, van der Ley P, van der Ark A, FH D (2014) Acellular pertussis vaccine based on outer membrane vesicles capable of conferring both long-lasting immunity and protection against different strain genotypes. Vaccine 32(8):931–937. https://doi.org/10.1016/j.vaccine.2013.12.048

Geurtsen J, Banus HA, Gremmer ER, Ferguson H, de la Fonteyne-Blankestijn LJ, Vermeulen JP, Dormans JA, Tommassen J, van der Ley P, Mooi FR, Vandebriel RJ (2007) Lipopolysaccharide analogs improve efficacy of acellular pertussis vaccine and reduce type I hypersensitivity in mice. Clin Vaccine Immunol 14 (7):821–829. https://doi.org/10.1128/CVI.00074-07

Gzyl A, Augustynowicz E, Gniadek G, Rabczenko D, Dulny G, Slusarczyk J (2004) Sequence variation in pertussis S1 subunit toxin and pertussis genes in Bordetella pertussis strains used for the whole-cell pertussis vaccine produced in Poland since 1960: efficiency of the DTwP vaccine-induced immunity against currently circulating B. pertussis isolates. Vaccine 22 (17–18):2122–2128. https://doi.org/10.1016/j.vaccine.2003.12.006

He Q, Makinen J, Berbers G, Mooi FR, Viljanen MK, Arvilommi H, Mertsola J (2003) Bordetella pertussis protein pertactin induces type-specific antibodies: one possible explanation for the emergence of antigenic variants? J Infect Dis 187(8):1200–1205. https://doi.org/10.1086/368412

Hegerle N, Dore G, Guiso N (2014) Pertactin deficient Bordetella pertussis present a better fitness in mice immunized with an acellular pertussis vaccine. Vaccine 32(49):6597–6600. https://doi.org/10.1016/j.vaccine.2014.09.068

Hozbor DF (2016) Outer membrane vesicles: an attractive candidate for pertussis vaccines. Expert Rev Vaccines 16:1–4. https://doi.org/10.1080/14760584.2017.1276832

Hozbor D, Rodriguez ME, Fernandez J, Lagares A, Guiso N, Yantorno O (1999) Release of outer membrane vesicles from Bordetella pertussis. Curr Microbiol 38(5):273–278

Kallonen T, Mertsola J, Mooi FR, He Q (2012) Rapid detection of the recently emerged Bordetella pertussis strains with the ptxP3 pertussis toxin promoter allele by real-time PCR. Clin Microbiol Infect 18(10):E377–E379. https://doi.org/10.1111/j.1469-0691.2012.04000.x

Kendrick P (1936) Progress report on pertussis immunization. Am J Public Health Nations Health 26:8–12

King AJ, Berbers G, van Oirschot HF, Hoogerhout P, Knipping K, Mooi FR (2001) Role of the polymorphic region 1 of the Bordetella pertussis protein pertactin in immunity. Microbiology 147. (Pt 11:2885–2895

Klein NP (2014) Licensed pertussis vaccines in the United States. History and current state. Hum Vaccin Immunother 10(9):2684–2690. https://doi.org/10.4161/hv.29576

Klein NP, Bartlett J, Fireman B, Rowhani-Rahbar A, Baxter R (2013) Comparative effectiveness of acellular versus whole-cell pertussis vaccines in teenagers. Pediatrics 131(6):e1716–e1722. https://doi.org/10.1542/peds.2012-3836

Koepke R, Eickhoff JC, Ayele RA, Petit AB, Schauer SL, Hopfensperger DJ, Conway JH, Davis JP (2014) Estimating the Effectiveness of Tdap Vaccine for Preventing Pertussis: Evidence of Rapidly Waning Immunity and Differences in Effectiveness by Tdap Brand. J Infect Dis 210:942–953. https://doi.org/10.1093/infdis/jiu322

Kulenkampff M, Schwartzman JS, Wilson J (1974) Neurological complications of pertussis inoculation. Arch Dis Child 49(1):46–49

Lam C, Octavia S, Ricafort L, Sintchenko V, Gilbert GL, Wood N, McIntyre P, Marshall H, Guiso N, Keil AD, Lawrence A, Robson J, Hogg G, Lan R (2014) Rapid Increase in Pertactin-deficient Bordetella pertussis Isolates, Australia. Emerg Infect Dis 20(4):626–633. https://doi.org/10.3201/eid2004.131478

Locht CMN (2014) Live attenuated vaccines against pertussis. Expert Rev Vaccines 13(9):1147–1158. https://doi.org/10.1586/14760584.2014.942222

Locht C (2016) Live pertussis vaccines: will they protect against carriage and spread of pertussis? Clin Microbiol Infect 22(Suppl 5):S96–S102. https://doi.org/10.1016/j.cmi.2016.05.029

Locht C, Papin JF, Lecher S, Debrie AS, Thalen M, Solovay K, Rubin K, Mielcarek N (2017) Live Attenuated Pertussis Vaccine BPZE1 Protects Baboons Against Bordetella pertussis Disease and Infection. J Infect Dis 216(1):117–124. https://doi.org/10.1093/infdis/jix254

Madsen T (1933) Vaccination against whooping cough. JAMA 101(3):187–188

Mäkelä PH (2000) Vaccines, coming of age after 200 years. FEMS Microbiol Rev 24(1):9–20

Marr N, Oliver DC, Laurent V, Poolman J, Denoel P, Fernandez RC (2008) Protective activity of the Bordetella pertussis BrkA autotransporter in the murine lung colonization model. Vaccine 26 (34):4306–4311. https://doi.org/10.1016/j.vaccine.2008.06.017

Martin SW, Pawloski L, Williams M, Weening K, DeBolt C, Qin X, Reynolds L, Kenyon C, Giambrone G, Kudish K, Miller L, Selvage D, Lee A, Skoff TH, Kamiya H, Cassiday PK, Tondella ML, Clark TA (2015) Pertactin-negative Bordetella pertussis strains: evidence for a possible selective advantage. Clin Infect Dis 60(2):223–227. https://doi.org/10.1093/cid/ciu788

McGirr A, Fisman DN (2015) Duration of pertussis immunity after DTaP immunization: a meta-analysis. Pediatrics 135:331–343. https://doi.org/10.1542/peds.2014-1729

McGirr AA, Tuite AR, Fisman DN (2013) Estimation of the underlying burden of pertussis in adolescents and adults in Southern Ontario, Canada. PLoS One 8(12): e83850. https://doi.org/10.1371/journal.pone.0083850

Meeting of the Strategic Advisory Group of Experts on immunization, April 2015: conclusions and recommendations (2015) Releve epidemiologique hebdomadaire / Section d'hygiene du Secretariat de la

Societe des Nations = Weekly epidemiological record / Health Section of the Secretariat of the League of Nations 90 (22):261–278

Mielcarek N, Debrie AS, Raze D, Bertout J, Rouanet C, Younes AB, Creusy C, Engle J, Goldman WE, Locht C (2006) Live attenuated B. pertussis as a single-dose nasal vaccine against whooping cough. PLoS Pathog 2 (7):e65. https://doi.org/10.1371/journal.ppat.0020065

Mielcarek N, Debrie AS, Mahieux S, Locht C (2010) Dose response of attenuated Bordetella pertussis BPZE1-induced protection in mice. Clin Vaccine Immunol 17 (3):317–324. https://doi.org/10.1128/CVI.00322-09

Mills KH, Barnard A, Watkins J, Redhead K (1993) Cell-mediated immunity to Bordetella pertussis: role of Th1 cells in bacterial clearance in a murine respiratory infection model. Infect Immun 61(2):399–410

Mills KH, Ross PJ, Allen AC, Wilk MM (2014) Do we need a new vaccine to control the re-emergence of pertussis? Trends Microbiol 22(2):49–52. https://doi.org/10.1016/j.tim.2013.11.007

Mooi FR, van Oirschot H, Heuvelman K, van der Heide HG, Gaastra W, Willems RJ (1998) Polymorphism in the Bordetella pertussis virulence factors P.69/pertactin and pertussis toxin in The Netherlands: temporal trends and evidence for vaccine-driven evolution. Infect Immun 66(2):670–675

Mooi FR, van Loo IH, King AJ (2001) Adaptation of Bordetella pertussis to vaccination: a cause for its reemergence? Emerg Infect Dis 7(3 Suppl):526–528

Olin P, Rasmussen F, Gustafsson L, Hallander HO, Heijbel H (1997) Randomised controlled trial of two-component, three-component, and five-component acellular pertussis vaccines compared with whole-cell pertussis vaccine. Ad Hoc Group for the Study of Pertussis Vaccines. Lancet 350 (9091):1569–1577

Ormazabal M, Bartel E, Gaillard ME, Bottero D, Errea A, Zurita ME, Moreno G, Rumbo M, Castuma C, Flores D, Martin MJ, Hozbor D (2014) Characterization of the key antigenic components of pertussis vaccine based on outer membrane vesicles. Vaccine 32 (46):6084–6090. https://doi.org/10.1016/j.vaccine.2014.08.084

Podda A, Carapella De Luca E, Titone L, Casadei AM, Cascio A, Bartalini M, Volpini G, Peppoloni S, Marsili I, Nencioni L et al (1993) Immunogenicity of an acellular pertussis vaccine composed of genetically inactivated pertussis toxin combined with filamentous hemagglutinin and pertactin in infants and children. J Pediatr 123(1):81–84

Raeven RH, Brummelman J, Pennings JL, Nijst OE, Kuipers B, Blok LE, Helm K, van Riet E, Jiskoot W, van Els CA, Han WG, Kersten GF, Metz B (2014) Molecular signatures of the evolving immune response in mice following a Bordetella pertussis infection. PLoS One 9(8):e104548. https://doi.org/10.1371/journal.pone.0104548

Rappuoli R (1999) The vaccine containing recombinant pertussis toxin induces early and long-lasting protection. Biologicals 27(2):99–102. https://doi.org/10.1006/biol.1999.0189

Roberts R, Moreno G, Bottero D, Gaillard ME, Fingermann M, Graieb A, Rumbo M, Hozbor D (2008) Outer membrane vesicles as acellular vaccine against pertussis. Vaccine 26(36):4639–4646. https://doi.org/10.1016/j.vaccine.2008.07.004

Romanus V, Jonsell R, Bergquist SO (1987) Pertussis in Sweden after the cessation of general immunization in 1979. Pediatr Infect Dis J 6(4):364–371

Ross PJ, Sutton CE, Higgins S, Allen AC, Walsh K, Misiak A, Lavelle EC, McLoughlin RM, Mills KH (2013) Relative contribution of Th1 and Th17 cells in adaptive immunity to Bordetella pertussis: towards the rational design of an improved acellular pertussis vaccine. PLoS Pathog 9(4):e1003264. https://doi.org/10.1371/journal.ppat.1003264

Ryan M, Murphy G, Gothefors L, Nilsson L, Storsaeter J, Mills KH (1997) Bordetella pertussis respiratory infection in children is associated with preferential activation of type 1 T helper cells. J Infect Dis 175 (5):1246–1250

Ryan M, Murphy G, Ryan E, Nilsson L, Shackley F, Gothefors L, Oymar K, Miller E, Storsaeter J, Mills KH (1998) Distinct T-cell subtypes induced with whole cell and acellular pertussis vaccines in children. Immunology 93(1):1–10

Safarchi A, Octavia S, Luu LD, Tay CY, Sintchenko V, Wood N, Marshall H, McIntyre P, Lan R (2015) Pertactin negative Bordetella pertussis demonstrates higher fitness under vaccine selection pressure in a mixed infection model. Vaccine 33(46):6277–6281. https://doi.org/10.1016/j.vaccine.2015.09.064

Sato H, Sato Y (1985) Protective antigens of Bordetella pertussis mouse-protection test against intracerebral and aerosol challenge of B. pertussis. Dev Biol Stand 61:461–467

Sato Y, Kimura M, Fukumi H (1984) Development of a pertussis component vaccine in Japan. Lancet 1 (8369):122–126

Sauer LW (1948) Simultaneous immunization against diphtheria, tetanus and pertussis; a preliminary report. Q Bull Northwest Univ Med Sch 22(3):281–285

Seubert A, D'Oro U, Scarselli M, Pizza M (2014) Genetically detoxified pertussis toxin (PT-9K/129G): implications for immunization and vaccines. Expert Rev Vaccines 13(10):1191–1204. https://doi.org/10.1586/14760584.2014.942641

Sheridan SL, Ware RS, Grimwood K, Lambert SB (2012) Number and order of whole cell pertussis vaccines in infancy and disease protection. JAMA 308 (5):454–456. https://doi.org/10.1001/jama.2012.6364

Sirivichayakul C, Chanthavanich P, Limkittikul K, Siegrist CA, Wijagkanalan W, Chinwangso P, Petre J, Hong Thai P, Chauhan M, Viviani S (2016) Safety and immunogenicity of a combined Tetanus, Diphtheria, recombinant acellular Pertussis vaccine (TdaP) in healthy Thai adults. Hum Vaccin Immunother 13(1):36–143

Skerry CM, Cassidy JP, English K, Feunou-Feunou P, Locht C, Mahon BP (2009) A live attenuated Bordetella pertussis candidate vaccine does not cause disseminating infection in gamma interferon receptor knockout mice. Clin Vaccine Immunol 16 (9):1344–1351. https://doi.org/10.1128/CVI.00082-09

Solans L, Locht C (2018) The Role of Mucosal Immunity in Pertussis. Front Immunol 9:3068. https://doi.org/10.3389/fimmu.2018.03068

Stefanelli P, Buttinelli G, Vacca P, Tozzi AE, Midulla F, Carsetti R, Fedele G, Villani A, Concato C (2017) Severe pertussis infection in infants less than 6 months of age: Clinical manifestations and molecular characterization. Hum Vaccin Immunother 13(5):1073–1077. https://doi.org/10.1080/21645515.2016.1276139

Tan T, Dalby T, Forsyth K, Halperin SA, Heininger U, Hozbor D, Plotkin S, Ulloa-Gutierrez R, von Konig CH (2015) Pertussis across the globe: recent epidemiologic trends from 2000–2013. Pediatr Infect Dis J 34 (9):e222–e232. https://doi.org/10.1097/INF.0000000000000795

Tartof SY, Lewis M, Kenyon C, White K, Osborn A, Liko J, Zell E, Martin S, Messonnier NE, Clark TA, Skoff TH (2013) Waning immunity to pertussis following 5 doses of DTaP. Pediatrics 131(4):e1047–e1052. https://doi.org/10.1542/peds.2012-1928

Thorstensson R, Trollfors B, Al-Tawil N, Jahnmatz M, Bergstrom J, Ljungman M, Torner A, Wehlin L, Van Broekhoven A, Bosman F, Debrie AS, Mielcarek N, Locht C (2014) A phase I clinical study of a live attenuated Bordetella pertussis vaccine--BPZE1; a single centre, double-blind, placebo-controlled, dose-escalating study of BPZE1 given intranasally to healthy adult male volunteers. PLoS One 9(1): e83449. https://doi.org/10.1371/journal.pone.0083449

Warfel JM, Merkel TJ (2013) Bordetella pertussis infection induces a mucosal IL-17 response and long-lived Th17 and Th1 immune memory cells in nonhuman primates. Mucosal Immunol 6(4):787–796. https://doi.org/10.1038/mi.2012.117

WHO (2016) Pertussis vaccines: WHO position paper, August 2015—Recommendations. Vaccine 34 (12):1423–1425. https://doi.org/10.1016/j.vaccine.2015.10.136

Wilk MM, Mills KHG (2018) CD4 TRM Cells Following Infection and Immunization: Implications for More Effective Vaccine Design. Front Immunol 9:1860. https://doi.org/10.3389/fimmu.2018.01860

Witt MA, Katz PH, Witt DJ (2012) Unexpectedly limited durability of immunity following acellular pertussis vaccination in preadolescents in a North American outbreak. Clin Infect Dis 54(12):1730–1735. https://doi.org/10.1093/cid/cis287

Witt MA, Arias L, Katz PH, Truong ET, Witt DJ (2013) Reduced risk of pertussis among persons ever vaccinated with whole cell pertussis vaccine compared to recipients of acellular pertussis vaccines in a large US cohort. Clin Infect Dis 56(9):1248–1254. https://doi.org/10.1093/cid/cit046

Yeung KHT, Duclos P, Nelson EAS, Hutubessy RCW (2017) An update of the global burden of pertussis in children younger than 5 years: a modelling study. Lancet Infect Dis 17(9):974–980. https://doi.org/10.1016/S1473-3099(17)30390-0

Adv Exp Med Biol - Advances in Microbiology, Infectious Diseases and Public Health (2019) 1183: 127–136
https://doi.org/10.1007/5584_2019_408
© Springer Nature Switzerland AG 2019
Published online: 19 July 2019

Pertussis: Identification, Prevention and Control

Paola Stefanelli

Abstract

Pertussis is a vaccine-preventable disease. Despite the high vaccination coverage among children, pertussis is considered a re-emerging disease for which identification, prevention and control strategies need to be improved. To control pertussis it is important to maintain a high vaccination coverage to protect the age groups considered at high risk for the disease. Laboratory confirmation of *Bordetella pertussis* infection together with a differential diagnostic test for other *Bordetellae* are prerequisite for a correct and timely diagnosis of pertussis. Moreover, investigations of antimicrobial susceptibility and whole genome sequencing may permit to monitor the circulation of antimicrobials resistant and/or vaccine-escape strains. Finally, the preventive framework should no longer consider pertussis exclusively as a childhood infectious disease, since adults may play a role in transmission events.

Keywords

Bordetella pertussis · Surveillance system · Vaccination · Vaccine · Whole genome sequencing

P. Stefanelli (✉)
Department of Infectious Diseases, Istituto Superiore di Sanità, Rome, Italy
e-mail: paola.stefanelli@iss.it

1 Pertussis Identification

Pertussis has French births, since it was recognized for the first time in Paris in 1578. Guillaume de Baillou (Ballonius), 1538–1616, described the first epidemic (Cherry et al. 1988). He was an important figure in medicine and for the development of pediatrics, and one of the first influencers in leading the University of Paris from Galenism into the new paths of learning.

However, we will have to wait until 1906 to discover the cause of the disease, when Jules Bordet and Octave Gengou (Bordet and Gengou 1906) isolated the bacterium in Bordet-Gengou agar plates.

Bordetella pertussis is a Gram-negative, aerobic coccobacillus ranking to the genus *Bordetella*. Phylogenetic analysis revealed nine different *Bordetella* species. Five of them are known to cause respiratory tract infections in humans: *B. pertussis, Bordetella parapertussis, Bordetella bronchiseptica, Bordetella holmesii,* and *Bordetella petrii* (Guiso and Hegerle 2014). Within the species Bordetella, *B. pertussis, B. bronchiseptica*, and *B. parapertussis* are closely related pathogens. *B. bronchiseptica* causes a mild or chronic respiratory infection in a large range of mammalian hosts. In humans, it causes respiratory tract infections mostly in immunocompromised patients (Woolfrey and Moody 1991). Regarding *B. parapertussis,* two distinct hosts have been identified: humans and sheep (Brinig et al. 2006; Porter et al. 1994).

B. holmesii is part of a different genetic lineage within the Bordetella genus causing either pertussis-like symptoms or invasive infections (e.g., septicemia, pneumonia, meningitis, arthritis, etc.) (Pittet and Posfay-Barbe 2015). *B. petrii* was isolated in patients with cystic fibrosis and in some cases of long-lasting respiratory tract infections (Le Coustumier et al. 2011).

To this regard, an extremely important issue is not only to correctly confirm the presence of *B. pertussis* in the clinical specimens but also to distinguish *B. pertussis* from other *Bordetellae* species.

Unfortunately, despite the European (EU) case definition (Commission Implementing Decision 2018) which includes and underlines the need of a laboratory identification test to confirm a suspected pertussis case, the diagnosis of pertussis still relies on clinical symptoms and laboratory confirmation is rarely performed in countries like Italy (Stefanelli et al. 2017). Clinical case definition requires, at first, one or more typical clinical symptoms, such as paroxysmal cough for at least 2 weeks. Moreover, it should be considered that the specificity of the case definition might be negatively influenced by the time between infection and diagnosis by previous vaccination, and by increasing age of patients. In fact, pertussis is rarely identified in adults due to mitigated signs of the disease and to the fact that pertussis is considered a disease of childhood. Consequently, the risk of disease

increases in the population with consequent impact on babies (Fig. 1).

After the incubation period, pertussis begins with a catarrhal phase. This phase lasts 1–2 weeks, during which the contagiousness is very high. As the catarrhal stage progresses, the cough increases in frequency and severity. The subsequent paroxysmal phase, which lasts 3–6 weeks, is characterized by cough with the characteristic whoop, vomiting, cyanosis, and apnea. The symptoms gradually decrease in severity during the convalescent phase, which can last up to several months (Cherry 1996).

Therefore, the diagnosis of pertussis must take into account the timing of symptoms onset and the vaccination status. The most suitable test for pertussis diagnosis depends on the temporal phase of the disease. In the European Centre for Disease Prevention and Control (ECDC) guidelines in the presence of cough lasting up to 2 weeks, the simultaneous use of culture and molecular tests is recommended (ECDC 2012a). In the presence of cough lasting from 2 to 4 weeks, molecular tests and serum dosage of IgG immunoglobulins against the pertussis toxin (PT) are also recommended (ECDC 2012b) (Fig. 2).

For this purpose, a number of microbiological techniques are readily available in routine diagnostic laboratories and, more recently, Next Generation Sequencing (NGS) is becoming widely available in pertussis reference laboratories allowing a complete genomic characterization.

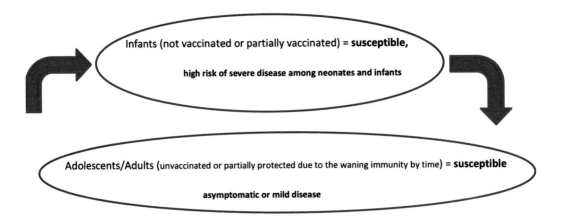

Fig. 1 Transmission cycle of pertussis

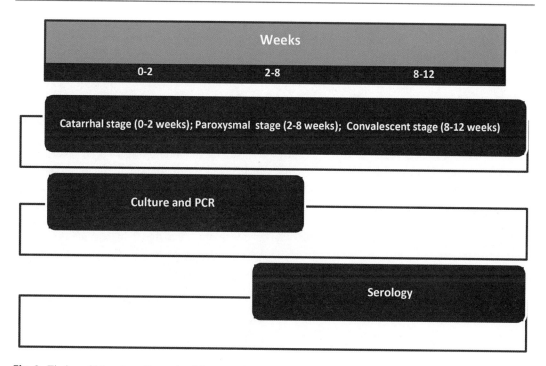

Fig. 2 Timing of laboratory diagnosis of *B. pertussis*

Although culture remains the gold standard, it has low sensitivity (Guillot et al. 2014; Lee et al. 2018). Serology is not an appropriate method to diagnose pertussis in pediatric populations, since it provides results that are difficult to interpret in immunized individuals and, ideally, requires measuring antibody titers in the acute and convalescent phases of the disease, thus delaying the diagnosis (Fedele et al. 2018).

Rapid, sensitive, and specific molecular assays are being increasingly implemented to overcome the limitations of culture and serology and are currently recommended for a rapid management of infected patients (Fry et al. 2009).

In addition to the timescale for the use of the most successful laboratory tests, there is the need to an accurate identification at the species level, as nowadays recognized. This is not only important from a clinical point of view, to select the most appropriate antibiotic treatment, but also for health public purposes, since misdiagnosis of Bordetella species can lead to an incorrect assessment of pertussis vaccine effectiveness and disease burden (Mattoo and Cherry 2005).

ECDC is addressing the harmonization and improvement of pertussis diagnosis for surveillance and outbreak detection in order to assure quality and comparability of data among EU Member States. In 2011, ECDC launched the project 'Coordination of activities for laboratory surveillance of whooping cough in Member States and EEA countries' as part of the EUpert-Labnet surveillance network. One of the main aim of this project was to produce a laboratory guidance for the use of real-time PCR on clinical specimens of patients with suspected whooping cough. Several amplification targets are considered for Bordetella PCR diagnostics. However, most of them are present in multiple Bordetella species. The interpretation of the results following the data obtained using different target genes is also included in the ECDC Technical Document (ECDC 2012a). Recently, ECDC has published an algorithm to correctly interpret the diagnosis by serology (ECDC 2012b).

The development of matrix-assisted laser desorption ionization time-of-flight mass spectrometry (MALDI-TOF MS) device has revolutionized the routine identification of microorganisms in clinical microbiology laboratories by introducing an easy, rapid, high throughput, low-cost, and efficient identification technique (Seng et al. 2009). This technology has the potential to replace and/or complement conventional bacterial identification techniques. A short-term activity to increase the identification pertussis rate by MALDI-TOF has been recently described (Zintgraff et al. 2018), but further evaluation is needed in order to ensure the capability of the assay to distinguish correctly among different Bordetella species and to measure the sensitivity in patients with previous antibiotic treatment.

2   Pertussis Prevention

The most important way to prevent pertussis is to evoke protective immunity through a complete immunization and to maintain it over time. Generally speaking, the vaccination shows a double effect:

(a) a direct effect: the vaccination protects vaccinated individuals, the target population;

(b) an indirect effect: the vaccination plays a major role in community protection afforded by a vaccination program, (the so-called "herd immunity "or "herd effect").

Both of them are fundamental to contribute to a rapid decrease of childhood mortality due to pertussis and to reduce the circulation of the pathogen in the population.

Nevertheless, there is still a debate concerning the influence of herd immunity on the evolution of the bacterial population, what is called "vaccine-driven selection". On the other hand, there are some publications suggesting how the genetic diversity of B. pertussis is low, decreasing with time, leading to a clonal expansion with the cycles of the disease (Barkoff et al. 2018).

An apparent paradox concerning this disease is that the reported incidence of pertussis has been increasing in several countries despite high vaccination pertussis coverage. While the changing epidemiology of pertussis and several factors that contribute to its resurgence have been extensively described, few data exist on the current burden and characteristics of the disease, especially among older age groups. These factors increase the difficulty to define a correct prevention strategy in a country.

Waning immunity has been increasingly recognized as an important factor in pertussis resurgence (Martinón-Torres et al. 2018; Burdin et al. 2017). Multiple observational studies found that acquired protective immunological memory is relatively short-lived, increasing the risk of pertussis (Koepke et al. 2014). Waning specific antibody levels against B. pertussis, after infection and vaccination, indicate the absence of durable effector mechanisms in the humoral compartment. The role of the evoked antibodies was clearly demonstrated as a key to prevent the attachment of B. pertussis to the cells of the upper and lower respiratory tract; in addition, cell-mediated immunity of the proper CD4+ T helper cell type is also implied, either by its own effector mechanism or by helping the antibody response (Plotkin and Gilbert 2012).

Overall, the above findings are relevant since pertussis affects all age groups. From this assumption, the prevention strategy should involve not only infants (primary vaccination) but also need of booster doses through the entire life should be considered, including maternal immunization. Health care professionals are another important target for vaccination.

One of the main issues in Europe is the lack of a high level of immunization coverage against pertussis, due to several reasons, including vaccine hesitancy. Primary vaccination coverage varies among countries, despite EU and WHO required goals (Sheikh et al. 2018). Although many schedules routinely include booster vaccinations for preschool children, recommendations for adolescents and adults are less common and immunization uptake is low in these groups. In addition to low vaccine uptake, a substantial gap in data on vaccination coverage in the European Region has to be underlined. Efforts to increase vaccination uptake should be accompanied by establishing systems for monitoring vaccination coverage, which is a key indicator of program performance, and essential for understanding gaps and trends in coverage and assessing the impact of the program itself.

To date, the only proven strategy to protect babies immediately after birth is maternal immunization during pregnancy. Maternal antibodies, in particular the IgG isotype, cross the placental barrier and then contribute to the newborn's passive immunity. This helps to protect infants during the first vulnerable weeks of life, when the immune system is still immature and active immunity has not been yet acquired. After birth, antibodies are transferred via breastmilk (Pandolfi et al. 2017); the colostrum contains ten times more pertussis-specific IgA antibodies than IgG after tetanus-diphtheria-pertussis (Tdap) immunization during pregnancy (Faucette et al. 2015).

The first successful maternal immunization experience, that drove a new concept in preventing pertussis infection in newborns, occurred in 2011, in the USA, the first country to recommend pertussis vaccination to pregnant women in the third trimester (CDC 2011); in October 2012, this advice was updated to recommend vaccination in every pregnancy (CDC 2013). Argentina introduced universal maternal pertussis vaccination in February 2012, from 20 weeks of pregnancy (Vizzotti et al. 2015). Then, in the UK, following the national outbreak of pertussis in 2012, a temporary antenatal vaccination program began in October 2012, offering vaccine to pregnant women, ideally at 28–32 weeks of gestation (Gkentzi et al. 2017; Campbell et al. 2018). More recently, several countries, including Italy, introduced maternal immunization against pertussis as recommendation to prevent the disease in newborns (Piano Nazionale Prevenzione Vaccinale 2017–2019). A scientific debate, starting with the study of Halperin et al. (2018), was focused on if and how high levels of transplacentally acquired antibodies might blunt the immune response to active immunization in the newborns (Niewiesk 2014). The inhibitory effect is, however, difficult to ascertain and seems not to affect the pertussis toxin antibody response to DTaP in infants (Englund et al. 1995; Rice et al. 2019). Unlike other studies, a recent study by Saul et al. (2018) showed evidence that the protective effect is stronger in mitigating disease severity than preventing pertussis disease. The overall vaccine effectiveness (VE) reported in this study was lower than in prior studies and suggests that maternal vaccination, while an effective strategy in preventing severe pertussis, is less effective in protecting against infection or mild disease.

The immunization strategies might be improved as follows: in the short term, younger age at first dose and maternal immunization may be effective; in the long term, maintaining high vaccine coverage adopting the current vaccination strategies among adolescents, adults, health workers, and improving disease surveillance (including bacterial surveillance by genomic evaluation and comparison) may contribute to pertussis prevention and control.

3   Pertussis Control

There is an urgent need to improve surveillance and assessment of the disease burden and the impact of infant immunization, with particular focus on fatality among infants <1 year of age, where it is very high. However, pertussis surveillance and control are greatly hampered by the under-notification of cases affecting adults, which usually present mild symptoms and, then, are rarely diagnosed and confirmed, contributing to the spread of B. pertussis. Moreover, especially in children, pertussis symptoms, may be caused by other Bordetellae and/or by viruses (i.e. Respiratory Syncytial Virus, RSV). For all the above-mentioned reasons, it is not easy to avoid pertussis under-notification. A more precise estimate of the number of pertussis cases, which are missed and/or not laboratory confirmed, might be obtained using different data sources.

The objectives of pertussis surveillance, as for other communicable disease surveillance systems, are to:

1. monitor disease burden and the impact of the pertussis vaccination programs, with a special focus on morbidity and mortality among children <5 years of age

2. generate data to update immunization schedules and strategies to increase the impact of vaccination
3. detect pertussis outbreaks

The recommended minimal standard surveillance is a case-based surveillance with laboratory confirmation, in agreement with EU pertussis case definition (Commission implementing decision (EU) 2018/945) and laboratory confirmation of a suspected pertussis case that comprises the use of cultivation and/or molecular methods to identify *B. pertussis* and to differentiate this species from other *Bordetellae* (ECDC Technical document 2012a, b). After 2 weeks from the onset of the disease serological methods are also considered even though the vaccination history of the case need to be considered for the final interpretation of the results.

Among *B. pertussis* strains, an increasing number of pertactin deficient (PRN-) strains have been reported worldwide and recently also in Europe (Barkoff et al. 2019). In that study, a collection of *B. pertussis* isolates (1998–2015) from several EU countries, with different vaccines and vaccination strategies, was investigated. The main finding was an increase of the prevalence of PRN-deficient *B. pertussis* isolates in Europe, which was associated with the time since the introduction of acellular pertussis vaccines (ACVs), as a possible consequence of the selective advantage of the strains (Safarchi et al. 2015; Hegerle and Guiso 2014). However, a long-term surveillance on antigen expression from *B. pertussis*, especially for those included in acellular pertussis vaccine formulation, should be improved since the circulation of deficient strains may have an impact in the efficacy of the vaccines currently in use. Other studies demonstrated how PNR- strains already circulated in the pre-vaccine era, but their circulation was increasing in regions where ACVs were introduced between 10 and 25 years ago (Zeddeman et al. 2014). No significant difference between groups of infants infected by *prn* negative strains for clinical symptoms have been described (Stefanelli et al. 2017).

Macrolides, such as erythromycin and azithromycin, are the first-line drugs of choice for the treatment of *B. pertussis* infection (Kilgore et al. 2016). *B. pertussis* isolates resistant to macrolides have been sporadically identified in Asia and then in America and Europe (Wang et al. 2014; CDC 1994; Guillot et al. 2012). Until now, there have been no standardized screening method or interpretative standards for susceptibility testing of *B. pertussis*. Macrolide resistant *B. pertussis* strains carries an A-to-G transition at nucleotide position 2047 of the 23S rRNA gene (Bartkus et al. 2003) in a region critical for macrolide binding. Sensitivity testing for *B. pertussis* may continue to be warranted to monitor the macrolide effect and to provide early warning of the emergence of resistant strains.

Genomic evaluation of circulating strains is, nowadays, of fundamental importance to monitor virulent strains and the emergence of vaccine-escape variants. Whole genome sequencing (WGS) investigates the evolution of the pathogen even within geographically or temporally defined epidemics. What clearly emerged from several publications is that the *B. pertussis* chromosome includes a large number of repetitive mobile genetic elements facilitating rearrangements that alter essential protein-encoding genes and the gene order. Some of these rearrangements might involve changes in gene expression positively selected (Weigand et al. 2017).

Over the recent years, pertussis outbreaks have been described mainly among adolescents (Hara et al. 2015). Prompt recognition of an outbreak through rapid diagnostic measures is crucial for outbreak containment, together with the actions to reinforce the monitoring of pertussis cases, and evaluate the vaccination status of the population, since waning immunity is one of the main cause of persisting circulation of the pathogen. Moreover, the decision about when and if administering the Post-exposure prophylaxis (PEP) to close contacts plays a crucial part of the disease outbreak control. PEP of close contacts has to be administered considering the severity of the disease, the duration of exposure and the immune-competence of the contact

(Tiwari et al. 2005). In the Cochrane review, the benefit of PEP treatment of pertussis contacts (Altunaiji et al. 2007) was difficult to determine due to insufficient evidence in the general practices.

Several PEP guidelines have been recently available, including those of Public Health Management of Pertussis in England (PHE 2018), which define the contacts as priority groups for public health action. In these guidelines, "vulnerable" contacts (Group 1) are: a) unimmunized infants (born after 32 weeks) less than 2 months of age whose mothers did not receive pertussis vaccine after 16 weeks of pregnancy and at least 2 weeks prior to delivery; b) unimmunized infants (born <32 weeks) less than 2 months of age regardless of maternal vaccine status; c) unimmunized and partially immunized infants (less than 3 doses of vaccine) aged 2 months and above regardless of maternal vaccine status as well as those at risk of transmitting the infection to others at risk of severe disease (Group 2). Group 2 includes Individuals at increased risk of transmitting to 'vulnerable' individuals in 'group 1', who have not received a pertussis containing vaccine more than 1 week and less than 5 years ago: a) pregnant women (>32 weeks gestation); b) healthcare workers working with infants and pregnant women; c) people whose work involves regular, close or prolonged contact with infants too young to be fully vaccinated; d) people who share a household with an infant too young to be fully vaccinated Immunization should be considered for those who have been offered PEP.

Given the limited benefit of PEP, PHE guidelines recommend antibiotic prophylaxis only to close contacts when the onset of disease in the index case is within the preceding 21 days and there is a close contact in one of the priority groups as defined above. Close contacts of a confirmed case (regardless of age and previous immunization history) should be offered PEP with an antibiotic dose as for the treatment of cases (PHE 2018).

WHO guidelines (WHO 2013, 2018) recommends to administer early treatment with macrolide antibiotics to close contacts who are infants <6 months of age, who develop symptoms of a respiratory infection. In addition to early treatment of young infants; some countries (Australian Government Department of Health. Communicable Disease Network of Australia 2015; PHE 2018; NICD 2017) have chosen to administer PEP to asymptomatic high-risk contacts even when symptoms are not present.

In conclusion:

- Pertussis cases and death are reported annually worldwide.
- Pertussis under-reporting and under-diagnosis, especially among adolescents and adults, is still one of the main issue to successfully prevent the disease.
- Although national immunization strategies have traditionally focused on infants and children, policymakers are moving towards a life-course immunization approach.
- Pertussis prevention strategies and vaccine coverage vary across Europe.
- Harmonization in data collection and improving lab-based surveillance are required.
- Genomic studies to monitor vaccine-escape circulating *B. pertussis* strains is needed.

References

Altunaiji S, Kukuruzovic R, Curtis N, Massie J (2007) Antibiotics for whooping cough (pertussis). Cochrane Database Syst Rev (3):CD004404

Australian Government Department of Health (2015) Communicable Disease Network of Australia. CDNA national guidelines for public health units: pertussis [website]. Australian Government Department of Health, Canberra. http://www.health.gov.au/internet/main/publishing.nsf/content/cdna-song-pertussis.htm

Barkoff AM, Mertsola J, Pierard D, Dalby T, Hoegh SV, Guillot S, Stefanelli P, van Gent M, Berbers G, Vestrheim DF, Greve-Isdahl M, Wehlin L, Ljungman M, Fry NK, Markey K, Auranen K, He Q (2018) Surveillance of circulating *Bordetella pertussis* strains in Europe during 1998 to 2015. J Clin Microbiol 56(5). pii: e01998-17

Barkoff AM, Mertsola J, Pierard D, Dalby T, Hoegh SV, Guillot S, Stefanelli P, van Gent M, Berbers G, Vestrheim D, Greve-Isdahl M, Wehlin L,

Ljungman M, Fry NK, Markey K, He Q (2019) Pertactin-deficient *Bordetella pertussis* isolates: evidence of increased circulation in Europe, 1998 to 2015. Euro Surveill 24(7). https://doi.org/10.2807/1560-7917.ES.2019.24.7.1700832

Bartkus JM, Juni BA, Ehresmann K, Miller CA, Sanden GN, Cassiday PK et al (2003) Identification of a mutation associated with erythromycin resistance in *Bordetella pertussis*: implications for surveillance of antimicrobial resistance. J Clin Microbiol 41:1167–1172

Bordet J, Gengou O (1906) Le microbe de la coqueluche. Ann Inst Pasteur (Paris) 20:48–68

Brinig MM, Register KB, Ackermann MR, Relman DA (2006) Genomic features of *Bordetella parapertussis* clades with distinct host species specificity. Genome Biol 7:R81

Burdin N, Handy LK, Plotkin SA (2017) What is wrong with pertussis vaccine immunity? The problem of waning effectiveness of pertussis vaccines. Cold Spring Harb Perspect Biol 9(12):pii: a029454

Campbell H, Gupta S, Dolan GP, Kapadia SJ, Kumar Singh A, Andrews N, Amirthalingam G (2018) Review of vaccination in pregnancy to prevent pertussis in early infancy. J Med Microbiol 67 (10):1426–1456

Centers for Disease Control and Prevention (CDC) (1994) Erythromycin-resistant *Bordetella pertussis*—Yuma County, Arizona, May–October 1994. MMWR Morb Mortal Wkly Rep 43:807–810

Centers for Disease Control and Prevention (CDC) (2011) Updated recommendations for use of tetanus toxoid, reduced diphtheria toxoid and acellular pertussis (Tdap) vaccine from the Advisory Committee on Immunization Practices, 2010. MMWR Morb Mortal Wkly Rep 60:1424–1426

Centers for Disease Control and Prevention (CDC) (2013) Updated recommendations for use of tetanus toxoid, reduced diphtheria toxoid, and acellular pertussis vaccine (Tdap) in pregnant women-Advisory Committee on Immunization Practices (ACIP), 2012. MMWR Morb Mortal Wkly Rep 62:131–135

Cherry JD (1996) Historical review of pertussis and the classical vaccine. J Infect Dis 174:259–263

Cherry JD, Brunell PA, Golden GS, Karzon DT (1988) Report of the task force on pertussis and pertussis immunization—1988. Pediatrics 81:939–984

Commission Implementing Decision (EU) 2018/945 of 22 June 2018 on the communicable diseases and related special health issues to be covered by epidemiological surveillance as well as relevant case definitions

Englund JA, Anderson EL, Reed GF, Decker MD, Edwards KM et al (1995) The effect of maternal antibody on the serologic response and the incidence of adverse reactions after primary immunization with acellular and whole-cell pertussis vaccines combined with diphtheria and tetanus toxoids. Pediatrics 96:580–584

European Center for Disease Prevention and Control (2012a) Guidance and protocol for the use of realtime PCR in laboratory diagnosis of human infection with *Bordetella pertussis* or *Bordetella parapertussis*. https://ecdc.europa.eu/sites/portal/files/media/en/publications/Publications/Guidance-protocol-PCR-laboratory-diagnosis-bordatella-pertussis-parapertussis.pdf

European Center for Disease Prevention and Control (2012b) Guidance and protocol for the serological diagnosis of human infection with *Bordetella pertussis*. https://ecdc.europa.eu/sites/portal/files/media/en/publications/Publications/bordetella-pertussis-guidance-protocol-serological-diagnosis.pdf

Faucette AN, Pawlitz MD, Pei B, Yao F, Chen K (2015) Immunization of pregnant women: future of early infant protection. Hum Vaccin Immunother 11 (11):2549–2555

Fedele G, Leone P, Bellino S, Schiavoni I, Pavia C, Lazzarotto T, Stefanelli P (2018) Diagnostic performance of commercial serological assays measuring *Bordetella pertussis* IgG antibodies. Diagn Microbiol Infect Dis 90(3):157–162

Fry NK, Duncan J, Wagner K, Tzivra O, Doshi N, Litt DJ, Crowcroft N, Miller E, George RC, Harrison TG (2009) Role of PCR in the diagnosis of pertussis infection in infants: 5 years' experience of provision of a same-day real-time PCR service in England and Wales from 2002 to 2007. J Med Microbiol 58:1023–1029

Gkentzi D, Katsakiori P, Marangos M, Hsia Y, Amirthalingam G, Heath PT, Ladhani S (2017) Maternal vaccination against pertussis: a systematic review of the recent literature. Arch Dis Child Fetal Neonatal Ed 102(5):F456–F463

Guillot S, Descours G, Gillet Y, Etienne J, Floret D, Guiso N (2012) Macrolide-resistant *Bordetella pertussis* infection in newborn girl, France. Emerg Infect Dis 18:966–968

Guillot S, Guiso N, Riffelmann M, Wirsing Von König CH (2014) Laboratory manual for the diagnosis of whooping cough caused by *Bordetella pertussis/Bordetella parapertussis*, update 2014. World Health Organization, Geneva

Guiso N, Hegerle N (2014) Other *Bordetellas*, lessons for and from pertussis vaccines. Expert Rev Vaccines 13 (9):1125–1133

Halperin SA, Langley JM, Ye L, MacKinnon-Cameron D, Elsherif M, Allen VM, Smith B, Halperin BA, McNeil SA, Vanderkooi OG, Dwinnell S, Wilson RD, Tapiero B, Boucher M, Le Saux N, Gruslin A, Vaudry W, Chandra S, Dobson S, Money D (2018) A randomized controlled trial of the safety and immunogenicity of tetanus, diphtheria, and acellular pertussis vaccine immunization during pregnancy and subsequent infant immune response. Clin Infect Dis 67(7):1063–1071

Hara M, Fukuoka M, Tashiro K et al (2015) Pertussis outbreak in university students and evaluation of acellular pertussis vaccine effectiveness in Japan. BMC Infect Dis 15:45

Hegerle N, Guiso N (2014) *Bordetella pertussis* and pertactin-deficient clinical isolates: lessons for pertussis vaccines. Expert Rev Vaccines 13(9):1135–1146

Kilgore PE, Salim AM, Zervos MJ, Schmitt HJ (2016) Pertussis: microbiology, disease, treatment, and prevention. Clin Microbiol Rev 29:449–486

Koepke R, Eickhoff JC, Ayele RA et al (2014) Estimating the effectiveness of tetanus, diphtheria, acellular pertussis vaccine (Tdap) for preventing pertussis: evidence of rapidly waning immunity and difference in effectiveness by Tdap brand. J Infect Dis 210 (6):942–953

Le Coustumier A, Njamkepo E, Cattoir V, Guillot S (2011) Guiso N. *Bordetella petrii* infection with long-lasting persistence in human. Emerg Infect Dis 17 (4):612–618

Lee AD, Cassiday PK, Pawloski LC, Tatti KM, Martin MD, Briere EC, Lucia Tondella M, Martin SW (2018) Clinical evaluation and validation of laboratory methods for the diagnosis of *Bordetella pertussis* infection: culture, polymerase chain reaction (PCR) and anti-pertussis toxin IgG serology (IgG-PT). PLoS One 13:e0195979

Martinón-Torres F, Heininger U, Thomson A, Wirsing von König CH (2018) Controlling pertussis: how can we do it? A focus on immunization. Expert Rev Vaccines 17(4):289–297

Mattoo S, Cherry JD (2005) Molecular pathogenesis, epidemiology, and clinical manifestations of respiratory infections due to *Bordetella pertussis* and other Bordetella subspecies. Clin Microbiol Rev 18:326–382

National Institute for Communicable Diseases (2017) Pertussis: NICD recommendations for diagnosis, management and public health response. National Institute for Communicable Diseases, Johannesburg. http://www.nicd.ac.za/wp-content/uploads/2017/03/Guidelines_pertussis_v1_20-December-2017_Final.pdf

Niewiesk S (2014) Maternal antibodies: clinical significance, mechanism of interference with immune responses, and possible vaccination strategies. Front Immunol 5:446

Pandolfi El, Gesualdo F, Carloni E, Villani A, Midulla F, Carsetti R, Stefanelli P, Fedele G, Tozzi AE (2017) Pertussis study group. Does breastfeeding protect young infants from pertussis? Case-control study and immunologic evaluation. Pediatr Infect Dis J 36(3): e48–e53

Piano Nazionale Prevenzione Vaccinale, PNPV 2017–2019. http://www.salute.gov.it/imgs/C_17_pubblicazioni_2571_allegato.pdf

Pittet LF, Posfay-Barbe KM (2015) *Bordetella holmesii* infection: current knowledge and a vision for future research. Expert Rev Anti-Infect Ther 13(8):965–971

Plotkin SA, Gilbert PB (2012) Nomenclature for immune correlates of protection after vaccination. Clin Infect Dis 54:1615–1617

Porter JF, Connor K, Donachie W (1994) Isolation and characterization of *Bordetella parapertussis*-like bacteria from ovine lungs. Microbiology 140 (Pt 2):255–261

Public Health England (2018) Guidelines for the public health management of pertussis in England. Public Health England, London. https://www.gov.uk/government/uploads/system/uploads/attachment_data/file/576061/Guidelines_for_the_Public_Health_Management_of_Pertussis_in_England.pdf

Rice TF, Diavatopoulos DA, Smits GP, van Gageldonk PGM, Berbers GAM, van der Klis FR, Vamvakas G, Donaldson B, Bouqueau M, Holder B, Kampmann B (2019) Antibody responses to *Bordetella pertussis* and other childhood vaccines in infants born to mothers who received pertussis vaccine in pregnancy – a prospective, observational cohort study from the United Kingdom. Clin Exp Immunol. https://doi.org/10.1111/cei.13275

Safarchi A, Octavia S, Luu LD, Tay CY, Sintchenko V, Wood N, Marshall H, McIntyre P, Lan R (2015) Pertactin negative *Bordetella pertussis* demonstrates higher fitness under vaccine selection pressure in a mixed infection model. Vaccine 33(46):6277–6281

Saul N, Wang K, Bag S, Baldwin H, Alexander K, Chandra M, Thomas J, Quinn H, Sheppeard V, Conaty S (2018) Effectiveness of maternal pertussis vaccination in preventing infection and disease in infants: The NSW Public Health Network case-control study. Vaccine 36:1887–1892

Seng P, Drancourt M, Gouriet F, La Scola B, Fournier PE et al (2009) Ongoing revolution in bacteriology: routine identification of bacteria by matrix-assisted laser desorption ionization time-of-flight mass spectrometry. Clin Infect Dis 49(4):543–551

Sheikh S, Biundo E, Courcier S, Damm O, Launay O, Maes E, Marcos C, Matthews S, Meijer C, Poscia A, Postma M, Saka O, Szucs T, Begg N (2018) A report on the status of vaccination in Europe. Vaccine 36 (33):4979–4992

Stefanelli P, Gabriele B, Paola V, Tozzi AE, Fabio M, Rita C, Giorgio F, Alberto V, Carlo C, and The Pertussis Study Group (2017) Severe pertussis infection in infants less than 6 months of age: clinical manifestations and molecular characterization. Hum Vaccin Immunother 13(5):1073–1077

Tiwari T, Murphy TV, Moran J (2005) Recommended antimicrobial agents for the treatment and postexposure prophylaxis of pertussis: 2005 CDC Guidelines. MMWR Recomm Rep 54:1–16

Vizzotti C, Neyro S, Katz N, Juárez MV, Perez Carrega ME et al (2015) Maternal immunization in Argentina: a storyline from the prospective of a middle income country. Vaccine 33:6413–6419

Wang Z, Cui Z, Li Y, Hou T, Liu X, Xi Y et al (2014) High prevalence of erythromycin-resistant *Bordetella pertussis* in Xi'an, China. Clin Microbiol Infect 20: O825–O830

Weigand MR, Peng Y, Loparev V, Batra D, Bowden KE, Burroughs M, Cassiday PK, Davis JK, Johnson T, Juieng P, Knipe K, Mathis MH, Pruitt AM, Rowe L, Sheth M, Tondella ML, Williams MM (2017) The history of *Bordetella pertussis* genome evolution includes structural rearrangement. J Bacteriol 199(8): pii: e00806-16

Woolfrey BF, Moody JA (1991) Human infections associated with *Bordetella bronchiseptica*. Clin Microbiol Rev 4(3):243–245

World Health Organization (WHO) (2013) Pocket book for hospital care of children: guidelines for the management of common childhood illnesses, 2nd edn. World Health Organization, Geneva. http://apps.who.int/iris/bitstream/10665/81170/1/9789241548373_eng.pdf?ua=1

World Health Organization (WHO) (2018) Vaccine preventable diseases surveillance standards. Pertussis. Last updated September 5, 2018. https://www.who.int/immunization/monitoring_surveillance/burden/vpd/standards/en/

Zeddeman A, van Gent M, Heuvelman CJ, van der Heide HG, Bart MJ, Advani A, Hallander HO, Wirsing von Konig CH, Riffelman M, Storsaeter J, Vestrheim DF, Dalby T, Krogfelt KA, Fry NK, Barkoff AM, Mertsola J, He Q, Mooi F (2014) Investigations into the emergence of pertactin-deficient *Bordetella pertussis* isolates in six European countries, 1996 to 2012. Euro Surveill 19(33):pii: 20881

Zintgraff J, Irazu L, Lara CS, Rodriguez M, Santos M (2018) The classical *Bordetella* species and MALDI-TOF technology: a brief experience. J Med Microbiol 67(12):1737–1742

Adv Exp Med Biol - Advances in Microbiology, Infectious Diseases and Public Health (2019) 1183: 137–149
https://doi.org/10.1007/5584_2019_409
© Springer Nature Switzerland AG 2019
Published online: 25 July 2019

Pertussis in Low and Medium Income Countries: A Pragmatic Approach

Nicole Guiso and Fabien Taieb

Abstract

Pertussis is a vaccine preventable disease since late 1940s. However, it is still endemic in all countries and occurs in epidemic cycles. The number of cases/deaths has decreased during the last decade but a high number of deaths persists in Low and Medium Income Countries (LMIC). The epidemiological situation in LMIC is not precisely known due to lack of surveillance and specific diagnostic tools. A pragmatic approach in these countries should be to establish; (i) a hospital-based surveillance in the largest cities of the country with clinicians and nurses trained to detect clinical symptoms, to obtain biological samples for specific analysis and diagnosis; (ii) a reference laboratory as part of an international network of reference laboratories, under quality assurance, and able to perform at least PCR diagnosis and if possible detection of antibiotic resistance. This surveillance network would allow specific diagnosis of pertussis and facilitate the reporting of cases at national level, thereby improving awareness of the disease at clinician, population and decision maker levels. This network could allow a better evaluation of vaccine coverage, timely vaccination and impacts of modification of national vaccine strategy or type of pertussis vaccine used.

Collaboration between this network and basic scientists should be strengthened through translational research projects in order to improve fundamental knowledge on pertussis in LMIC and help clinicians' access to specific diagnostic tools.

Keywords

Diagnosis · Low and medium income countries · Pertussis · Surveillance · Vaccines

1 Introduction

Whooping cough, often called pertussis, is a highly infectious respiratory disease particularly for infants and persons at risk such as pregnant women and elderly. It is a vaccine preventable disease. Vaccine is available since late 1940's in North America, late 1950's in Europe and since 1974, with the Expanded Programme on Immunization (EPI), in the rest of the world. Despite high vaccination coverage, the disease is still endemic in all countries and occurs in epidemic cycles every 3–5 years, whatever the type of pertussis vaccines used. The intensity of the cycles varies among regions mostly due to demographic differences, different vaccine coverage and differences in the surveillance of the disease, when it exists (Broutin et al. 2010; Fine and Clarkson 1982). Before vaccines became widely available, pertussis was one of the most common paediatric diseases worldwide. It is a recent

N. Guiso (✉) and F. Taieb
Institut Pasteur, Centre of Translational research, Paris, France
e-mail: nicole.guiso@pasteur.fr

disease since the first description of the disease as an epidemic was made in Persia around one century before the one described by G de Baillou in 1578 in Paris (Aslanabadi et al. 2015). The unique features of the Persian epidemics, suggest that whooping cough emerged as a pandemic (and later as an endemic) disease 500–600 years ago, estimation which correlates with the studies performed on the agent of the disease *Bordetella pertussis* (Bart et al. 2014). The disease was probably then introduced into Europe from Western Asia.

During the pre-vaccination period according to few historical descriptions cases occurred mainly in children <5 years of age and less in adults (WHO 2015). The introduction of vaccination induced a dramatic decline in the incidence of the disease with a decline in mortality. In 2014, according to WHO estimates, pertussis was still causing 160,000 deaths and the number of cases in children <5 years of age at 24 million, a significant reduction as compared to the 1999 estimations (Yeung et al. 2017). There is, however, considerable uncertainty over these estimates since the sensitivity analysis ranges the number deaths between 38,000 and 670,000 among which the African region had the greatest share (von Koenig and Guiso 2017).

Around three decades after the introduction of vaccination, of infants and young children only, since the disease was considered as paediatric, a shift in the age distribution of pertussis towards older age groups (adolescents and young adults) has been first reported in High Income Countries (HIC) such as in United States of America (USA) and then in Europe (Baron et al. 1998; Bass and Stephenson 1987; Zepp et al. 2011). This shift induced an increase of the incidence in non-vaccinated infants less than 2 months of age contaminated by an adolescent or an adult. This change was observed in the late 1980's after the use of the whole cell pertussis vaccines (wPV) and since a few years also after the use of acellular pertussis vaccine (aPV) (WHO 2015; Sweden PHAo 2018). The age shift may, in part, be due to more sensitive diagnosis and surveillance covering not only young children but also adolescents and adults, change of wPV to aPV,

adaptation of the agent of the disease to vaccine pressure. However, one of the major cause is probably the waning of immunity induced by infant vaccination combined with less natural boosters in the population since the circulation of the agent of the disease decreased after the introduction of vaccination. Furthermore, low vaccine coverage as well as timeliness routine immunization can also participate to the non-control of pertussis around the world. In response to the waning immunity induced by wPV or aPV, several HIC countries introduced vaccine boosters for older children, adolescents, adults and recently maternal immunization with aPV, since it is not possible to use wPV after 6 years of age due to its reactogenicity (WHO 2015). In most of the Low and Medium Income Countries (LMIC) wPV are still used, the surveillance is mostly clinical and the vaccine strategy includes either only primary vaccination or primary vaccination and one vaccine booster.

The objective of this review is Pertussis in LMIC and what could be a pragmatic approach to increase the control of the disease in these countries.

2 Disease

Whooping cough is a strictly Human respiratory disease occurring after transmission of the bacteria from person- to-person in airborne droplets. The bacteria are highly infectious. It is not only a childhood disease, it can affect a person whatever her age. It is dramatic for neonates and infants less than 3 months of age but can also be very severe for children and adults.

In non-immune patients, the classical disease follows three phases. The first, called the *catarrhal phase*, starts after an incubation period of 7–10 days with non-specific symptoms, such as rhinorrhoea, sneezing and non-specific coughs. Typically, the patients do not develop fever. The second phase is called the *paroxysmal phase* and is characterized by the specific cough, inspiratory whooping, vomiting. The third phase is the *convalescent phase* with a decrease of the cough (WHO 2014a; Cherry 2016; Cherry et al. 2012).

Cases in neonates and unvaccinated young infants often present with non-specific coughs and apnoea as the only symptom.

In older, vaccinated schoolchildren, adolescents and adults, the symptoms can vary widely. Adult pertussis is often associated with a long illness and the persistent cough is often paroxysmal and has a mean duration of approximately 6 weeks. It is frequently accompanied by choking, vomiting and by whooping (Cherry et al. 2012; WHO 2014a). *B. pertussis*, the agent of the disease circulates all around the world whatever the vaccine strategy and reinfection can occur throughout a person life.

Major complications of pertussis in infants and children are pulmonary, neurologic (acute pertussis encephalopathy) and nutritional. In LMIC the average Case Fatality Rate (CFR) for pertussis has been estimated at almost 4% in infants aged <1 year and at 1% in children aged 1–4 years (WHO 2015).

3   Agents of the Disease

Bordetella pertussis is a strictly human pathogen. Pertussis infection is mainly due to this Gram negative bacterium but can also be due to another species, *Bordetella parapertussis* to a less extent (Guiso and Hegerle 2014). The infection begins with the attachment of the bacteria to the ciliated epithelium of the respiratory tract; the subsequent manifestations are thought to be due to a team of virulence factors including toxins and adhesins. These virulence factors include pertussis toxin or PT, adenylate cyclase-hemolysin or AC-Hly, filamentous haemagglutinin or FHA, pertactin or PRN, fimbrial proteins or FIM.

The genomes of *B. pertussis* and *B. parapertussis* reference strains and of several clinical isolates circulating around the world as well as of some other *Bordetella* spp. have been sequenced and are publicly available (Bouchez and Guiso 2013; Gross et al. 2010; Linz et al. 2016; Parkhill et al. 2003; Sebaihia et al. 2006).

Compared to other human pathogens, isolates of *B. pertussis* show only small genomic heterogeneity, suggesting a more recent development as

a human pathogen, but the population structure of *B. pertussis* is constantly evolving (Bart et al. 2014; Linz et al. 2016).

It is now well established that *B. pertussis* circulating during the pre-vaccine era and the post-wPV era are different (Bart et al. 2014; Bouchez and Guiso 2015). Vaccination with wPV controlled the vaccine strains type but not all isolates. However, as mentioned by WHO, there is no proof that the circulation of these isolates decreased the effectiveness of the wPV (WHO 2014b, 2015).

Since the end of 2000's *B. pertussis* isolates non producing PRN are circulating in regions using aPV. These isolates are as virulent as those producing PRN in infants. Recently, it was shown that the effectiveness of aPV didn't change in regions where 90% of the isolates are not producing PRN (Breakwell et al. 2016; Hegerle and Guiso 2014).

4   Pertussis Vaccines

4.1   Whole Cell Pertussis Vaccines

All wPV are suspensions of killed *B. pertussis* which are prepared by cultures that favour the expression of the virulence factors. All wPV are combined with diphtheria toxoid and tetanus toxoid and often now also with other vaccines routinely administered during infancy, such as *Haemophilus influenzae* type b (Hib) and hepatitis B (HepB). The reactogenicity of wPV was considered too high for use in older children, adolescents and adults. For this reason wPV are not licensed for routine use in persons older than 6 years.

Although the production process of wPV appears to be simple, significant differences have been observed in the immunogenicity, efficacy and effectiveness of these vaccines from different producers (WHO 2017). The vaccine efficacy estimates varied from 98% to 96% with a German vaccine, 94–96% in France and in Senegal with the French vaccine, 36–48% for the American vaccine in Sweden and Italy. The effectiveness of the Dutch vaccine was estimated to be 51% and that of the Canadian one 57% (WHO 2017). Estimates for effectiveness of these products in LMIC are lacking. In 2014, WHO has developed a set of recommendations for quality (production and lot-release), safety and efficacy of wP vaccines (WHO 2014c). However, there are still divergences of regulatory requirements and processes in LMIC which contribute to delay access to high-quality, safe and efficacious vaccines for their respective populations. wPV may have different antigenic content, ways of production and controls, leading to variations in post-vaccination immune response (Dellepiane et al. 2018). Better coordination of evaluation procedures are urgently needed.

4.2 Acellular Pertussis Vaccines

Characterization of the roles of PT, FHA, PRN and FIM during the disease and concerns about the safety of wPV conducted to the development of aPV. All aPV are associated with significantly fewer and less serious side-effects, and thus the replacement of the wPV was mainly driven by the safety profile of these vaccines (WHO 2017). The other important advantage of the aPV is the reproducible production process with its use of purified antigens. All aPV contained PT inactivated chemically or genetically (aPV1). Some of them contain only PT, others associate with FHA (aPV2) or with FHA and PRN (aPV3) or with FHA, PRN and FIMs (aPV5). Due to their safety profile, aPV can be used for vaccine boosters in older children, adolescents and adults (WHO 2015).

4.3 Correlates of Protection

No correlate of protection for humoral or cell-mediated immunities against the different pertussis antigens has been determined so far. Overall, it seems most probable that no single correlate of protection exists and that antibodies to many antigens in conjunction with cell-mediated immunity, confer protection (Plotkin 2013).

5 Vaccine Strategy Recommended by WHO

5.1 Children Less Than 5 Years of Age

Epidemiological observations suggest that the efficacy of pertussis vaccine is high only for a limited period and wanes after immunization. In 2014, WHO concluded that there is increasing and consistent evidence, both from observational and analytical studies, from a number of countries using wPV and aPV showing that a single dose of either vaccine have an effectiveness in preventing severe disease, hospitalization and death and that two doses of either pertussis vaccines offers higher protection (83–87%) (WHO 2014b). However, a study in rural Kenya was conducted before and after mass wP vaccination. For almost 5 years after vaccination, pertussis incidence was reduced in both DTP2 (two doses of diphtheria–tetanus–pertussis) and DTP3 (three doses of DTP) vaccinated groups but waning of immunity was much more rapid in those vaccinated with DTP2 (Muloiwa et al. 2018).

The WHO recommendation include that "*all children worldwide, including HIV-positive individuals, should be immunized against pertussis. Every country should seek to achieve early and timely vaccination initiated at 6 weeks and no later than 8 weeks of age, and maintain high coverage (≥90%) with at least 3 doses of assured quality pertussis vaccine This will ensure high levels of protection in children in the <5 year age group. Any reduction in overall coverage can lead to an increase in cases of pertussis. The reactogenicity of wP vaccines is significantly reduced when given in early short timeframe schedules*" (WHO 2015). However, the duration of protection following primary immunization varies considerably depending upon factors such as local epidemiology, vaccine manufacturer, vaccine coverage, vaccination schedule and timely vaccination. Therefore, WHO recommends also a booster dose for children aged 1–6 years, preferably during the second year of life ≥6 months after last primary dose (WHO 2015).

5.2   Vaccination of Pregnant Women

Vaccination of pregnant women, after 27th weeks of gestation, is likely to be the most cost-effective additional strategy for preventing at least severe disease in infants too young to be vaccinated (Saul et al. 2018). National programmes may consider the vaccination of pregnant women with one dose of aPV as a strategy additional to routine primary infant pertussis vaccination in countries or settings with high morbidity/mortality from pertussis (Forsyth et al. 2018; WHO 2015). In the recent study of Halperin et al., it is shown that maternal immunization is safe and induces high level of pertussis antibodies that are transmitted to the foetus (Halperin et al. 2018). However, this study confirms that there is a blunting of the immune response to primary series immunization in infants of women immunized with aPV during pregnancy (This phenomenom is described in details in the "Immune response to pertussis vaccination" in the present book). This blunting was significant for several pertussis vaccine components before and after a vaccine booster at 12 months. According to the authors since all deaths from pertussis occur in infants <3 months of age, maternal immunization is achieving its goal of preventing pertussis deaths in the first months of life. However, this blunting, could lead to an increased burden of pertussis disease in the second half of the first year of life or afterward. This could be a particular concern in LMIC where children are receiving only a primary immunization with wPV. In addition, the study was unable to assess the effect of the combined vaccine during pregnancy on the infants' response to diphtheria and tetanus immunizations or to vaccines using a tetanus or diphtheria toxoid–based conjugate which could also be a major problem in LMIC.

Furthermore, introducing this strategy will need the knowledge, attitudes, and beliefs around vaccines which are key determinants to vaccine acceptance in LMIC (Sobanjo-Ter Meulen et al. 2016).

5.3   Vaccine Coverage and Timely Vaccination

The vaccine coverage has increased in the past decades through the Global Immunization Vision and Strategy and the Global Vaccine Action Plan (Muloiwa et al. 2018). However, this coverage remains frequently under 90% whereas WHO is recommending a coverage above this threshold (WHO 2015). The barriers include funding's, training of public health authorities and health care workers, vaccine hesitancy and delivery chain.

A recent survey on public health priorities conducted in six African nations (Ghana, Kenya, Nigeria, Senegal, South Africa, and Uganda) revealed that 76% of responders considered increasing child vaccination coverage to be less important than improving hospitals (Muloiwa et al. 2018).

In Africa, childhood immunization coverage remains low (74% for DTP3 in 2016) and poor disease awareness among the general population was a major reason for this (Muloiwa et al. 2018).

Furthermore, in a recent study in Taiwan, it was shown that approximately up to one-fifth pertussis cases in children aged 3–35 months could have been prevented with on-time vaccination (Huang et al. 2017). On time vaccination is rarely taking into account in LMIC as well as in HIC (Kurova et al. 2018).

LMIC continue to have a substantial burden of vaccine-preventable diseases, and their vaccination coverage is lagging behind that of HIC. As mentioned recently, the extent of the challenges facing vaccination programmes in LMIC is beginning to be recognised. WHO has convened a Task Force to develop a coordinated strategy to enhance sustainable access to vaccines in these countries (Turner et al. 2018). However, a lot remains to do.

6   Surveillance of the Disease

6.1   Awareness of the Disease

Under-recognition, under-reporting, misdiagnosis, lack of disease awareness (by the population and the health care workers), and other health priorities are major problems in LMIC.

The decrease of the incidence of the disease after introduction of vaccination might have decreased its awareness all around the world. It is often observed by the number of times parents have seen a physician before whooping cough was diagnosed. Moreover, in a clinician point of view, the non-specific clinical signs, especially at the beginning of the disease (*catarrhal phase*), makes identification of the disease complicated in a setting with no specific biological diagnostic tools. In a patient point of view, in a context of fee-for-service health care, only the most serious cases can be referred to the hospital.

In most of LMIC there is no surveillance of pertussis since (i) in general national surveillance systems are complicated to set up sustainably and especially for specific disease and not syndromic surveillance (ii) the disease is not really perceived as a dangerous one (iii) the disease is often classified with pneumonia or other respiratory diseases and notifications of pertussis deaths or cases are not always reported (iv) there is often no possibility to confirm the diagnosis (Guiso and Wirsing von Konig 2016).

Since the end of 2000's some countries are trying to implement biological diagnosis in order to confirm the diagnosis. There are some progresses in Asia and in Africa as reflected by recent reviews or studies (Muloiwa et al. 2018; Son et al. 2019).

The type of surveillance to be established in a region depends on the question being asked (incidence, effectiveness, duration of protection) but will always need a specific and sensitive biological diagnosis.

6.2   Laboratory Diagnosis

The WHO laboratory diagnosis manual of whooping cough was updated in 2014 describing the two types of approaches to diagnosis: direct and indirect (WHO 2014a).

Direct diagnosis consists of identifying the microorganism responsible for the disease, either by culture or by real-time polymerase chain reaction (RT-PCR). Indirect diagnosis is, essentially, by serology and consists of detecting specific anti-pertussis toxin antibodies, these antibodies being the only specific ones (Guiso et al. 2011).

For direct diagnosis, nasopharyngeal aspirates (NPA) or swabs (NPS) are required and this technique needs training as it was mentioned in several publications (WHO 2014a). It was previously shown that a 15% gain in the isolation rate is obtained using NPA compared to NPS in new-borns and infants (WHO 2014a). However, NPA collection is found sometimes uncomfortable and requires skilled personnel and often NPS is now used for infants. For older subjects induced sputum can be used (Nunes et al. 2016).

Culture is the most specific diagnosis but is not very sensitive since samples must be collected within the two first weeks after the beginning of the cough. The percentage of success is generally no higher than 60% even in neonates (WHO 2014a). It is a long process. However, it is important to continue to culture in order, at least, to perform surveillance of the resistance to macrolides which seems to increase in some regions where the delivery of antibiotics is free (Wang et al. 2014).

The other direct method now widely used is Real Time-PCR which is more sensitive than bacterial culture and faster. However, this technique needs expensive equipment and experimented technicians, expensive reagents, regular quality controls...

Indirect diagnosis (serology) consists of detecting specific anti-PT antibodies in the serum of infected individuals after 3 weeks of the cough. However, serology can not be used in infants, as their immune system is immature and liable to interference of maternal antibodies, or in patients vaccinated within 1 year since antibodies after vaccination or infection can't be differentiated. The presence of a high level of anti-PT antibodies in the serum of a non-vaccinated individual indicates infection. However, the choice of the diagnostic kits is important since it required purified antigen and validation (Guiso et al. 2011).

6.3 Type of Surveillance

Different types of surveillance can be established (Guiso and Wirsing von Konig 2016). Since WHO recommends to estimate the burden of pertussis in children less than 5 years, the first and cheapest approach to establish is the hospital-based surveillance which can give an estimate of severe and life-threatening disease. A questionnaire of household contacts as well as sampling is easier in a hospital setting, and the hospital laboratory can perform culture and PCR for the infant and PCR or serology for the household contacts.

Latter on, when the hospital paediatricians and biologists are trained and able to establish quickly a diagnosis, paediatricians and general practioners of the region can be trained and a Sentinel surveillance can be insured. Once this type of surveillance is established, during an outbreak, cases can be confirmed as soon as possible so that the appropriate control measures can be taken.

National sero-surveys serves as a marker allowing to estimate the circulation of *B. pertussis* in a population at a given time point, irrespective of any clinical symptoms. As pertussis is a cyclical disease, sero-surveys do only reflect the circulation of the bacteria in the population at a given time point, and the differences between through years and peak years of a cycle can be huge (Guiso and Wirsing von Konig 2016).

7 Incidence of the Disease in 2019

Surveillance of pertussis incidence and mortality is of high importance in LMIC as adequate vaccine strategy must be based on data (WHO 2015). In 2015, more than half of all estimated pertussis deaths of children <5 years of age occurred in Africa (Sobanjo-Ter Meulen et al. 2016). In 2016, Kampmann and Mackenzie wrote *"Due to the lack of systematic and targeted surveillance with laboratory confirmation of B. pertussis infection, we cannot definitively conclude that pertussis disease is well controlled in West Africa. However, based on observations by clinicians and ongoing demographic surveillance systems that capture morbidity and mortality data in general terms, currently there is no evidence that pertussis causes a significant burden of disease in young children in West Africa"* (Kampmann and Mackenzie 2016). However, since the 2010's there is some hope. Surveillance is beginning to be established in some LMIC such as North and South Africa, China, Malaysia, Iran, Korea, Taiwan, Thailand...

As reviewed recently by Global Pertussis Initiative (GPI), hospital-based surveillance was established in North and South Africa. In South Africa case fatality rate was shown to vary between 1.8 and 5% and to be higher in HIV infected children (7.5% vs 21.8% for non-exposed) (Muloiwa et al. 2018). Among children less than 5 years old who were hospitalized with severe pneumonia as part of the Pneumonia Etiology Research for Child Health (PERCH) study between 2011 and 2014, the pertussis detection rate was 4.0% in children <6 months old and pertussis was identified in 3.7% of 137 in-hospital deaths among African cases in this age group (Barger-Kamate et al. 2016).

In Algeria, Morocco and Tunisia hospital-based surveillance was also set up in the biggest cities of the countries and are still on-going (Muloiwa et al. 2018). The last hospital-based surveillance data are showing that pertussis is still present as a cyclical disease in Tunisia, despite high primo-vaccination coverage with wPV, and the estimated pertussis incidence in the Tunis area is 134/100,000 in children aged less than 5 years (Ben Fraj et al. 2018). Furthermore, a cross sectional study to determine the sero-prevalence of health care workers in the Tunis hospital shows that 11.4% were positive and sero-prevalence was higher in the 21–31 years of age group (Ben Fraj et al. 2019). Further efforts are now required to develop and expand the surveillance in Tunisia.

Sero-surveys were recently performed in several LMIC. A large study was performed in Asian children and adolescents confirming the circulation of *B. pertussis* with 5% of individuals having evidence of recent infection (Son et al. 2019). Similar results were obtained in Gambia (Scott et al. 2015) and China (Chen et al. 2016; Yao et al. 2018).

8   The Pragmatic Approach

It is now well established that infections or primary vaccinations of infants don't grant a lifelong protection and it is now well established that the source of transmission is now mainly adults and adolescents. The latter contaminate unvaccinated or partially vaccinated infants leading to the most severe cases and deaths. To protect infants and, in particular, those younger than 2 months of age, HIC have decided to introduce vaccine boosters using aPV for school children, adolescents, adults and have introduced cocooning and maternal strategies.

In LMIC, where the majority of cases and deaths occur, the strategy is based on: i) a primary vaccination with wPV only ii) primary vaccination and one booster, rarely two boosters (due to financial limits regarding to the cost of aPV). Furthermore, whereas primary vaccination coverage seems relatively high (although not always >90%), neither the vaccine coverage at the age of the vaccine boosters is known nor if these boosters are given on time.

Then, what can be the pragmatic approach to change this situation?

8.1   Establishment of a Hospital-Based Surveillance

As previously mentioned development of a hospital-based surveillance is the cheapest and most cost-effective way to establish a surveillance. When a patient is hospitalized with a pertussis suspicion, based on WHO criteria, a quick questionnaire can be filled up, a nasopharyngeal sampling can be obtained and a specific biological diagnosis can be performed.

However, the first objective should be primarily to educate, at the regional level, Health Care Personnel in the hospital to detect the clinical symptoms in infants and in young children and to be able to perform the nasopharyngeal sampling on the patient. This training could be performed by experts using video conferences. Means of transport of the biological sample are now available allowing a transport in less than 4–6 h for culture or 24–48 h for PCR to the reference laboratory.

The establishment of this type of surveillance network would:

– give access to specific pertussis diagnosis to the patients and clinicians
– facilitate the reporting of cases at national level
– evaluate efficacy and impacts of modification of national vaccine strategy or type of pertussis vaccine used

8.2   Access to Specific Diagnosis

Access to specific diagnosis as discussed previously seems to be extremely important, in particular due to the lack of specificity of the clinical symptoms. This would allow an increase of the awareness of the disease in the population in general as well as in the medical and decision-makers community. The increase of the awareness might improve (i) at the population level, the compliance and comprehension towards vaccination (ii) at the medical community level, the early recognition of the disease allowing the establishment of respiratory isolation procedures preventing further contamination especially at hospital where shared rooms are common. Moreover, specific procedures could be established to treat all contact cases in order to stop transmission of the disease at community as well as hospital level.

8.3 Reference Laboratory

Establishment of a network of regional laboratories is crucial and should be linked to a reference laboratory in the biggest hospital of the country. Culture is a long, complex and expensive process. However, it has always been considered as the gold standard and can usefully be employed to analyse the evolution of the clinical isolates and their resistance to macrolides which is increasing in some regions (Wang et al. 2014). In the near future sequencing will replace culture; in this respect it has already been shown that analysis of the evolution of the isolates can be made directly on DNA extracted from NPA or NPS (Moriuchi et al. 2017). PCR or sequencing request skilled staff, expensive equipment and reagents as well as appropriated locations. This diagnostic tools can't be available immediately in all hospital of LMIC. Moreover, it is important to keep in mind that he price of PCR materials and reagents is higher in LMIC than in HIC since the numbers of kits bought is likely to be low. WHO should recommend the manufacturers of equipment and kits to decrease the prices when a reference laboratory is built up in a LMIC since this will be followed by a surveillance of the disease in the country. Basic science should also be enhanced to develop in the near future point of care tests (POCTs).

It is also important to stress the fact that the main role of a reference laboratory, in addition to national surveillance, confirmation of complex cases and bacterial resistance testing (as mentioned above), would be to ensure quality control of the results reported by the regional laboratories through samples and quality control cross-checking, sample panels and training of laboratory and clinicians staff.

Thus, the establishment of such a network would make it possible to evaluate the effectiveness of the vaccines used, the duration of the vaccine induced protection and the vaccine strategies used.

8.4 Quality of wPV

The recent analysis undertaken by the global registration experts highlighted the divergence of regulatory requirements for registration of vaccines worldwide (Dellepiane et al. 2018). It was agreed that vaccine registration processes should be streamlined and redundancies removed towards enabling faster access to vaccines in LMIC.

It should also be important to establish pharmacovigilance in all countries following changes of combination vaccines or introduction of a new type of vaccine or a new strategy such as maternal immunization.

8.5 Vaccine Strategy, Vaccine Coverage and Timely Vaccination

As recommended by WHO, before any amendment of national vaccine strategy is introduced, description and analysis of the national, current epidemiologic situation at country level are needed, including vaccine coverage estimation and surveillance of timely vaccination (WHO 2015).

The first objective of a country, in a context a limited financial resources, should be the decrease of deaths and severe cases, e.g. the protection of infants too young to be vaccinated or partially vaccinated. In this context, primary-vaccination and 18-months old booster should be prioritized with high-coverage surveillance. Whenever outbreaks or sudden increase of cases is taking place, maternal immunization should also be introduced. However, introduction of maternal immunization (i) needs acceptance (ii) needs high coverage (iii) should be followed by long term surveillance. Indeed, the following questions are still unanswered: (a) will maternal immunization induce a decrease in mortality of

Table 1 Implementation of a Pertussis surveillance in LMIC

What to implement?	Which impact expected?
An hospital-based surveillance after training of clinicians about the clinical presentation and the severity of the disease in the different age groups of the population	Facilitate the reporting of the cases coming to hospital
	Improve the quality of the estimation of incidence and burden of the disease
	Document the impact of local vaccine strategy modification
A National Centre of Reference under quality assurance and able to perform recommended pertussis biological diagnosis	Give access to specific diagnosis to the clinicians and patients
	Improve the awareness of the Health Care Workers and the population
	Help to stop transmission of the disease at hospital and community levels
	Implement Research studies (antibiotic resistance, evaluation of new diagnostic tests, etc.)
	Increase of the number of local experts
	Implement an international network
A regular control of vaccination coverage and of timely vaccination	Reach the goal of >90% of vaccine coverage
	Improve immune protection through timely vaccination without any additional cost
Pharmacovigilance after introduction of new type or combined vaccines	Generate data on the characteristics of the wPV used in order to compare with a new vaccine candidate in term of safety, immunogenicity and efficacy
	Evaluate the acceptance of the vaccine and the vaccine strategy by the population

new-borns? (b) will the blunting of the immune response of new-borns from immunized mothers (receiving only a primary vaccination) be monitored on a long term?

Propositions for pragmatic approach are summarized in Table 1.

9 Conclusions

It is urgent to have a pragmatic approach to the surveillance of whooping cough in LMIC (Table 1). This surveillance should integrate not only the demography, the vaccine composition and the vaccine strategies used but also **a perennial surveillance of** (i) quality of wPV used, (ii) vaccine coverage >90%, (iii) timely vaccination according to recommendations, (iv) duration of vaccine induced protection (v), impact of macrolide resistance of circulating isolates. The priority is to build up a reference laboratory as part of an international network of reference laboratories, to be complemented by few regional laboratories under supervision of the former, in

order to give access to patients and clinicians to specific diagnostic tools.

Pertussis is still killing new-born in LMIC, and the first objective should be high vaccination coverage, timely vaccination, increase of the awareness of the disease and biological surveillance of the disease at least using PCR.

References

Aslanabadi A, Ghabili K, Shad K, Khalili M, Sajadi MM (2015) Emergence of whooping cough: notes from three early epidemics in Persia. Lancet Infect Dis 15 (12):1480–1484. https://doi.org/10.1016/S1473-3099(15)00292-3

Barger-Kamate B, Deloria Knoll M, Kagucia EW, Prosperi C, Baggett HC, Brooks WA, Feikin DR, Hammitt LL, Howie SR, Levine OS, Madhi SA, Scott JA, Thea DM, Amornintapichet T, Anderson TP, Awori JO, Baillie VL, Chipeta J, DeLuca AN, Driscoll AJ, Goswami D, Higdon MM, Hossain L, Karron RA, Maloney S, Moore DP, Morpeth SC, Mwananyanda L, Ofordile O, Olutunde E, Park DE, Sow SO, Tapia MD, Murdoch DR, O'Brien KL, Kotloff KL, Pneumonia Etiology Research for Child Health Study G (2016) Pertussis-associated pneumonia in infants and children from low- and middle-income countries participating in the PERCH Study. Clin Infect Dis 63(suppl 4):S187–S196. https://doi.org/10.1093/cid/ciw546

Baron S, Njamkepo E, Grimprel E, Begue P, Desenclos JC, Drucker J, Guiso N (1998) Epidemiology of pertussis in French hospitals in 1993 and 1994: thirty years after a routine use of vaccination. Pediatr Infect Dis J 17(5):412–418

Bart MJ, Harris SR, Advani A, Arakawa Y, Bottero D, Bouchez V, Cassiday PK, Chiang CS, Dalby T, Fry NK, Gaillard ME, van Gent M, Guiso N, Hallander HO, Harvill ET, He QS, van der Heide HGJ, Heuvelman K, Hozbor DF, Kamachi K, Karataev GI, Lan RT, Lutynska A, Maharjan RP, Mertsola J, Miyamura T, Octavia S, Preston A, Quail MA, Sintchenko V, Stefanelli P, Tondella ML, Tsang RSW, Xu YH, Yao SM, Zhang SM, Parkhill J, Mooi FR (2014) Global population structure and evolution of Bordetella pertussis and their relationship with vaccination. MBio 5(2):e01074-14. https://doi.org/10.1128/mBio.01074-14

Bass JW, Stephenson SR (1987) The return of pertussis. Pediatr Infect Dis J 6(2):141–144

Ben Fraj I, Kechrid A, Guillot S, Bouchez V, Brisse S, Guiso N, Smaoui H (2018) Pertussis epidemiology in Tunisian infants and children and characterization of Bordetella pertussis isolates: results of a 9-year surveillance study, 2007 to 2016. J Med Microbiol. https://doi.org/10.1099/jmm.0.000892

Ben Fraj I, Smaoui H, Zghal M, Sassi O, Guiso N, Kechrid A (2019) Seroprevalence of pertussis among healthcare workers: a cross-sectional study from Tunisia. Vaccine 37(1):109–112. https://doi.org/10.1016/j.vaccine.2018.11.023

Bouchez V, Guiso N (2013) Bordetella holmesii: comparison of two isolates from blood and a respiratory sample. Adv Infect Dis 3:123–133. https://doi.org/10.4236/aid.2013.32020

Bouchez V, Guiso N (2015) Bordetella pertussis, B. parapertussis, vaccines and cycles of whooping cough. Pathog Dis 73(7):ftv055. https://doi.org/10.1093/femspd/ftv055

Breakwell L, Kelso P, Finley C, Schoenfeld S, Goode B, Misegades LK, Martin SW, Acosta AM (2016) Pertussis vaccine effectiveness in the setting of Pertactin-deficient pertussis. Pediatrics 137(5):e20153973. https://doi.org/10.1542/peds.2015-3973

Broutin H, Viboud C, Grenfell BT, Miller MA, Rohani P (2010) Impact of vaccination and birth rate on the epidemiology of pertussis: a comparative study in 64 countries. Proc Biol Sci 277(1698):3239–3245. https://doi.org/10.1098/rspb.2010.0994

Chen ZY, Zhang J, Cao LN, Zhang N, Zhu JP, Ping GL, Zhao JH, Li SM, He QS (2016) Seroprevalence of pertussis among adults in China where whole cell vaccines have been used for 50 years. J Infect 73 (1):38–44. https://doi.org/10.1016/j.jinf.2016.04.004

Cherry JD (2016) Pertussis in young infants throughout the world. Clin Infect Dis 63(suppl 4):S119–S122. https://doi.org/10.1093/cid/ciw550

Cherry JD, Tan T, Wirsing von Konig CH, Forsyth KD, Thisyakorn U, Greenberg D, Johnson D, Marchant C, Plotkin S (2012) Clinical definitions of pertussis: summary of a global pertussis initiative roundtable meeting, February 2011. Clin Infect Dis 54 (12):1756–1764. https://doi.org/10.1093/cid/cis302

Dellepiane N, Pagliusi S, Registration Experts Working G (2018) Challenges for the registration of vaccines in emerging countries: differences in dossier requirements, application and evaluation processes. Vaccine 36(24):3389–3396. https://doi.org/10.1016/j.vaccine.2018.03.049

Fine PE, Clarkson JA (1982) The recurrence of whooping cough: possible implications for assessment of vaccine efficacy. Lancet 1(8273):666–669

Forsyth KD, Tan T, von Konig CW, Heininger U, Chitkara AJ, Plotkin S (2018) Recommendations to control pertussis prioritized relative to economies: a global pertussis initiative update. Vaccine 36 (48):7270–7275. https://doi.org/10.1016/j.vaccine.2018.10.028

Gross R, Keidel K, Schmitt K (2010) Resemblance and divergence: the "new" members of the genus Bordetella. Med Microbiol Immunol 199(3):155–163. https://doi.org/10.1007/s00430-010-0148-z

Guiso N, Hegerle N (2014) Other Bordetellas, lessons for and from pertussis vaccines. Expert Rev Vaccines 13 (9):1125–1133. https://doi.org/10.1586/14760584.2014.942221

Guiso N, Wirsing von Konig CH (2016) Surveillance of pertussis: methods and implementation. Expert Rev Anti-Infect Ther 14(7):657–667. https://doi.org/10.1080/14787210.2016.1190272

Guiso N, Berbers G, Fry NK, He Q, Riffelmann M, Wirsing von Konig CH, group EUP (2011) What to do and what not to do in serological diagnosis of pertussis: recommendations from EU reference laboratories. Eur J Clin Microbiol Infect Dis 30 (3):307–312. https://doi.org/10.1007/s10096-010-1104-y

Halperin SA, Langley JM, Ye L, MacKinnon-Cameron D, Elsherif M, Allen VM, Smith B, Halperin BA, McNeil SA, Vanderkooi OG, Dwinnell S, Wilson RD, Tapiero B, Boucher M, Le Saux N, Gruslin A, Vaudry W, Chandra S, Dobson S, Money D (2018) A randomized controlled trial of the safety and immunogenicity of tetanus, diphtheria, and acellular pertussis vaccine immunization during pregnancy and subsequent infant immune response. Clin Infect Dis 67(7):1063–1071. https://doi.org/10.1093/cid/ciy244

Hegerle N, Guiso N (2014) Bordetella pertussis and pertactin-deficient clinical isolates: lessons for pertussis vaccines. Expert Rev Vaccines 13(9):1135–1146. https://doi.org/10.1586/14760584.2014.932254

Huang WT, Lin HC, Yang CH (2017) Undervaccination with diphtheria, tetanus, and pertussis vaccine: national trends and association with pertussis risk in young children. Hum Vaccin Immunother 13(4):757–761. https://doi.org/10.1080/21645515.2016.1249552

Kampmann B, Mackenzie G (2016) Morbidity and mortality due to Bordetella pertussis: a significant pathogen in West Africa? Clin Infect Dis 63:S142–S147. https://doi.org/10.1093/cid/ciw560

Kurova N, Timofeeva EV, Guiso N, Macina D (2018) A cross-sectional study of Bordetella pertussis seroprevalence and estimated duration of vaccine protection against pertussis in St. Petersburg, Russia. Vaccine 36 (52):7936–7942. https://doi.org/10.1016/j.vaccine.2018.11.007

Linz B, Ivanov YV, Preston A, Brinkac L, Parkhill J, Kim M, Harris SR, Goodfield LL, Fry NK, Gorringe AR, Nicholson TL, Register KB, Losada L, Harvill ET (2016) Acquisition and loss of virulence-associated factors during genome evolution and speciation in three clades of Bordetella species. BMC Genomics 17 (1):767. https://doi.org/10.1186/s12864-016-3112-5

Moriuchi K et al (2017) Molecular epidemiology of Bordetella pertussis in Cambodia determined by direct genotyping of clinical specimens. Int J Infect Dis 62:56–58. https://doi.org/10.1016/j.ijid.2017.07.015

Muloiwa R, Wolter N, Mupere E, Tan T, Chitkara AJ, Forsyth KD, von Konig CW, Hussey G (2018) Pertussis in Africa: findings and recommendations of the Global Pertussis Initiative (GPI). Vaccine 36 (18):2385–2393. https://doi.org/10.1016/j.vaccine.2018.03.025

Nunes MC, Soofie N, Downs S, Tebeila N, Mudau A, de Gouveia L, Madhi SA (2016) Comparing the yield of nasopharyngeal swabs, nasal aspirates, and induced sputum for detection of Bordetella pertussis in hospitalized infants. Clin Infect Dis 63(suppl 4): S181–S186. https://doi.org/10.1093/cid/ciw521

Parkhill J, Sebaihia M, Preston A, Murphy LD, Thomson N, Harris DE, Holden MT, Churcher CM, Bentley SD, Mungall KL, Cerdeno-Tarraga AM, Temple L, James K, Harris B, Quail MA, Achtman M, Atkin R, Baker S, Basham D, Bason N, Cherevach I, Chillingworth T, Collins M, Cronin A, Davis P, Doggett J, Feltwell T, Goble A, Hamlin N, Hauser H, Holroyd S, Jagels K, Leather S, Moule S, Norberczak H, O'Neil S, Ormond D, Price C, Rabbinowitsch E, Rutter S, Sanders M, Saunders D, Seeger K, Sharp S, Simmonds M, Skelton J, Squares R, Squares S, Stevens K, Unwin L, Whitehead S, Barrell BG, Maskell DJ (2003) Comparative analysis of the genome sequences of Bordetella pertussis, Bordetella parapertussis and Bordetella bronchiseptica. Nat Genet 35(1):32–40. https://doi.org/10.1038/ng1227

Plotkin SA (2013) Complex correlates of protection after vaccination. Clin Infect Dis 56(10):1458–1465. https://doi.org/10.1093/cid/cit048

Saul N, Wang K, Bag S, Baldwin H, Alexander K, Chandra M, Thomas J, Quinn H, Sheppeard V, Conaty S (2018) Effectiveness of maternal pertussis vaccination in preventing infection and disease in infants: the NSW public health network case-control study. Vaccine 36(14):1887–1892. https://doi.org/10.1016/j.vaccine.2018.02.047

Scott S, van der Sande M, Faye-Joof T, Mendy M, Sanneh B, Barry Jallow F, de Melker H, van der Klis F, van Gageldonk P, Mooi F, Kampmann B (2015) Seroprevalence of pertussis in the Gambia: evidence for continued circulation of bordetella pertussis despite high vaccination rates. Pediatr Infect Dis J 34(4):333–338. https://doi.org/10.1097/INF.0000000000000576

Sebaihia M, Preston A, Maskell DJ, Kuzmiak H, Connell TD, King ND, Orndorff PE, Miyamoto DM, Thomson NR, Harris D, Goble A, Lord A, Murphy L, Quail MA, Rutter S, Squares R, Squares S, Woodward J, Parkhill J, Temple LM (2006) Comparison of the genome sequence of the poultry pathogen Bordetella avium with those of B-bronchiseptica, B-pertussis, and B-parapertussis reveals extensive diversity in surface structures associated with host interaction. J Bacteriol 188(16):6002–6015. https://doi.org/10.1128/Jb.01927-05

Sobanjo-Ter Meulen A, Duclos P, McIntyre P, Lewis KD, Van Damme P, O'Brien KL, Klugman KP (2016) Assessing the evidence for maternal pertussis immunization: a report from the Bill & Melinda Gates Foundation Symposium on pertussis infant disease burden in low- and lower-middle-income countries. Clin Infect Dis 63(suppl 4):S123–S133. https://doi.org/10.1093/cid/ciw530

Son S, Thamlikitkul V, Chokephaibulkit K, Perera J, Jayatilleke K, Hsueh PR, Lu CY, Balaji V, Moriuchi H, Nakashima Y, Lu M, Yang Y, Yao K, Kim SH, Song JH, Kim S, Kim MJ, Heininger U, Chiu CH, Kim YJ (2019) Prospective multinational serosurveillance study of Bordetella pertussis infection among 10- to 18-year-old Asian children and adolescents. Clin Microbiol Infect 25(2):250 e251–250 e257. https://doi.org/10.1016/j.cmi.2018.04.013

Sweden PHAo (2018) Pertussis surveillance in Sweden. Nineteen-year report. Available at https://www.folkhalsomyndigheten.se/contentassets/65ed8f6dbdab4999bc358fcd9b657e77/pertussis-sweden-nineteen-year-report.pdf. Accessed on Feb 2019

Turner HC, Thwaites GE, Clapham HE (2018) Vaccine-preventable diseases in lower-middle-income countries. Lancet Infect Dis 18(9):937–939. https://doi.org/10.1016/S1473-3099(18)30478-X

von Koenig CHW, Guiso N (2017) Global burden of pertussis: signs of hope but need for accurate data. Lancet Infect Dis 17(9):889–890. https://doi.org/10.1016/S1473-3099(17)30357-2

Wang Z, Cui Z, Li Y, Hou T, Liu X, Xi Y, Liu Y, Li H, He Q (2014) High prevalence of erythromycin-resistant Bordetella pertussis in Xi'an, China. Clin Microbiol

Infect 20(11):O825–O830. https://doi.org/10.1111/1469-0691.12671

WHO (2014a) Laboratory Manual for the diagnosis of whooping cough caused by bordetella pertussis/bordetella parapertussis: update 2014. Available at http://www.who.int/iris/handle/10665/127891. Accessed on Feb 2019

WHO (2014b) Report from the SAGE working group on pertussis vaccines, 26–27 august 2014 meeting, Geneva. Available at https://www.who.int/immunization/sage/meetings/2015/april/1_Pertussis_report_final.pdf?ua=1. Accessed on Feb 2019

WHO (2014c) WHO expert committee on biological standardization. sixty-third report. Recommendations to assure the quality, safety and efficacy of DT-based combined vaccines Available at https://www.who.int/biologicals/vaccines/Combined_Vaccines_TRS_980_Annex_6.pdf?ua=1. Accessed on Feb 2019

WHO (2015) Pertussis vaccines: WHO position paper. Available at https://www.who.int/wer/2015/wer9035.pdf. Accessed on Feb 2019

WHO (2017) The immunological basis for immunization series: module 4: pertussis, update 2017. Available at https://apps.who.int/iris/bitstream/handle/10665/259388/9789241513173-eng.pdf?sequence=1&isAllowed=y. Accessed on Feb 2019

Yao N, Zeng Q, Wang Q (2018) Seroepidemiology of diphtheria and pertussis in Chongqing, China: serology-based evidence of Bordetella pertussis infection. Public Health 156:60–66. https://doi.org/10.1016/j.puhe.2017.12.009

Yeung KHT, Duclos P, Nelson EAS, Hutubessy RCW (2017) An update of the global burden of pertussis in children younger than 5 years: a modelling study. Lancet Infect Dis 17(9):974–980. https://doi.org/10.1016/S1473-3099(17)30390-0

Zepp F, Heininger U, Mertsola J, Bernatowska E, Guiso N, Roord J, Tozzi AE, Van Damme P (2011) Rationale for pertussis booster vaccination throughout life in Europe. Lancet Infect Dis 11(7):557–570. https://doi.org/10.1016/S1473-3099(11)70007-X

Adv Exp Med Biol - Advances in Microbiology, Infectious Diseases and Public Health (2019) 1183: 151–160
https://doi.org/10.1007/5584_2019_410
© Springer Nature Switzerland AG 2019
Published online: 30 July 2019

Clinical Findings and Management of Pertussis

Ilaria Polinori and Susanna Esposito

Abstract

Pertussis is an endemic highly infectious vaccine-preventable disease. The disease is a major cause of childhood morbidity and mortality. In the most recent years, the re-emergence of pertussis occurred, and many efforts were done to identify the possible causes. Certainly, more effective laboratory methods have a role in making the diagnosis easier. However, sub-optimal efficacy of available vaccines as well as their limited duration of protection could explain the resurgence of the disease. Many forms and clinical features of the disease, ranging from the most classical to atypical and very nuanced forms, have been reported. There are many aspects that influence the clinical features of the pathology, such as a previous immunization or infection, patient's age, gender and antibiotic treatment. A prompt suspect and a rapid diagnosis of pertussis is fundamental for an appropriate clinical management and for preventing pertussis complications, especially in children. However, under a clinical point of view, pertussis is often difficult to be diagnosed. A prompt treatment may decrease the duration and severity of cough; the cornerstone drugs are the macrolides. Although prompt diagnosis and effective therapy are important for pertussis control, only with a broad vaccination coverage will be possible to reduce circulation of *Bordetella pertussis*.

Keywords

Antibiotic treatment · Paroxysmal cough · Pertussis · Post-exposure prophylaxis · Whooping cough pathology

1 Introduction

Whooping cough is an endemic highly infectious vaccine-preventable disease caused by *Bordetella pertussis*, a Gram-negative, aerobic coccobacillus that lives in upper respiratory tract, which was isolated in the laboratory in 1906. The easy spreading of the disease through the cough and sneezing droplets makes itself a relatively common clinical problem in pediatric population (Nieves and Heininger 2016). Fomites are not a factor in transmission, and there is no known animal reservoir for pertussis (de Greeff et al. 2012). The disease is a major cause of childhood morbidity and mortality: there were estimated 24.1 million pertussis cases and 160,700 deaths from pertussis in children <5 years of age in 2014 (Yeung et al. 2017). In 2013, the Global Burden of Disease Study estimated mortality due to pertussis in the first year of life to be approximately 400 per million live births, or approximately 56,000 deaths (Liu et al. 2015).

I. Polinori and S. Esposito (✉)
Pediatric Clinic, Department of Surgical and Biomedical Sciences, Università degli Studi di Perugia, Perugia, Italy
e-mail: susanna.esposito@unipg.it

Due to the severity of this disease, there were developed many preventive strategies (Esposito 2018). The first vaccine introduced was whole-cell-pertussis vaccine (wPV) and some years later to overcome the frequency of local and systemic adverse event due to the wPV, acellular pertussis vaccine (aPV) was developed.

In the most recent years, the re-emergence of pertussis occurred, and many efforts was done to identify the possible causes (Esposito 2018; Tan et al. 2015). Certainly, more effective laboratory methods have a role in making the diagnosis easier. But, there is also another reason that could explain the resurgence of the pathology: the emergence of strains with a mutation, i.e. the loss of pertactin (PRN), that could reduce the vaccine efficacy (Otsuka et al. 2012). The WHO (2014) demonstrated a true resurgence of the pertussis, probably linked with the use of aPV; indeed differences in induction of immune response exist between wPV and aPV, with a lower immune response induced, in terms of quality and duration, by the aPV (Klein et al. 2016); moreover, aPV vaccination do not prevent colonization (Warfel et al. 2014; Polak et al. 2018). For these reasons and since neither natural illness nor vaccination will ensure a permanent immunity, the circulation of *B. pertussis* is unavoidable (Esposito and Principi 2016; Wendelboe et al. 2005).

All the factors mentioned above (i.e., the re-emergence of the disease, the role of immunization, advances in diagnosis) allow to explain the very wide clinical spectrum of the disease.

2 What Could Influence Clinical Features of Pertussis

Pertussis has been reported to cause mild to serious upper and lower respiratory disease with different degrees of severity, that can lead to death. There are many aspects that influence the clinical features of the pathology, such as a previous immunization or infection, patient's age, gender and antibiotic treatment (Wang and Harnden 2011). A prompt treatment can reduce the severity of *B. pertussis* infection using an adequate anti-microbial therapy in the early catarrhal phase of the disease (Nguyen and Simon 2018).

Vaccination is one of the most relevant factors that changes the clinical presentation, either in infants or in older children (Briand et al. 2007; Esposito and Principi 2016). Some authors have analyzed 788 cases of pertussis, all laboratory-confirmed, and demonstrated that the duration of cough in unvaccinated was two times longer than in vaccinated children (Tozzi et al. 2003). Moreover, unvaccinated children have more frequently a broad spectrum of symptoms, as demonstrated by Heininger et al. (1997).

The presence of transplacentally acquired specific antibodies appeared efficacious against infant deaths (Amirthalingam et al. 2016; Healy et al. 2018; Winter et al. 2017). Other factors with an impact on pertussis severity are low birth weight, low gestational age, young age at time of cough onset, and high peak white blood cells (WBC) and lymphocyte counts (Winter et al. 2015; Scanlon et al. 2017). The number of WBCs appeared the best predictor of fatal infection (Mikelova et al. 2003). Both apnea and pneumonia are common in fatality cases, even if they are not a risk factor for death. Moreover, early onset of pneumonia is more likely linked to death than delayed onset pneumonia (Cherry 2018). Furthermore, seizures, that are mainly a consequence of hypoxia deriving from apnea, are strongly correlated to mortality (Winter et al. 2015).

3 Clinical Presentation and Complications

3.1 Typical Presentation

The incubation period usually lasts 7–10 days, but incubation periods as long as 4 weeks have been observed (Heininger et al. 1998).

Table 1 summarizes clinical manifestations of pertussis among infants. In the typical presentation observed in unvaccinated children, there are three stages of the disease: catarrhal, paroxysmal, and convalescent. The first phase, the milder one, of 1–2 weeks in duration is characterized by

Table 1 Clinical manifestations of pertussis among infants

Clinical sign	Frequency (%) in *B. pertussis* positive	Reference[s]
Paroxysmal cough, cyanosis, and apnea	61%, 42%, 55%	Stefanelli et al. (2017)
Apnea	68.7%	Raymond et al. (2007)
Paroxysmal cough	98.1% inpatients vs 95.9% outpatients	Mbayei et al. (2018)
Post-tussive vomiting	61.2% inpatients vs 46.6% outpatients	
Whoop	48.4% inpatients vs 28.9% outpatients	
Cyanosis	44.1% inpatients vs 1.7% outpatiens	
Apnea	35.7% inpatients vs 23.4% outpatients	

malaise, mild sore throat, mild cough, nasal congestion, and lacrimation with conjunctival injection (Kilgore et al. 2016). Usually, fever is absent. Due to its similarity with common cold, pertussis is often not suspected during this early phase. Asking for prolonged cough in other family members increases the chance to diagnose the disease, and to avoid its transmission to contacts, and other family members (Sali et al. 2015). In the worsening stage, that usually lasts 2–6 weeks, there are series of 5–10 or more paroxysmal cough that can lead to the characteristic whoop (Mattoo and Cherry 2005). Paroxysm consists of a series of rapid, forced expirations, followed by gasping inhalation, producing the typical whooping sound (Hewlett and Edwards 2005). In severe cases, patients may have 30 or more paroxysms per day, and the child may be exhausted after a paroxysm (Hartzell and Blaylock 2014).

Between attacks, the patient appears relatively well and asymptomatic. Weight loss can be seen as a result of frequent vomiting or the refusal of the child to eat. Crying, laughing, and eating can trigger paroxysms that occur with a high frequency during the night. Possible complications, typical of this phase, are intracranial hemorrhage, stroke, vertebral artery dissection, syncope, subconjunctival hemorrhages, rib fractures, urinary incontinence, and hernias (Irwin 2006; McEniery et al. 2004).

The convalescent stage is characterized by less frequent and less severe coughing. This phase represents the slow decline of the pathology. The patient is most infectious during the catarrhal phase, and infectivity decreases during the paroxysmal phase.

As already mentioned, vaccination plays a relevant role in the severity of the disease. In a recent prospective study, Stefanelli et al. investigated the clinical manifestations in unvaccinated infants aged less than 6 months, and found that about 50% of infants with respiratory symptoms were positive for *B. pertussis*. Moreover, paroxysmal cough, cyanosis, and apnea were strongly associated with pertussis positivity. Interestingly, data showed that 12% of the patients in this study received one dose of aPV and only 4% had been received 2 doses of aPV, meaning that also a partial immunization reduces the severity of symptoms (Stefanelli et al. 2017).

The Centers for Disease Control and Prevention (CDC) reported than almost all younger children of less than 12 months of age have complications, and about a half of them needed hospitalization. CDC data about complications reported that 23% infants hospitalized for pertussis get pneumonia lung infection, 1.1% had convulsions, 61% had apnea, 0.3% had encephalopathy, and 1% died (CDC 2017). One reason explaining why children are more prone to develop such complications is due to their respiratory airway dimension and more accentuated bronchial hyperreactivity (Tanaka et al. 2003). In addition, they have a weak and not completed developed immune system with a lower adaptive response which leads to a diminished capacity to clear bacteria (Winter et al. 2015). In a recent study conducted in California, among 17 infants aged less than 3 months who were admitted to pediatric intensive care units with severe pertussis, six infants were diagnosed with pulmonary hypertension and four of those six died (Murray et al. 2013).

Raymond et al. conducted a study that underlined the differences in clinical manifestations between children and adults and the role of asymptomatic or poorly symptomatic carriers in spreading the disease (Raymond et al. 2007). The study included children <4 months of age, hospitalized with symptoms compatible with pertussis. The most frequent symptoms in patients with a polymerase chain reaction (PCR)-confirmed *B. pertussis* infection was apnea, reported in 68.7% of cases. In six patients there was a RSV co-infection.

In one study, Mbayei et al. investigated the incidence and the characteristics of patients with severe pertussis infections in United States from January 2011 to 31 December 2015. Among 15,942 cases, 515 (3.2%) were hospitalized. The risk of hospitalization (95% confidence intervals [CI], 22.7–30) was higher in patients aged <2 months and was of 26.2 times, whereas the risk of intensive care unit (ICU) admission (95% CI, 50.3–103.0) was of 72 times compared to all other age groups combined (Mbayei et al. 2018). Furthermore, the same authors in multivariate analysis evaluated factors associated with hospitalization divided for age: mother vaccination with Tdap during the third trimester of pregnancy was a protective factor in children of less than 2 months of age and in children aged 2 months to 11 years aPV was associated with a lower risk of hospitalization. Data about effect of mother's vaccination in reducing children hospitalization were concordant with another case-control study among pertussis patients age < 2 months (Skoff et al. 2017).

3.2 Atypical Presentation in Adolescents and Immunised Patients

The introduction of aPV into the immunization schedule of infants and children has changed the epidemiology of the disease. Indeed, nowadays pertussis is common in adolescents and adults but with a different and wide spectrum of symptoms (Table 2). Adolescents have usually milder symptoms (Paisley et al. 2012). Strebel et

el. found that only 27%, 7%, 15%, 27% have respectively paroxysmal cough, whooping, post-tussive vomiting, post-tussive gagging in a cohort of patients with a median age of 35 years (Strebel et al. 2001). Similar data were found in a prospective study among adults who consulted general practitioners for a persistent cough without an evident diagnosis, conducted in a French Area with very high wPV (Gilberg et al. 2002).

In school-age children persistent cough could be the principal pertussis manifestation. It was reported that up to one-third of illnesses with prolonged cough are caused by Bordetella infections (von Konig et al. 2002). These data were concordant with those coming from a study recently conducted in Italy among 96 children with a long-lasting cough (from 2 to 8 weeks) and found evidence of pertussis infection in 18 (18.7%; 95% CI, 11.5–28.0) cases (Principi et al. 2017).

A recent systematic review and meta-analysis on diagnostic accuracy of pertussis clinical symptoms in adults and children showed that typical pertussis presentation, consisting in paroxysmal cough and absence of fever, have a high sensitivity (93.2% [CI, 83.2–97.4] and 81.8% [CI, 72.2–88.7], respectively, and low specificity (20.6% [CI, 14.7–28.1], 18.8% [CI, 8.1–37.9]), whereas post-tussive vomiting and whooping have low sensitivity (32.5% [CI, 24.5–41.6], 29.8% [CI, 8.0–45.2]) and high specificity (77.7% [CI, 73.1–81.7] and 79.5% [CI, 69.4–86.9]) (Moore et al. 2017). Post-tussive vomiting in children is moderately sensitive (60.0% [CI, 40.3–77.0]) and specific (66.0% [CI, 52.5–77.3]) (Moore et al. 2017).

Among older patients, isolated cases of serious respiratory complications, including respiratory distress or failure, and hypoxia, are reported (Rutledge and Keen 2012; Zycinska et al. 2017).

4 Clinical Definition Cases of *B. pertussis* Infections

A prompt suspect and diagnosis of pertussis is fundamental for an appropriate clinical management and for preventing pertussis complications,

Table 2 Clinical manifestations of pertussis among adolescents

Clinical sign	Frequency (%) in *B. pertussis* positive	Reference[s]
Paroxysmal cough, whooping, post-tussive vomiting, post-tussive gagging	27%, 7%, 15%, 27%	Strebel et al. (2001)
Whoop, post-tussive vomiting	16%, 25%	Gilberg et al. (2002)

especially in children. However, under a clinical point of view, pertussis is often difficult to be diagnosed. This complexity has emerged some years ago in the context of the Global Pertussis Initiative (GPI) (Cherry et al. 2012).

Throughout the world there are different definitions of pertussis cases. The last CDC definition comprises clinical criteria, laboratory criteria and epidemiologic linkage. Based on these aspects the diagnosis is qualified as confirmed or probable. The clinical criteria include as main symptom, in absence of a more likely diagnosis, a cough illness lasting ≥2 weeks, and with at least one the following signs or symptoms: paroxysms of coughing, or inspiratory whoop, or post-tussive vomiting, or apnea (for infants aged <1 year only). The laboratory test, with the isolation of *B. pertussis* or PCR positivity, but also in presence of contact with a laboratory-confirmed case of pertussis, make the diagnosis confirmed (CDC 2014). The European CDC pertussis definition is very similar, including also the pertussis toxin specific antibody response as adjunctive laboratory criteria (ECDC 2018).

Also European CDC considered clinical, laboratory and epidemiologic criteria (ECDC 2018). Clinical criteria include any person with a cough lasting at least 2 weeks and at least one of the following three: (i) paroxysms of coughing, (ii) inspiratory 'whooping', (iii) post-tussive vomiting or any person diagnosed as pertussis by a physician or apnoeic episodes in infants. Laboratory criteria include at least one of the following three: (i) isolation of *B. pertussis* from a clinical specimen, (ii) detection of *B. pertussis* nucleic acid in a clinical specimen, (iii) *B. pertussis* specific antibody response. Epidemiological criteria include an epidemiological link by human to human transmission.

The World Health Organization (WHO) definition has in common the requirement of 2 weeks of cough, and at least 1 additional symptom among paroxysms of coughing, inspiratory whooping, and post-tussive vomiting (WHO 2018).

In a recent meta-analysis on the accuracy of signs and symptoms for the diagnosis of pertussis, Ebell et al. (2017) reported that some clinical aspects, paroxysmal and whooping cough, and vomiting are more accurate to diagnose pertussis in children than adults. Moreover, the typical signs and symptoms resulted to be more sensitive but less specific in vaccinated people than in the unvaccinated population. A fundamental issue is that the clinician's overall impression is the most accurate way to determine the likelihood of *B. pertussis* infection (Ebell et al. 2017; Cornia et al. 2010).

By the way, new criteria have been developed, dividing patients in three different cohorts: 0–3 months, 4 months–9 years, and ≥ 10 years. In young infants, the association of cough that is increasing in frequency and severity and a coryza which does not became purulent in afebrile illness are elements suggestive of pertussis. Moreover, the addition of apnea, seizures, cyanosis, emesis, or pneumonia increases both sensitivity and specificity. About laboratory role, the elevation of white blood cell count (≥20,000 cells/μL) with absolute lymphocytosis is potentially diagnostic. In children aged between 4 months and 9 years, the same clinical setting of afebrile child with no purulent coryza with the main symptom of non-producing long lasting (≥7 days) cough is suggestive for pertussis; the same clinical aspects that increase sensitivity and specificity (i.e., whoop, apnea, and post-tussive emesis) are applicable in this age range. In older children, the same criteria reported above could be used, but in this case sweating episodes between paroxysms should be added (Cherry et al. 2012).

5 Management of *B. pertussis* Cases and Contacts

5.1 Postexposure Prophylaxis (PP)

Anyone who has had face-to-face exposure or other type of close contact with a patient infected by *B. pertussis* is recommended to receive a post-exposure prophylaxis (PP) taking into account the vaccination status (Tiwari et al. 2005). The effectivity of PP depends on the time of initiation, i.e. the earlier starts the treatment the greater is the benefit. The optimal time for initiation is within 21 days from the contact. The cornerstone drugs are the macrolides. There is no agreement about duration of PP, but most recent data support the notion that shorter treatment are as efficacious as longer ones (Altunaiji et al. 2007).

5.2 Antibiotic Treatment

The CDC recommends to begin the treatment in the early stage of the disease during the catarrhal phase, indeed a prompt treatment may decrease the duration and severity of cough (Tiwari et al. 2005). Because no proven treatments exist that reduce disease severity and frequency of symptoms during subsequent phases, it is suggested to begin a therapy even before laboratory results. Anyway, the contagiousness can last for a long time, even more than a month from the infection, and for this reason it is always recommended to administer antibiotics in the course of the disease to reduce the spreading of the infection.

The preferred antibiotics for treatment of *B. pertussis* infections are macrolides (i.e., erythromycin, clarithromycin, or azithromycin), but also other options are available as trimethoprim-sulfamethoxasole (Table 3). The choice between macrolides depends on drug safety and tolerability related to individual patients' factors, including the age.

In children aged less than a month, two risks must be taken into consideration: the one related to disease's complications and the one deriving from the use of macrolides. Hypertrophic pyloric stenosis (IHPS) is related to erythromycin exposure. A low risk of IHPS seems to exist also with azithromycin, as demonstrated in a recent work in which of seven premature infants treated with azithromycin for pertussis two developed IHPS (Morrison 2007).

For children and adults, erythromycin should be administered in 3–4 divided daily doses for 14 days. Shorter regimens are associated with relapses (Tiwari et al. 2005). The dose in infants aged >1 month and older children is 40–50 mg/kg per day (maximum: 2 g per day) in 3–4 divided doses for 14 days. The most common adverse

Table 3 Treatment of pertussis

Agent	Infant (<1 mo)	Infants (1–5 mo)	Children (>6 mo)	Adults
Azithromycin	10 mg/kg once daily for 3 days	10 mg/kg once daily for 3 days	Day 1: 10 mg/kg (max, 500 mg) once daily for 3 days	Day 1: 500 mg once daily for 3 days
Erythromycin*	Not recommended	40–50 mg/kg per day (max, 2 g/d) divided 3–4 times daily for 14 days	40–50 mg/kg per day (max, 2 g/day) divided 3–4 times daily for 14 days	2 g/day divided 3–4 times daily for 14 days
Clarithromycin	Not recommended	15 mg/kg per day (max, 1 g/day) divided twice a day for 7–10 days	15 mg/kg per day (max, 1 g/day) divided twice a day for 7–10 days	1 g/day divided twice a day for 7–10 days
TMP-SMX*	Not recommended	TMP 8 mg/kg per day (max, 320 mg/day); SMX 40 mg/kg per day divided twice a day for 14 days *Contraindicated in infants <2 mo of age	TMP 8 mg/kg per day (max, 320 mg/d); SMX 40 mg/kg per day divided twice a day for 14 days	TMP 320 mg; SMX 1600 mg/d divided twice a day for 14 days

*Is related to the controindication of erythromycin and TMP-SMX

effects involve the gastrointestinal system they include nausea, vomiting, and diarrhea. Other effects are hypersensitive reactions such as skin rashes or eosinofilia, but anaphylaxis is rare. The most severe adverse events are cardiovascular effects (i.e., electrocardiographic QT/QTc interval prolongation, cardiac arrest, other arhythmias), they are rare related to co-administration with other drugs (i.e., cisapride, pimazole, or terfenadine). Due its inhibition on the cytochrome P450 enzyme system (CYP3A subclass), co-administration with other families drugs who act on this metabolic system is not allowed to avoid toxicity resulting from an elevations in drug concentrations.

Clarithromycin is administered at the dose of 15 mg/kg per day (maximum: 1 g per day) in 2 divided doses each day for 7–10 days. Adults recommended dose is 1 g per day in two divided doses for 7–10 days. Clarithromycin has a good tolerability, although adverse events involving the gastrointestinal tract are possible, and no co-administration with other drugs acting on CYP3A enzyme is allowed.

Azithromycin is administered at 10 mg/kg (maximum: 500 mg) per day for 3 days.

An alternative is trimethoprim-sulfamethoxazole (TMP-SMX), although it is contraindicated in infants <2 months of age due to the risk of kernicterus.

5.3   Supportive Therapies and Novel Potential Drugs

The main target of supportive therapies is to prevent life-threatening complications. Neonates with pertussis should be admitted to the hospital with continuous cardiorespiratory monitoring. The most important symptoms to be recorded are the episodes of vomiting and the ability to feed to promptly establish an intravenous hydration and avoid dehydration.

Oxygen therapy and suction of secretions are the principal supportive therapy during the critical phases. In addition, mechanical ventilation has to be considered in case of frequent and long lasting apnea or severe pneumonia. When this measure is required, admission in ICU is needed (Berti et al. 2014).

Cases of severe pertussis characterized by refractory hypoxemia, cardiogenic shock, and pneumonia associated with extreme leukocytosis can be treated with exchange transfusion (ET) with the purpose of leukoreduction and the decrease of circulating toxins. Data about beneficial effects of ET demonstrated an improving in cardiorespiratory function, with a reduction in mortality from 45% (4/9) to 10% (1/10) (Kuperman et al. 2014; Rowlands et al. 2010).

No benefits were observed with administration of bronchodilators or steroids and for this reason they are not recommended as supported therapy (Scanlon et al. 2015; Wang et al. 2014; Winter et al. 2015).

Nowadays, a fundamental issue is to find novel therapies for severe cases. Starting from animal models in which pendrin knockout mice exhibited lower levels of lung inflammation, some authors supposed a certain role of acetazolamide. Indeed, pendrin is an excreter of bicarbonates and allows an alkalization of pH to optimal levels for inflammatory mediator activity (Scanlon et al. 2014); for this reason, acetazolamide, a carbonic anhydrase inhibitor, could reduce inflammation level and could be a potential novel drug (Carbonetti 2016). Another possible therapeutic target is Sphingosine-1-phosphate (S1P) pathway, already involved in many pathological processes (Spiegel and Milstien 2011). In a recent work using the S1P receptor agonist, was shown a reduction in pertussis lung inflammation (Skerry et al. 2017). However, further studies are needed on these novel therapeutic approaches.

6   Conclusive Remarks

Pertussis is still a pathology of global importance and of great relevance. Its peculiarity lies in the evolution that has endured over time. In fact, contrary to many other infectious diseases, pertussis has evolved on a clinical level. Today, in fact, we can recognize many forms and clinical features of the disease, ranging from the most classical to atypical and very nuanced forms. Its evolution has also involved aspects not only exclusively clinical but also immunological. Although prompt diagnosis and effective therapy are important for pertussis control, broad vaccination coverage is the only way to reduce circulation of *B. pertussis*.

References

Altunaiji S, Kukuruzovic R, Curtis N et al (2007) Antibiotics for whooping cough (pertussis). Cochrane Database Syst Rev (3):Cd004404. https://doi.org/10.1002/14651858.CD004404.pub3

Amirthalingam G, Campbell H, Ribeiro S et al (2016) Sustained effectiveness of the maternal pertussis immunization program in England 3 years following introduction. Clin Infect Dis 63(Suppl 4):S236–s243. https://doi.org/10.1093/cid/ciw559

Berti E, Venturini E, Galli L et al (2014) Management and prevention of pertussis infection in neonates. Expert Rev Anti-Infect Ther 12(12):1515–1531. https://doi.org/10.1586/14787210.2014.979156

Briand V, Bonmarin I, Levy-Bruhl D (2007) Study of the risk factors for severe childhood pertussis based on hospital surveillance data. Vaccine 25(41):7224–7232. https://doi.org/10.1016/j.vaccine.2007.07.020

Carbonetti NH (2016) *Bordetella pertussis*: new concepts in pathogenesis and treatment. Curr Opin Infect Dis 29(3):287–294. https://doi.org/10.1097/qco.0000000000000264

Centers for Disease Control and Prevention. Pertussis (whooping cough): surveillance and reporting – 2014 case definition. Available at: https://wwwn.cdc.gov/nndss/conditions/pertussis/case-definition/2014/. Accessed on 10 Feb 2019

CDC (2017) Centers for Disease Control and Prevention. Pertussis (whooping cough): Complications 2017. Available at: https://www.cdc.gov/pertussis/about/complications.html. Accessed on 10 Feb 2019

Cherry JD (2018) Treatment of pertussis-2017. J Pediatr Infect Dis Soc 7(3):e123–e125. https://doi.org/10.1093/jpids/pix044

Cherry JD, Tan T, Wirsing von Konig CH et al (2012) Clinical definitions of pertussis: summary of a global pertussis initiative roundtable meeting, February 2011. Clin Infect Dis 54(12):1756–1764. https://doi.org/10.1093/cid/cis302

Cornia PB, Hersh AL, Lipsky BA et al (2010) Does this coughing adolescent or adult patient have pertussis? JAMA 304(8):890–896. https://doi.org/10.1001/jama.2010.1181

de Greeff SC, de Melker HE, Westerhof A et al (2012) Estimation of household transmission rates of pertussis and the effect of cocooning vaccination strategies on infant pertussis. Epidemiology 23(6):852–860. https://doi.org/10.1097/EDE.0b013e31826c2b9e

Ebell MH, Marchello C, Callahan M (2017) Clinical diagnosis of *Bordetella pertussis* infection: a systematic review. J Am Board Fam Med 30(3):308–319. https://doi.org/10.3122/jabfm.2017.03.160330

European Centre for Disease Control and Prevention. EU definitions (2018) Pertussis. Available at: https://ecdc.europa.eu/en/surveillance-and-disease-data/eu-case-definitions. Accessed on 31 Dec 2018

Esposito S (2018) Prevention of pertussis: from clinical trials to real world evidence. J Prev Med Hyg 59(3):e177–e186. https://doi.org/10.15167/2421-4248/jpmh2018.59.3.1041

Esposito S, Principi N (2016) Immunization against pertussis in adolescents and adults. Clin Microbiol Infect 22(Suppl 5):S89–s95. https://doi.org/10.1016/j.cmi.2016.01.003

Gilberg S, Njamkepo E, Du Chatelet IP et al (2002) Evidence of *Bordetella pertussis* infection in adults presenting with persistent cough in a french area with very high whole-cell vaccine coverage. J Infect Dis 186(3):415–418. https://doi.org/10.1086/341511

Hartzell JD, Blaylock JM (2014) Whooping cough in 2014 and beyond: an update and review. Chest 146(1):205–214. https://doi.org/10.1378/chest.13-2942

Healy CM, Rench MA, Swaim LS et al (2018) Association between third-trimester Tdap immunization and neonatal pertussis antibody concentration. JAMA 320(14):1464–1470. https://doi.org/10.1001/jama.2018.14298

Heininger U, Klich K, Stehr K et al (1997) Clinical findings in *Bordetella pertussis* infections: results of a prospective multicenter surveillance study. Pediatrics 100(6):E10

Heininger U, Cherry JD, Stehr K et al (1998) Comparative efficacy of the Lederle/Takeda acellular pertussis component DTP (DTaP) vaccine and Lederle whole-cell component DTP vaccine in German children after household exposure. Pertussis Vaccine Study Group. Pediatrics 102(3 Pt 1):546–553

Hewlett EL, Edwards KM (2005) Clinical practice. Pertussis--not just for kids. N Engl J Med 352(12):1215–1222. https://doi.org/10.1056/NEJMcp041025

Irwin RS (2006) Introduction to the diagnosis and management of cough: ACCP evidence-based clinical practice guidelines. Chest 129(1 Suppl):25s–27s. https://doi.org/10.1378/chest.129.1_suppl.25S

Kilgore PE, Salim AM, Zervos MJ et al (2016) Pertussis: microbiology, disease, treatment, and prevention. Clin Microbiol Rev 29(3):449–486. https://doi.org/10.1128/cmr.00083-15

Klein NP, Bartlett J, Fireman B et al (2016) Waning Tdap effectiveness in adolescents. Pediatrics 137(3):e20153326. https://doi.org/10.1542/peds.2015-3326

Kuperman A, Hoffmann Y, Glikman D et al (2014) Severe pertussis and hyperleukocytosis: is it time to change for exchange? Transfusion 54(6):1630–1633. https://doi.org/10.1111/trf.12519

Liu L, Oza S, Hogan D et al (2015) Global, regional, and national causes of child mortality in 2000-13, with projections to inform post-2015 priorities: an updated systematic analysis. Lancet 385(9966):430–440. https://doi.org/10.1016/s0140-6736(14)61698-6

Mattoo S, Cherry JD (2005) Molecular pathogenesis, epidemiology, and clinical manifestations of respiratory infections due to *Bordetella pertussis* and other *Bordetella* subspecies. Clin Microbiol Rev 18 (2):326–382. https://doi.org/10.1128/cmr.18.2.326-382.2005

Mbayei SA, Faulkner A, Miner C et al (2018) Severe pertussis infections in the United States, 2011–2015. Clin Infect Dis. https://doi.org/10.1093/cid/ciy889

McEniery JA, Delbridge RG, Reith DM (2004) Infant pertussis deaths and the management of cardiovascular compromise. J Paediatr Child Health 40(4):230–232

Mikelova LK, Halperin SA, Scheifele D et al (2003) Predictors of death in infants hospitalized with pertussis: a case-control study of 16 pertussis deaths in Canada. J Pediatr 143(5):576–581. https://doi.org/10.1067/s0022-3476(03)00365-2

Moore A, Ashdown HF, Shinkins B et al (2017) Clinical characteristics of pertussis-associated cough in adults and children: a diagnostic systematic review and meta-analysis. Chest 152(2):353–367. https://doi.org/10.1016/j.chest.2017.04.186

Morrison W (2007) Infantile hypertrophic pyloric stenosis in infants treated with azithromycin. Pediatr Infect Dis J 26(2):186–188. https://doi.org/10.1097/01.inf.0000253063.87338.60

Murray EL, Nieves D, Bradley JS et al (2013) Characteristics of severe *Bordetella pertussis* infection among infants </=90 days of age admitted to pediatric intensive care units – Southern California, September 2009-June 2011. J Pediatr Infect Dis Soc 2(1):1–6. https://doi.org/10.1093/jpids/pis105

Nguyen VTN, Simon L (2018) Pertussis: the whooping cough. Prim Care 45(3):423–431. https://doi.org/10.1016/j.pop.2018.05.003

Nieves DJ, Heininger U (2016) Bordetella pertussis. Microbiol Spectr 4(3). https://doi.org/10.1128/microbiolspec.EI10-0008-2015

Otsuka N, Han HJ, Toyoizumi-Ajisaka H et al (2012) Prevalence and genetic characterization of pertactin-deficient Bordetella pertussis in Japan. PLoS One 7 (2):e31985. https://doi.org/10.1371/journal.pone.0031985

Paisley RD, Blaylock J, Hartzell JD (2012) Whooping cough in adults: an update on a reemerging infection. Am J Med 125(2):141–143. https://doi.org/10.1016/j.amjmed.2011.05.008

Polak M, Zasada AA, Mosiej E et al (2018) Pertactin-deficient *Bordetella pertussis* isolates in Poland-a country with whole-cell pertussis primary vaccination. Microbes Infect. https://doi.org/10.1016/j.micinf.2018.12.001

Principi N, Litt D, Terranova L et al (2017) Pertussis-associated persistent cough in previously vaccinated children. J Med Microbiol 66(11):1699–1702. https://doi.org/10.1099/jmm.0.000607

Raymond J, Armengaud JB, Cosnes-Lambe C et al (2007) Pertussis in young infants: apnoea and intra-familial infection. Clin Microbiol Infect 13(2):172–175. https://doi.org/10.1111/j.1469-0691.2006.01616.x

Rowlands HE, Goldman AP, Harrington K et al (2010) Impact of rapid leukodepletion on the outcome of severe clinical pertussis in young infants. Pediatrics 126(4): e816–e827. https://doi.org/10.1542/peds.2009-2860

Rutledge RK, Keen EC (2012) Images in clinical medicine. Whooping cough in an adult. N Engl J Med 366 (25):e39. https://doi.org/10.1056/NEJMicm1111819

Sali M, Buttinelli G, Fazio C et al (2015) Pertussis in infants less than 6 months of age and household contacts, Italy, April 2014. Hum Vaccin Immunother 11(5):1173–1174. https://doi.org/10.1080/21645515.2015.1019190

Scanlon KM, Skerry C, Carbonetti NH (2015) Novel therapies for the treatment of pertussis disease. Pathog Dis 73(8):ftv074. https://doi.org/10.1093/femspd/ftv074

Scanlon KM, Snyder YG, Skerry C et al (2017) Fatal pertussis in the neonatal mouse model is associated with pertussis toxin-mediated pathology beyond the airways. Infect Immun 85(11). https://doi.org/10.1128/iai.00355-17

Scanlon KM, Gau Y, Zhu J et al (2014) Epithelial anion transporter pendrin contributes to inflammatory lung pathology in mouse models of *Bordetella pertussis* infection. Infect Immun 82(10):4212–4221. https://doi.org/10.1128/iai.02222-14

Skerry C, Scanlon K, Ardanuy J et al (2017) Reduction of pertussis inflammatory pathology by therapeutic treatment with Sphingosine-1-phosphate receptor ligands by a pertussis toxin-insensitive mechanism. J Infect Dis 215(2):278–286. https://doi.org/10.1093/infdis/jiw536

Skoff TH, Blain AE, Watt J et al (2017) Impact of the US maternal tetanus, diphtheria, and acellular pertussis vaccination program on preventing pertussis in infants <2 months of age: a case-control evaluation. Clin Infect Dis 65(12):1977–1983. https://doi.org/10.1093/cid/cix724

Spiegel S, Milstien S (2011) The outs and the ins of sphingosine-1-phosphate in immunity. Nat Rev Immunol 11(6):403–415. https://doi.org/10.1038/nri2974

Stefanelli P, Buttinelli G, Vacca P et al (2017) Severe pertussis infection in infants less than 6 months of age: clinical manifestations and molecular characterization. Hum Vaccin Immunother 13(5):1073–1077. https://doi.org/10.1080/21645515.2016.1276139

Strebel P, Nordin J, Edwards K et al (2001) Population-based incidence of pertussis among adolescents and adults, Minnesota, 1995–1996. J Infect Dis 183 (9):1353–1359. https://doi.org/10.1086/319853

Tan T, Dalby T, Forsyth K et al (2015) Pertussis across the globe: recent epidemiologic trends from 2000 to 2013. Pediatr Infect Dis J 34(9):e222–e232. https://doi.org/10.1097/inf.0000000000000795

Tanaka M, Vitek CR, Pascual FB et al (2003) Trends in pertussis among infants in the United States, 1980–1999. JAMA 290(22):2968–2975. https://doi.org/10.1001/jama.290.22.2968

Tiwari T, Murphy TV, Moran J (2005) Recommended antimicrobial agents for the treatment and postexposure prophylaxis of pertussis: 2005 CDC guidelines. MMWR Recomm Rep 54(Rr-14):1–16

Tozzi AE, Rava L, Ciofi Degli Atti ML et al (2003) Clinical presentation of pertussis in unvaccinated and vaccinated children in the first six years of life. Pediatrics 112(5):1069–1075

von Konig CH, Halperin S, Riffelmann M et al (2002) Pertussis of adults and infants. Lancet Infect Dis 2(12):744–750

Wang K, Harnden A (2011) Pertussis-induced cough. Pulm Pharmacol Ther 24(3):304–307. https://doi.org/10.1016/j.pupt.2010.10.011

Wang K, Bettiol S, Thompson MJ et al (2014) Symptomatic treatment of the cough in whooping cough. Cochrane Database Syst Rev (9):Cd003257. https://doi.org/10.1002/14651858.CD003257.pub5

Warfel JM, Zimmerman LI, Merkel TJ (2014) Acellular pertussis vaccines protect against disease but fail to prevent infection and transmission in a nonhuman primate model. Proc Natl Acad Sci U S A 111(2):787–792. https://doi.org/10.1073/pnas.1314688110

Wendelboe AM, Van Rie A, Salmaso S et al (2005) Duration of immunity against pertussis after natural infection or vaccination. Pediatr Infect Dis J 24 (5 Suppl):S58–S61

Winter K, Cherry JD, Harriman K (2017) Effectiveness of prenatal tetanus, diphtheria, and acellular pertussis vaccination on pertussis severity in infants. Clin Infect Dis 64(1):9–14. https://doi.org/10.1093/cid/ciw633

Winter K, Zipprich J, Harriman K et al (2015) Risk factors associated with infant deaths from pertussis: a case-control study. Clin Infect Dis 61(7):1099–1106. https://doi.org/10.1093/cid/civ472

Yeung KHT, Duclos P, Nelson EAS et al (2017) An update of the global burden of pertussis in children younger than 5 years: a modelling study. Lancet Infect Dis 17(9):974–980. https://doi.org/10.1016/s1473-3099(17)30390-0

Zycinska K, Cieplak M, Chmielewska M et al (2017) Whooping cough in adults: a series of severe cases. Adv Exp Med Biol 955:47–50. https://doi.org/10.1007/5584_2016_167

World Health Organization (WHO) (2018) Immunization, vaccines, biologicals. Pertussis. Available at: https://www.who.int/immunization/monitoring_surveillance/burden/vpdsurveillance_type/passive/pertussis/en/. Accessed on 20 Jan 2019

World Health Organization. (WHO) (2014) SAGE Pertussis Working Group. Background paper. World Health Organization, Geneva. Available at: http://www.who.int/immunization/sage/meetings/2014/april/1_Pertussis_background_FINAL4_web.pdf.9 Accessed on 23 Jan 2019

Adv Exp Med Biol - Advances in Microbiology, Infectious Diseases and Public Health (2019) 1183: 161–167
https://doi.org/10.1007/5584_2019_411
© Springer Nature Switzerland AG 2019
Published online: 25 July 2019

Pertussis Vaccines and Vaccination Strategies. An Ever-Challenging Health Problem

Antonio Cassone

Abstract

Vaccines and vaccination against pertussis (whooping cough) have had one of the longest and most complex history, with alternating splendour and public disbelief, enthusiasm and concerns, overall resulting in changes in composition and replacement of vaccines, and associated vaccination strategies, including use of different vaccines in different countries, with no apparent equals for other bacterial vaccines. Of this both frustrating and exciting venue no end has been reached. In this note, I am shortly recapitulating the history of pertussis vaccines, from the inactivated, whole-cell vaccine to the acellular ones, with their merits and limitations, particularly concerning the debated issue of waning immunity, and a glimpse on a new vaccine proposal. Some reflections on the complexity and apparent peculiarity of this field are also made to the final scope of discussing aspects of the evolving strategies of disease control in a high-income country.

Keywords

Pertussis · Pertussis vaccines · Vaccination strategies · Whooping cough

A. Cassone (✉)
Polo d'innovazione della genomica, genetica e biologia,
University of Siena, Siena, Italy
e-mail: a.cassone@pologgb.com

1 The Vaccines Against Pertussis

Early during the 50's of the past century, when little to nothing was known about pathogenesis of, and immune responses to pertussis main causative agent, *Bordetella pertussis*, a vaccine composed of whole inactivated bacterial cells (wP) was generated and began to be used. At that time, pertussis was a worldwide rampant disease in terms of both morbidity and mortality, as illustrated elsewhere in this volume (see Ausiello et al. Chapter 6). As happened for other old vaccines, the contrast between the unknowns about the bacterium/ the disease and the effectiveness of the wP vaccine may appear now to us as a truly astonishing one: the mortality dropped to very low levels in all countries where vaccination was sustained by health authorities. This excellent profile in terms of disease control was probably due to the vaccine capacity of protecting both from the actual respiratory illness and, at least in part, to the disease-initiating, airways infection by the concurrent stimulation of Th1 and Th17 -cell mediated immunity and consequent generation of sufficient herd immunity (see below).

Unfortunately, the wP vaccine was highly reactogenic, that included such infrequent but alarming systemic disorders as encephalopathy and drowsiness (Edwards and Decker 2013; Plotkin 2014). As usually happens when a vaccine -preventable disease drastically falls in its incidence, the perception of the above side effects became more acute and lead most parents to abandon vaccinating their children. This particularly occurred in high-income countries where vaccine coverage against pertussis reached in the 70′-80s' of the past century such a low level as to cause disease resurgence, including vast epidemics (Plotkin 2014; Sizaire et al. 2014). Enhanced awareness of the need to replace this old vaccine with a new one was a consequence, with obvious emphasis on vaccine safety.

Conception and feasibility of a new pertussis vaccine was made possible by some remarkable biochemical and genetic progress in identifying, characterizing and purifying a number of virulence factors expressed by *B. pertussis*, in particular its potent pertussis toxin (Stein et al. 1994) that together with the extraordinary advances in knowledge of the mechanisms of antimicrobial immunity achieved in the aforementioned period, led manufactures to privilege the generation of subunit vaccines with highly purified components, the so-called acellular pertussis (aP) vaccines. Marked investments, public-private collaboration and collective enthusiasm in the formulation and testing of these new vaccines took the whole -90s' and more of the last century and happened to be a complex and rather unique story itself. More than twenty acellular vaccines were proposed for use and several of them underwent large clinical trials for efficacy and safety (Edwards et al. 1995; Pichichero et al. 1997; Gustafsson et al. 1996; Greco et al. 1996; Giuliano et al. 1998).

The antigenic formulation of aP vaccines, approved and some of them currently used in various high-income countries, is quite distinctive with respect to all other bacterial vaccines for human use. The absence of valid and universally accepted correlate(s) of vaccine protection has much complicated the decision as to which vaccines were most acceptable, hence aP vaccines have been approved for use which have largely different composition, from just one antigen, the chemically -or genetically-detoxified PT to a pentavalent vaccine containing FHA, pertactin and fimbriae 1 and 2 (Plotkin 2014; Cherry 1997). As discussed

elsewhere (Ausiello and Cassone 2014), vaccines with such different antigenic formulations could be anticipated conferring unequal levels of efficacy and long-term protection, and this has been confirmed in a recent Cochrane review (Zhang et al. 2014). On the other hand, the aP vaccines were all characterized by a marked reduction of those safety issues which led to misfortune of the wP vaccine which, nonetheless, is still largely and effectively used in low-income countries (Edwards and Decker 2013).

2 A Critical Difference in Immune Responses to Whole Cell and Acellular Pertussis Vaccines

Cell-mediated immunity (CMI) fine-tunes the balance between host-beneficial and host-damaging, inappropriate or exaggerated immune responses. Among other things, it regulates the type of help for antibody formation, as well B- and T-memory persistence (Kurup et al. 2019). The impact of these central functions on pertussis vaccine-induced immunity was poorly known before the advent of aP vaccines, thus the clinical trials of these vaccines provided a strong impetus to assess CMI responses. Since some trials included children receiving a wP vaccine to measure aP relative efficacy, CMI studies were able to show marked differences of intensity, quality and type of CMI between the recipients of wP or aP vaccines. In fact, these latter showed a mixed Th1-Th2 or pure Th2 rather than the Th1 (and, in subsequent studies, also Th17) cell priming (Zepp et al. 1996; Cassone et al. 1997; Ausiello et al. 1997; Ryan et al. 1998; Ross et al. 2013). These results were in good agreement with concurrent and subsequent studies in animal models which altogether suggested the different T cell priming had a critical impact on nature and duration of protection conferred by the two vaccines (Mills et al. 1998; Warfel et al. 2012a, b, 2014). As a likely consequence of the above differences, clinical and epidemiological data showed that the aP vaccines, including those containing factors affecting bacterial adherence to the pharyngeal epithelial tissue, did not protect from airways infection (Gill et al. 2017).

The above studies also provided for at least a partial explanation of an unexpected and quite disappointing event observed in most countries relatively soon after aP introduction in the real world of infant immunization, i.e. that the ability to protect from disease lasted significantly lesser than anticipated (or simply hoped?) by the results of clinical trials. This was mostly due to the decreased level of antibodies against the vaccine antigens and, as major consequence, a shift in disease prevalence toward the older age (adolescent and adults) was evident (CDC 2012). Collectively, these events were interpreted as due to some waning of immunity, and absence of sufficient herd immunity conferred by the aP vaccines. Immune-protection from disease also wanes after wP vaccination but it takes longer (Wendelboe et al. 2005). Accelerated waning of aP vaccine immunity has since been confirmed in many countries (with few apparent exceptions; see below) by extensive investigations and robust clinical and epidemiological evidence. The waning of long-term protection by the aP vaccines has probably multiple concurrent reasons, both of immunological, epidemiological and microbiological origin (Burns et al. 2014; Althouse and Scarpino 2015; Wendelboe et al. 2005). Whichever the contribution, and the unequal weight of the different factors above, the problem is now that pertussis continues to be a serious public health issue, with disease outbreaks in several countries with high vaccination coverage, involvement of adolescent and adult ages, possible presence of asymptomatic or paucy-symptomatic carriage in these subjects and transmission to children, high morbidity and non-neglectable mortality, particularly in areas with disparity of access to health services (reviewed by Kandeil et al. 2019).

Somewhat surprisingly, not all countries appear to have been equally affected by the waning of aP-induced immunity (Gill et al. 2017; Hallander et al. 2005; Salmaso et al. 2001). Of interest, Ausiello and collaborators (1999) had shown that the decay of antibody titers in four-years old recipient of aP vaccines primary schedule was in some children counteracted by acquisition of cell-mediated immunity (CMI) to vaccine components associated to protection against pertussis. Quite interestingly, this priming was of a pure Th1 type, similar to that associated with wP vaccination, hence different from that induced by the aP vaccine. The data were interpreted as due to the circulation of *B. pertussis* and asymptomatic or pauci-symptomatic infection, meaning that, under some epidemiological settings, waning of aP-induced *"immunity"* does not necessarily bring to waning of *"protection"*.

The important issue of the asymptomatic carriage of *B. pertussis* has been verified in the baboon model of pertussis infection and vaccination (Warfel et al. 2014) and discussed with wealth of epidemiological insight by Gill et al. (2017) and Althouse and Scarpino (2015).

For a true balance of merits and failures of aP vaccines, it is important to remind the intrinsic difficulties of this research field. All other, highly successful bacterial subunit vaccines are made by a single dominant antigen, the toxin for tetanus and diphtheria vaccines, and the anti-phagocytic capsule for those against Hemophilus, pneumococcus and meningococcus. The situation with the pertussis vaccines is much more complex because infection and disease caused by *B. pertussis* are due to a multiplicity (up to more than a dozen only among those known) of distinct virulence factors interplaying in a cascade of events from early colonization of the nasopharyngeal mucosa to the final inflammation and damage of the upper respiratory tract and systemic symptomatology, all due to the bacterial capacity to resist host defence (reviewed by Kilgore et al. 2016). By even admitting a minor pathogenicity role of some of the above factors, the disease setting remains of high unusual complexity, much closer to some vaccine-orphan diseases (those caused by *Staphylococcus aureus*, or *Escherichia.coli*, for instance) than to pneumonia caused by *S. pneumoniae* or meningitis. All this virulence armamentarium was only partially known by those who developed the aP vaccines. I posit nobody would now realistically propose for the approval by the health authorities a vaccine made up with only one or two *B. pertussis* antigens. An implicit conclusion is that a *B. pertussis* whole cell vaccine (inactivated or attenuated) should constitute a primary choice for disease control (discussed below).

3 How to Implement an Effective Vaccination Strategy with Current Vaccines?

So, what to do now? The pertussis vaccine immunology summarized above would suggest that rich countries take a bath of humblness, copy from the low-income ones and return using the wP vaccine. Since this is unlikely to occur, a re-shaping of the aP vaccination strategy is being implemented so as trying to push longer aP immune durability with new adjuvants, more appropriate vaccination schedules, new vaccination targets in the population with particular regard to adolescents and adults. A special target has become the pregnant women, given the critical need of protecting infants before they complete the primary vaccination course. This has become more relevant since immunization of household contacts of infants (cocooning strategy) has shown low or no effectiveness (Castagnini et al. 2012), In fact, the infants too young to receive the vaccine could die of whooping cough (more than 80% of all pertussis-related deaths occur in <3 months-old infants:),and infected (or simply asymptomatic carriers?) mothers account for 50% of this mortality burden (Fedele et al. 2017). Vaccination during pregnancy (at the end of the second trimester-third trimester) has been shown to be safe and is currently adopted in several countries (Campbell and Amirthalingam 2018; Agricola et al. 2016), with strong expectations but also some concerns regarding the possibility that antibodies transmitted to the newborn via the placenta could interact negatively with the vaccine first dose at 3 months of age (Barug et al.

2019), Delaying the primary course of vaccination to avoid this interference is one possibility, nonetheless, this should not contrast with the need of providing the earliest possible vaccination against other components of the primary vaccination schedule when polyvalent vaccines are used.

Studies on multi-boosters aP policy, particularly aimed at examining whether and to what extent they could overcome the expected waning of immunity previously discussed, as well as their apparently irreversible imprinting toward a Th2 response (da Silva Antunes et al. 2018) are recommended. Given the suggested expansion of aP recipients, the intriguing observation that TdaP vaccines could induce some form of immune-tolerance to unrelated antigens (Blok et al. 2019) should also be carefully evaluated by further studies.

In Italy, the current schedule of children, adolescents and adult immunization formulated by the National Plan of Vaccine Prevention 1917–1919 and approved by the governmental authorities (www.salute.gov.it), prescribes the usual three doses of aP vaccine (offered in a hexavalent vaccine combination with diphtheria, tetanus, hemophilusB, hepatitis B and polio vaccine) during the first year of age, one dose before school entrance (at 6 years of age) but also asks for additional booster doses at the adolescence with reduced content (a dTaP vaccine together with IPV) then another recall of dTaP every 10 years. It is anticipated a re-discussion of this issue when a novel Plan to be implemented for the next triennium is prepared. A re-assessment of effectiveness of the multi-boosters aP policy in adults, particularly in terms of contrasting the issue of possible waning of immunity previously discussed, is also expected. In order to implement an optimal vaccination strategy, studies should focus on the existence, duration and immune-effects of asymptomatic *B. pertussis* and cognate bacterial carriage, and establish with robust epidemiological and microbiological evidence whether this carriage confers natural boosters to vaccines, as also suggested by some of our past investigations (Ausiello et al. 1999). Has this carriage an impact on systemic immunity beyond airway mucosal tissue and is it training innate immunity? (Netea et al. 2011; Cassone 2018). An extended recommendation on how best to control pertussis by the present-day technical and economic resources has been updated by GPI experts (Forsyth et al. 2018).

4   A New Pertussis Vaccine

Whichever the currently accepted or newly proposed vaccination strategies, with their anticipated partial effectiveness, the generation of a new pertussis vaccine providing safe and long-term protection remains a proper need for an effective pertussis control. Given the strong commitments and huge costs required to generate and adopt a new pertussis vaccine, a consortium (Periscope 2018) gathering most active investigators of pertussis and pertussis vaccines and aimed at generating new knowledge and technical infrastructures for an improved pertussis vaccine has been established.

Any new pertussis vaccine should be able to represent the complexity of *B. pertussis* virulence armamentarium by either selecting a formulation with all major factors contributing to airways infection and disease, with an appropriate adjuvant helping promote the right CMI priming, or generating a whole cell vaccine that in turn could be a detoxified, old wP or a live, attenuated one. All of the above options are indeed being pursued.

Since attenuated, live vaccines usually confer the most pronounced and persistent protection (some of them for life), a vaccine that appears to be of major interest is the genetically attenuated BPZE1, formulated for intranasal delivery, that has been shown to be protective in mice and baboons and has recently completed a Phase 1 safety trial in humans (Locht 2018). While waiting for the necessary efficacy and confirmatory safety trials, one aspect of special interest of this vaccine resides in its potential to harness mucosal immunity capable of contrasting nasopharyngeal carriage of *B-pertussis* (Locht 2018). Another interest has been raised by the potent and persistent activation of innate immunity, as also reported for other attenuated, experimental and currently-used vaccines, and the consequent non-specific, off-target effects (innate immune memory or trained immunity; Netea et al. 2011; Bistoni et al. 1986; Quintin et al. 2012; Cauchi and Locht 2018; Li et al. 2010). Whether this or other new pertussis vaccines will one day be available is unknown. What can be safely anticipated is that preventing and controlling pertussis will continue to be as attractive as challenging for all those searching the best ways of fighting this disease by vaccination.

Acknowledgements The Author wishes to thank Clara Maria Ausiello and Giorgio Fedele for helpful suggestions regarding the content of this manuscript and overall editorial supervision.

References

Agricola E, Gesualdo F, Alimenti L, Pandolfi E, Carloni E, D'Ambrosio A, Russo L, Campagna I, Ferretti B, Tozzi AE (2016) Knowledge attitude and practice toward pertussis vaccination during pregnancy among pregnant and postpartum Italian women. Hum Vaccin Immunother 12:1982–1988

Althouse BM, Scarpino SV (2015) Asymptomatic transmission and the resurgence of Bordetella pertussis. BMC Med 13:146

Ausiello CM, Cassone A (2014) Acellular pertussis vaccines and pertussis resurgence: revise or replace? MBio 5(3):e01339–e01314. https://doi.org/10.1128/mBio.01339-14

Ausiello CM, Urbani F, la Sala A, Lande R, Cassone A (1997) Vaccine- and antigen-dependent type 1 and type 2 cytokine induction after primary vaccination of infants with whole-cell or acellular pertussis vaccines. Infect Immun 65(6):2168–2174

Ausiello CM, Lande R, Urbani F, La Sala A, Stefanelli P, Salmaso S, Mastarntonio P, Cassone A (1999) Cell-mediated immune responses in four- years-old hildren after primary immunization with cellular pertussis vaccines. Infect Immun 67:4064–4071

Barug D, Pronk I, van Houten MA, Versteegh FGA, Knol MJ, van de Kassteele J, Berbers GAM, Sanders EAM, Rots NY (2019) Maternal pertussis vaccination and its effects on the immune response of infants aged up to 12 months in the Netherlands: an open-label, parallel, randomised controlled trial. Lancet Infect Dis 19:392–401

Bistoni F, Vecchiarelli A, Cenci E, Puccetti P, Marconi P, Cassone A (1986) Evidence for macrophage-mediated protection against lethal Candida albicans infection. Infect Immun 51:668–674

Blok BA, de Bree LCJ, Diavatopoulos DA, Langereis JD, Joosten LAB, Aaby P, Crevel RV, Benn CS, Netea MG (2019) Interacting non-specific immunological effects of BCG and TdaP vaccinations: an explorative randomized. Clin Infect Dis Clin Infect Dis. https://doi.org/10.1093/cid/ciz246. [Epub ahead of print]

Burns DL, Meade BD, Messionnier NE (2014) Pertussis resurgence: perspectives from the working group meeting on pertussis on the causes, possible paths forward, and gaps in our knowledge. J Infect Dis 209 (Suppl 1):S32–S35

Campbell AN, Amirthalingam G (2018) Review of vaccination in pregnancy to prevent pertussis in early infancy. J Med Microbiol 67:1426–1456

Cassone A (2018) The case for an expanded concept of trained immunity. MBio 9(3):pii: e00570-18. https://doi.org/10.1128/mBio.00570-18

Cassone A, Ausiello CM, Urbani F, Lande R, Giuliano M, La Sala A, Piscitelli A, Salmaso S (1997) Cell-mediated and antibody responses to Bordetella pertussis antigens in children vaccinated with acellular or whole-cell pertussis vaccines. The Progetto Pertosse-CMI Working Group. Arch Pediatr Adolesc Med 151:283–289

Castagnini LA, Healy CM, Rench MA, Wootton SH, Munoz FM, Baker CJ (2012) Impact of maternal postpartum tetanus and diphtheria toxoids and acellular pertussis immunization on infant pertussis infection. Clin Infect Dis 54:78–84

Cauchi S, Locht C (2018) non-specific effects of live attenuated pertussis vaccine against heterologous infectious and inflammatory diseases. J Virol 9:7105–7113

Centers for Disease Control and Prevention (2012) Pertussis epidemic—Washington, 2012. MMWR 61:517–522

Cherry JD (1997) Comparative efficacy of acellular pertussis vaccines: an analysis of recent trials. Pediatr Infect Dis J 16(4 Suppl):S90–S96

da Silva Antunes R, Babor M, Carpenter C, Khalil N, Cortese M, Mentzer AJ, Seumois G, Petro CD, Purcell LA, Vijayanand P, Crotty S, Pulendran B, Peters B, Sette AJ (2018) Th1/Th17 polarization persists following whole-cell pertussis vaccination despite repeated acellular boosters. J Clin Invest 128(9):3853–3865

Edwards KM, Decker MD (2013) Pertussis vaccines. In: Plotkin SA, Orenstein W, Offit PA (eds) Vaccines, 6th edn. Elsevier Saunders, Philadelphia, pp 447–492

Edwards KM, Meade BD, Decker MD, Reed GF, Rennels RB, Steinhoff MC, Anderson EL, Enguland JA, Pichichero M, Deloria MA (1995) Comparison of thirteen acellular pertussis vaccines: overview and serological response. Pediatrics 96:548–557

Fedele G, Carollo M, Palazzo R, Stefanelli P, Pandolfi E, Gesualdo F, Tozzi AE, Carsetti R, Villani A, Nicolai A, Midulla F, Ausiello CM (2017) Pertussis Study Group Parents as source of pertussis transmission in hospitalized young infants. Infection 45 (2):171–178

Forsyth KD, Tan T, von König CW, Heininger U, Chitkara AJ, Plotkin S (2018) Recommendations to control pertussis prioritized relative to economies: a global pertussis initiative update. Vaccine 36:7270–7275

Gill C, Rohani P, Thea DM (2017) The relationship between mucosal immunity, nasopharyngeal carriage, asymptomatic transmission and the resurgence of Bordetella pertussis. F1000Res 6:1568–1574

Giuliano M, Mastrantonio P, Giammanco A, Piscitelli A, Salmaso S, Wassilak SGF (1998) Antibody responses and persistence in the 2 years following immunization with two acellular and one whole-cell vaccines against pertussis. J Pediatr 132:973–978

Greco D, Salmaso S, Mastrantonio P, Giuliano M, Tozzi AG, Anemona A, Ciofi Degli Atti ML, Giammanco A, Panei P, Blackwelder WC, Klein DL, Wassilak SGF, and the Progetto Pertosse Working Group (1996) A controlled trial of two acellular vaccines and one whole-cell vaccine against pertussis. N Engl J Med 334:341–348

Gustafsson L, Hallander HO, Olin P, Reizenstein E, Storsaeter J (1996) A controlled trial of a two-component acellular, a five-component acellular, and a whole-cell pertussis vaccine. N Engl J Med 334 (6):349–355

Hallander HO, Advani A, Donnelly D et al (2005) Shifts of Bordetella pertussis variants in Sweden from 1970 to 2003, during three periods marked by different vaccination programs. J Clin Microbiol 43:2856–2865

Kandeil W, Atanasov P, Avramioti D, Fu J, Demarteau N, Li X (2019) The burden of pertussis in older adults: what is the role of vaccination? A systematic literature review. Expert Rev Vaccines 19:1–17

Kilgore PE, Salim AM, Zervos MJ, Schmitt HJ (2016) Pertussis: microbiology, disease, treatment, and prevention. Clin Microbiol Rev 29:449–458

Kurup SP, Butler NS, Harty JT (2019) T cell-mediated immunity to malaria. Nat Rev Immunol. https://doi.org/10.1038/s41577-019-0158-z

Li R, Lim A, Phoon MC, Narasaraju T, Ng JK, Poh WP, Sim MK, Chow VT, Locht C, Alonso S (2010) Attenuated Bordetella pertussis protects against highly pathogenic influenza A viruses by dampening the cytokine storm. J Virol 84:7105–7113

Locht C (2018) Will we have new pertussis vaccines? Vaccine 36:5460–5469

Mills KHG, Ryan M, Ryan E, Mahon BP (1998) A murine model in which protection correlates with pertussis vaccine efficacy in children reveals complementary roles for humoral and cell-mediated immunity in protection against B. pertussis. Infect Immun 66:594–602

Netea MG, Quintin J, van der Meer JWM (2011) Trained immunity: a memory for innate host defense. Cell Host Microbe 9:355–361

PERISCOPE Consortium (2018) PERISCOPE: road towards effective control of pertussis. Lancet Infect Dis 19(5):e179–e186. https://doi.org/10.1016/S1473-3099(18)30646-7. [Epub ahead of print]

Pichichero ME, Deloria MA, Rennels MB, Anderson EL, Edwards KM, Decker MD, Englund JA, Steinhoff MC, Deforest A, Meade BD (1997) A safety and immunogenicity comparison of 12 acellular pertussis vaccines and one whole-cell pertussis vaccine given as a fourth dose in 15- to 20-month-old children. Pediatrics 100 (5):772–788

Plotkin SA (2014) The pertussis problem. Clin Infect Dis 58:830–833

Quintin J, Saed J, Martens JHA, Giamarellos Bourboulis EJ, Ifrim DC, Logie C, Jacobs L, Jansen T, Kullberg BJ, Wijmenga C, Joosten LAB, Xavier RJ, van der Meer JWM, Stunnenberg HG, Netea MG (2012) Candida albicans infection affords protection against reinfection via functional reprogramming of monocytes. Cell Host Microbe 12:223–232

Ross PJ, Sutton CE, Higgins S, Allen AC, Walsh K, Misiak A, Lavelle EC, McLoughlin RM, Mills KH (2013) Relative contribution of Th1 and Th17 cells in adaptive immunity to Bordetella pertussis: towards the rational design of an improved acellular pertussis Vaccine. PLoS Pathog 9:e1003264

Ryan M, Murphy G, Ryan E, Nilsson L, Shackley F, Gothefors L, Oymar K, Miller E, Storsaeter J, Mills KH (1998) Distinct T-cell subtypes induced with whole cell and acellular pertussis vaccines in children. Immunology 93:1–10

Salmaso S, Mastrantonio P, Tozzi AE, Stefanelli P, Anemona A, Ciofi Degli Atti ML, Giammanco A, Stage III Working Group (2001) Sustained efficacy during the first 6 years of life of 3-component acellular pertussis vaccines administered in infancy: the Italian experience. Pediatrics 108:E81

Sizaire V, Garrido-Estepa M, Masa-Calles J et al (2014) Increase of pertussis incidence in 2010 to 2012 after 12 years of low circulation in Spain. Euro Surveill 19 (32):pii: 20875

Stein PE, Boodhoo A, Armstrong GD, Heerze LD, Cockle SA, Klein MH, Read RJ (1994) Structure of a pertussis toxin-sugar complex as a model for receptor binding. Nat Struct Biol 1:591–596

Warfel JM, Beren J, Merkel TJ (2012a) Airborne transmission of Bordetella pertussis. J Infect Dis 206:902–906

Warfel JM, Beren J, Kelly VK, Lee G, Merkel TJ (2012b) Non-human primate model of pertussis. Infect Immun 80:1530–1536

Warfel JM, Zimmerman LI, Merkel TJ (2014) Acellular pertussis vaccines protect against disease but fail to prevent infection and transmission in a nonhuman primate model. Proc Natl Acad Sci U S A 111:787–792

Wendelboe AM, Van Rie A, Salmaso S, Englund JA (2005) Duration of immunity against pertussis after natural infection or vaccination. Pediatr Infect Dis J 24(5 Suppl):S58–S61

Zepp F, Knuf M, Habermehl P, Schmitt JH, Rebsch C, Schmidtke P, Clemens R, Slaoui M (1996) Pertussis-specific cell-mediated immunity in infants after vaccination with a tricomponent acellular pertussis vaccine. Infect Immun 64:4078–4084

Zhang L, Prietsch SO, Axelsson I, Halperin SA (2014) Acellular vaccines for preventing whooping cough in children. Cochrane Database Syst Rev 17(9): CD001478

Adv Exp Med Biol - Advances in Microbiology, Infectious Diseases and Public Health (2019) 1183: 169–170
https://doi.org/10.1007/978-3-030-33249-5
© Springer Nature Switzerland AG 2019

Index

A

Abrahams, J.S., 2–12
Acellular vaccines (ACV), 2–6, 11, 12, 19, 82, 84, 86,
    100, 116, 120, 121
Adenylate cyclase toxin-hemolysin (ACT/AC-Hly), v,
    38–41, 43, 57, 62, 68, 83, 84, 87, 139
Althouse, B.M., 163
Antibiotic treatment, 129, 152, 156–157
Anti-pertussis immunity, 100
Ausiello, C.M., 99–108, 163

B

Bagby, S., 2–12
de Baillou, G., 127, 138
Banus, S., 83
Barkoff, A.-M., 19–30
Bart, M.J., 5–9, 24
Bordetella pertussis (Bp), 2, 20, 35, 53, 81, 100, 115, 127,
    138, 151, 161
Bordetella toxins, v, 38, 56, 57
Bordet, J., 54, 99, 115, 127
Bouchez, V., 25, 27
Bowden, K.E., 7, 27
Brody, M., 116
Buisman, A.-M., 81–92

C

Campins-Martí, M., 103
Carbonetti, N., 35–44
Cassiday, P.K., 22, 23
Cassone, A., 161–165
Consortium, P., 70
Corbière, V., 99–108

D

Dalet, K., 10, 11
Diagnosis, vi, 2, 86, 128, 129, 133, 138, 142–144, 152,
    154, 155, 158
Diavatopoulos, D.A., 53–70
DNA sequencing, 7
Du, Q., 5

E

Ebell, M.H., 155

Edelman, K.J., 104
van Els, C.A.C.M., 81–92
Epidemiology, 2, 19–30, 53, 64, 82, 91, 92, 118,
    119, 130, 140, 145, 154, 155, 162, 163
Esposito, S., 151–158
Evolution, v, 1–12, 25, 29, 118, 130, 132, 145,
    157, 158

F

Fedele, G., 99–108
Fu, P., 28

G

Gengou, O., 99, 115, 127
Genomic variation, 11
Gill, C., 163
Gillard, J., 53–70
Guiso, N., 137–146

H

Hafler, J.P., 86
Hallander, H.O., 21
Halperin, S.A., 131, 141
Hamidou Soumana, I., 3
He, Q., 19–30
Hegerle, N., 27
Hovingh, E.S., 59
Hozbor, D., 115–122

I

Immune programming, 70
Immunization strategies, 131, 133
Innate immunity, v, 38, 54, 58, 59, 61–64, 165

K

Kendrick, P., 116
King, A.J., 8, 9

L

Lam, C., 27
Lambert, E.E., 81–92
Locht, C., 119
Low and medium income countries, vi,
    137–146

M
MacArthur, I., 2–12
Madsen, T., 116
Mascart, F., 86, 99–108
Mbayei, S.A., 154
Mechanism of protection, 54, 83, 89
Mills, K.H.G., 90

N
Natural infection, 54, 55, 60, 61, 65, 81–92, 104

O
Octavia, S., 27
Ostolaza, H., 39

P
Parkhill, J., 24
Paroxysmal cough, 84, 116, 154
Pawloski, L.C., 26
Pertactin (PRN), 2, 4–7, 19, 20, 24–29, 32, 58, 61, 83, 85, 86, 100–102, 116, 118–120, 122, 139, 140, 152, 162
Pertussis, v, vi, 1–12, 19–30, 35–44, 53–70, 81–92, 99–108, 115–122, 127–133, 137–146, 151–157, 161–165
Pertussis pathogenesis, 36, 43
Pertussis toxin (PTx), v, 2, 6, 19, 35–38, 43, 56, 57, 83, 100, 116, 120–121, 128, 131, 139, 142, 162
Pertussis vaccination, 85, 99–105, 107, 131
Pertussis vaccines, vi, 36, 40, 53, 54, 56, 57, 59, 64, 69, 70, 82, 83, 91, 92, 99–108, 115–122, 137–140, 161–165
Pohl-Koppe, A., 86
Polak, M., 25
Polinori, I., 151–158
Post-exposure prophylaxis (PP), 132, 156
Preston, A., 2–12

R
Raeven, R.H., 90
Raymond, J., 154
Rieber, N., 106
Ring, N., 2–12

S
Sauer, L.W., 116
Saul, N., 131
Scanlon, K., 35–44
Scarpino, S.V., 163
Schouls, L.M., 23, 24
van Schuppen, E., 53–70

Sealey, K.L., 24
da Silva Antunes, R., 86
Sirivichayakul, C., 120
Skerry, C., 35–44
Smits, K., 106
Sorley, R.G., 116
Stefanelli, P., 127–133
Strebel, P., 154
Surveillance, v, 2, 12, 20, 25, 53, 119, 129, 131–133, 137, 138, 141–146
Surveillance systems, 53, 131, 143

T
Taieb, F., 137–146
Th1 and Th17 polarized subsets, 55, 56, 59, 61–63, 85, 89, 104, 107
Tissue residency, v, 85–87
Tracheal cytotoxin (TCT), v, 41–43, 57, 65, 84
Tran Minh, N.N., 104
Tsang, R.S., 26, 27

V
Vaccination, v, vi, 2, 4–6, 20, 21, 24–27, 29, 31, 42, 53–70, 82, 83, 85, 86, 91, 100–108, 118, 120, 128, 130–132, 137–146, 152–156, 161–165
Vaccination strategies, 5, 25, 83, 132, 145, 146, 165
Vaccines, 2, 19, 36, 53, 82, 99, 115, 129, 137, 151, 161
van Gent, M., 23

W
Wang, X., 41
Weigand, M.R., 6, 7, 12, 27, 28
Weiss, A.A., 40
Whole-cell vaccines (WCV), vi, 2, 3, 5, 6, 10–12, 19, 40, 53, 54, 82, 86, 90, 99, 116, 118, 120, 138, 162–165
Whole genome sequencing (WGS), 6, 11, 12, 20, 24–25, 29, 30, 132
Whooping cough, v, 2, 4, 6–8, 11, 12, 19, 81, 82, 85, 115, 129, 137, 138, 142, 146, 151, 152, 155, 164
Whooping cough pathology, 153
Wilk, M.M., 84

X
Xu, Y., 5, 24

Z
Zeddeman, A., 27
Zomer, A., 5, 27